MORGAN COUNTY PUBLIC LIBRARY
110 SOUTH JEFFERSON ST
MARTINSVILLE, IN 46151

About the Author

Paul A. Tucci is an author of several books for Visible Ink Press, including *The Handy Geography Answer Book* (2nd edition), *The Handy Personal Finance Answer Book,* and *The Handy Investing Answer Book.* Over the course of his

910
TUC
Tucci, Paul A.,
1962-
The handy geography
answer book.

industry articles and
onal management at
versity, the University
versity, and Oakland
of a philanthropic organization and several privately held companies. As a former Global Publishing executive, he has managed, travelled, and conducted business throughout the world. Tucci brings all this experience and inspiration to this book. He resides in Michigan.

Also from Visible Ink Press

The Handy African American History Answer Book
by Jessie Carnie Smith
ISBN: 978-1-57859-452-8

The Handy American History Answer Book
by David L. Hudson Jr.
ISBN: 978-1-57859-471-9

The Handy Anatomy Answer Book, 2nd edition
by Patricia Barnes-Svarney and Thomas E. Svarney
ISBN: 978-1-57859-542-6

The Handy Answer Book for Kids (and Parents), 2nd edition
by Gina Misiroglu
ISBN: 978-1-57859-219-7

The Handy Art History Answer Book
by Madelynn Dickerson
ISBN: 978-1-57859-417-7

The Handy Astronomy Answer Book, 3rd edition
by Charles Liu
ISBN: 978-1-57859-190-9

The Handy Bible Answer Book
by Jennifer Rebecca Prince
ISBN: 978-1-57859-478-8

The Handy Biology Answer Book, 2nd edition
by Patricia Barnes Svarney and Thomas E. Svarney
ISBN: 978-1-57859-490-0

The Handy Boston Answer Book
by Samuel Willard Crompton
ISBN: 978-1-57859-593-8

The Handy California Answer Book
by Kevin S. Hile
ISBN: 978-1-57859-591-4

The Handy Chemistry Answer Book
by Ian C. Stewart and Justin P. Lamont
ISBN: 978-1-57859-374-3

The Handy Civil War Answer Book
by Samuel Willard Crompton
ISBN: 978-1-57859-476-4

The Handy Dinosaur Answer Book, 2nd edition
by Patricia Barnes-Svarney and Thomas E. Svarney
ISBN: 978-1-57859-218-0

The Handy English Grammar Answer Book
by Christine A. Hult, Ph.D.
ISBN: 978-1-57859-520-4

The Handy Geography Answer Book, 3rd edition
by Paul A. Tucci
ISBN: 978-1-57859-215-9

The Handy Geology Answer Book
by Patricia Barnes-Svarney and Thomas E. Svarney
ISBN: 978-1-57859-156-5

The Handy History Answer Book, 3rd edition
by David L. Hudson, Jr.
ISBN: 978-1-57859-372-9

The Handy Hockey Answer Book
by Stan Fischler
ISBN: 978-1-57859-513-6

The Handy Investing Answer Book
by Paul A. Tucci
ISBN: 978-1-57859-486-3

The Handy Islam Answer Book
by John Renard Ph.D.
ISBN: 978-1-57859-510-5

The Handy Law Answer Book
by David L. Hudson Jr.
ISBN: 978-1-57859-217-3

The Handy Math Answer Book, 2nd edition
by Patricia Barnes-Svarney and Thomas E. Svarney
ISBN: 978-1-57859-373-6

The Handy Military History Answer Book
by Samuel Willard Crompton
ISBN: 978-1-57859-509-9

The Handy Mythology Answer Book
by David A. Leeming, Ph.D.
ISBN: 978-1-57859-475-7

The Handy Nutrition Answer Book
by Patricia Barnes-Svarney and Thomas E. Svarney
ISBN: 978-1-57859-484-9

The Handy Ocean Answer Book
by Patricia Barnes-Svarney and Thomas E. Svarney
ISBN: 978-1-57859-063-6

The Handy Personal Finance Answer Book
by Paul A. Tucci
ISBN: 978-1-57859-322-4

The Handy Philosophy Answer Book
by Naomi Zack
ISBN: 978-1-57859-226-5

The Handy Physics Answer Book, 2nd edition
By Paul W. Zitzewitz, Ph.D.
ISBN: 978-1-57859-305-7

The Handy Politics Answer Book
by Gina Misiroglu
ISBN: 978-1-57859-139-8

The Handy Presidents Answer Book, 2nd edition
by David L. Hudson, Jr.
ISB N: 978-1-57859-317-0

The Handy Psychology Answer Book, 2nd edition
by Lisa J. Cohen
ISBN: 978-1-57859-508-2

The Handy Religion Answer Book, 2nd edition
by John Renard
ISBN: 978-1-57859-379-8

The Handy Science Answer Book, 4th edition
by The Carnegie Library of Pittsburgh
ISBN: 978-1-57859-321-7

The Handy Supreme Court Answer Book
by David L Hudson, Jr.
ISBN: 978-1-57859-196-1

The Handy Technology Answer Book
by Naomi Bobick and James Balaban
ISBN: 978-1-57859-563-1

The Handy Weather Answer Book, 2nd edition
by Kevin S. Hile
ISBN: 978-1-57859-221-0

Please visit the Handy Answers series website at www.handyanswers.com.

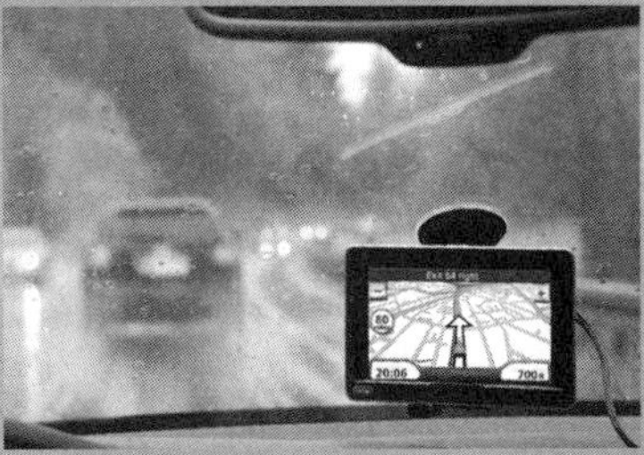

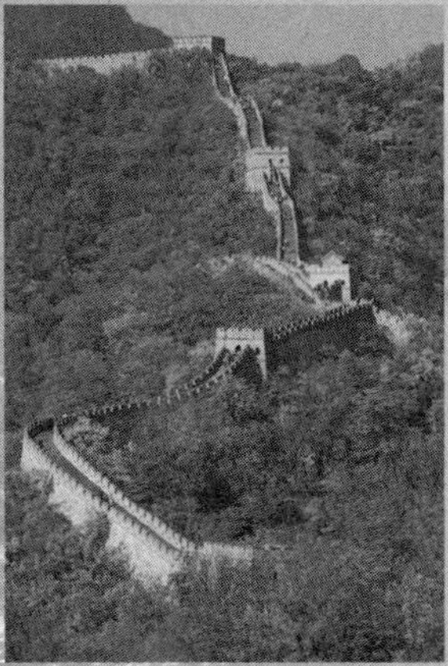

THE HANDY GEOGRAPHY ANSWER BOOK

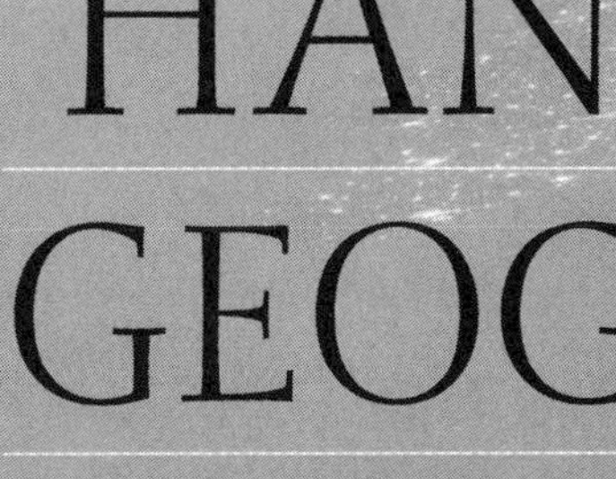

THIRD EDITION

Paul A. Tucci

Detroit

THE HANDY GEOGRAPHY ANSWER BOOK

Copyright © 2017 by Visible Ink Press®

This publication is a creative work fully protected by all applicable copyright laws, as well as by misappropriation, trade secret, unfair competition, and other applicable laws.

No part of this book may be reproduced in any form without permission in writing from the publisher, except by a reviewer who wishes to quote brief passages in connection with a review written for inclusion in a magazine, newspaper, or website.

All rights to this publication will be vigorously defended.

Visible Ink Press®
43311 Joy Rd., #414
Canton, MI 48187-2075

Visible Ink Press is a registered trademark of Visible Ink Press LLC.

Most Visible Ink Press books are available at special quantity discounts when purchased in bulk by corporations, organizations, or groups. Customized printings, special imprints, messages, and excerpts can be produced to meet your needs. For more information, contact Special Markets Director, Visible Ink Press, www.visibleinkpress.com, or 734-667-3211.

Managing Editor: Kevin S. Hile
Art Director: Mary Claire Krzewinski
Typesetting: Marco DiVita
Proofreaders: Larry Baker and Shoshana Hurwitz
Indexer: Shoshana Hurwitz

Cover images: Image of oil pump, Eric Kounce; all other images, Shutterstock.

Library of Congress Cataloging-in-Publication Data

Names: Tucci, Paul A., 1962- author.
Title: The handy geography answer book / Paul A. Tucci.
Description: Third edition. | New York : Visible Ink Press, [2017] |Description based on print version record and CIP data provided by publisher; resource not viewed.
Identifiers: LCCN 2016019256 (print) | LCCN 2016017255 (ebook) | ISBN 978157859627 (Kindl) | ISBN 9781578596263 (ePub) | ISBN 9781578596256 (Pdf) | ISBN 9781578595761 (paperback)
Subjects: LCSH: Geography--Miscellanea. | BISAC: SCIENCE / Earth Sciences / Geography.
Classification: LCC G131 (print) | LCC G131 .T83 2017 (ebook) | DDC 910--dc23
LC record available at https://lccn.loc.gov/2016019256

Printed in the United States of America

10 9 8 7 6 5 4 3 2 1

Contents

Acknowledgments

I wish to thank the following people whose great efforts help create this book: proofreaders Larry Baker and Shoshana Hurwitz (who also indexed this book), page and cover designer Mary Claire Krzewinski, typesetter Marco DiVita. Also, thanks to Matthew Rosenberg, who created the original first edition of this book many years ago. I also wish to thank the Visible Ink Editor Kevin Hile and Visible Ink Publisher Roger Jänecke for making this book possible.

Thank you also to my family and friends who encouraged me for over one year to write this book.

Photo Sources

Aamsse (Wikicommons): p. 190.
Amerikanisches Aussenministerium: p. 209.
Cavit (Wikicommons): p. 23.
CherryX (Wikicommons): p. 217.
CIA: p. 256.
Continentalis (Wikicommons): p. 154.
Glenmary Research Center: p. 177.
Hile, Kevin: p. 105.
Hosseini, Mahmoud: p. 269.
Kmusser (Wikicommons): p. 49.
Mareklug (Wikicommons): p. 167.
Maslowski, Wieslaw, at the U. S. Naval Postgraduate School, Monterey, California: p. 187.
Mutxamel (Wikicommons): p. 215.
NASA: pp. 45, 75, 245.
National Archives and Records Administration: p. 90.
National Maritime Museum: p. 289 (top).
National Park Service: p. 185.
NOAA: p. 93.
Shutterstock: pp. 4, 7, 9, 14, 21, 25, 27, 29, 31, 37, 38, 41, 44, 51, 54, 57, 58, 61, 64, 67, 69, 73, 78, 81, 86, 88, 97, 99, 108, 113, 116, 120, 133, 138, 143, 146, 148, 151, 157, 164, 171, 173, 175, 197, 199, 203, 212, 220, 223, 225, 228, 230, 234, 235, 238, 239, 242, 250, 252, 261, 265, 267, 270, 275, 280, 282, 284, 294, 296.
United Nations: p. 128.
U.S. Census: p. 130.
U.S. Geologic Survey: pp. 3, 36.
Public domain: pp. 10, 12, 17, 107, 136, 159, 161, 180, 192, 202, 246, 278, 289 (bottom), 293.

Introduction

What is so surprising about geography is how dynamic our world is and how it changes so quickly. It is profound to see just how much of the information contained within the previous edition has changed in just a few short years. Immutable facts of history are being challenged every day by scholars around the world. New artifacts are being discovered, and new interpretations of old texts are constantly being published. Technological advances have made measuring many things much more accurate and precise. We are now better able to see our world with greater clarity. Simultaneously, the enormous convergence of technology like software apps and powerful mobile devices, as well as the ubiquitous qualities of information in all its forms and formats, has brought access to geography and nearly all of its facets to billions of people. This is in sharp contrast to a much different world 500 years ago, when many people were convinced that the world is flat with edges not too far beyond the next pasture or village.

Things have gotten better around the world, but in some cases worse. Our previous list of the most polluted cities in the world changed in a surprising way. The people of the world became healthier, lived a bit longer, became more literate, donated more money, re-engineered rivers, and preserved forests; the list is endless. Centuries-old disputes over geography still spawn conflict and resolution around the world. A convergence of technology and political geography causes a country like China to alter the map of the world by building tiny islands in a nearby sea.

Interesting companies that are creating new ways to use maps and to visualize data are fueling our renewed interest in geographical concepts. Google's Street View lets users drill down to see mailboxes at specified GPS coordinates, or to see the front door of their next door neighbor's house, something that was inconceivable only a few years ago.

A friend recently gave me a geography textbook from 1890. The city of Los Angeles, which is one of the largest metropolitan areas in the world today, had a population of approximately 51,000 people. There were only three cities in the U.S. with populations greater than one million. Today, there are ten. People had access to very little information beyond what they garnered from this little geography textbook in their first year of high school. Many countries that exist today were not on the map at that time, and many colonies that had been established centuries before were still in existence. Great natural disasters occurred in the nineteenth century, but information about them would take months before they would be in the national consciousness. Today, billions of people are made aware of nearly any reported event, anywhere in the world, in seconds.

This access to information will hopefully translate into a future in which the study of geography will be revived at all levels of our educational system, and people will remember a time when a geography department was started at their university, instead of when it was closed, as in my case.

Our mission for this edition of *The Handy Geography Answer Book* is to provide you with a helpful grid on the field of geography, and populate the chapters of the book with really interesting questions and answers that make the field of geography come to life. Perhaps it will inspire an argument or two, or make you want to read more about the subject. The book will capture you no matter what your background or experience is.

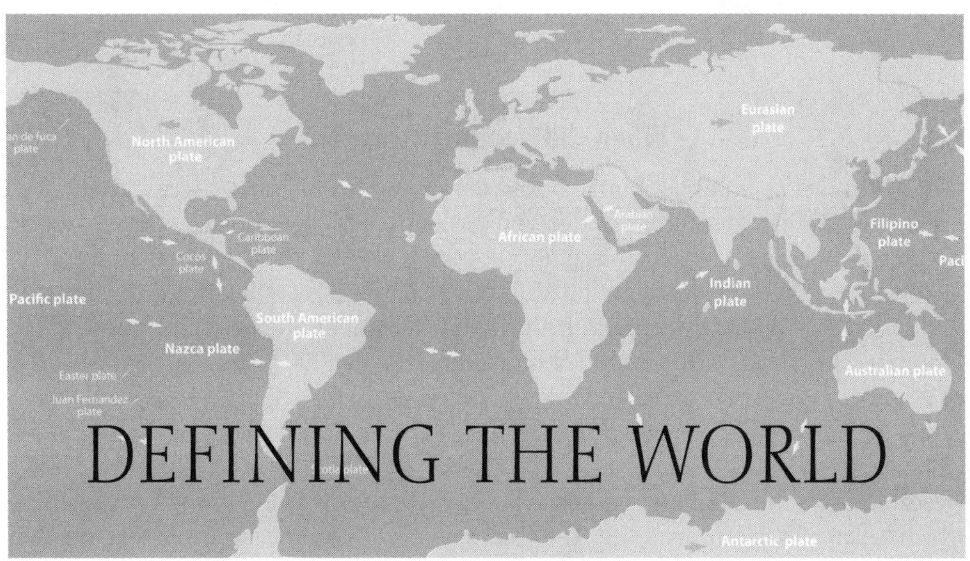

DEFINING THE WORLD

DEFINITIONS AND HISTORY

What is the origin of the word "geography"?

The word "geography" is from the Greek word *geographia* and can be divided into two parts, *geo*, meaning the Earth, and *graphy*, which refers to writing. So geography can be loosely translated as "writing about the Earth." Ancient geography was often filled with descriptions of remote places, but modern geography has become much more than writing about the Earth. Contemporary geographers have a difficult time defining the discipline. Some popular definitions include "the bridge between the human and the natural sciences," "the mother of all sciences," and "anything that can be mapped."

Who invented geography?

The Greek philosopher Thales was one of the first to argue about the shape of the world in the sixth century B.C.E. And Chinese texts of the fifth century B.C.E. describe the provinces of China in great detail. However, the Greek scholar Eratosthenes is credited with the first use of the word "geography" in the third century B.C.E. He is also known as the "father of geography" for his geographical writing and accomplishments, including the measurement of the circumference of the Earth, the tilt of the Earth's axis, the distance from the Earth to the sun, and the concept of "Leap Day."

What is geologic time?

Geologic time is a time scale that divides the history of the planet Earth into eras, periods, and epochs from the birth of the planet to the present. The oldest era is the Precambrian Era, which began 4.6 billion years ago and ended about 570 million years ago. Next came the Paleozoic Era, which lasted from 570 to 245 million years ago, followed by the

When did geography begin?

We must assume that humans have always wanted to know where sources of food are located, places to live, and the location of one's protective group. Geographic thought has been present for thousands of years. Maps drawn in the sand, etched in stone, or painted on the walls of caves, as well as explorations to distant lands, were made by the earliest civilizations. Geographic knowledge has been accumulating since the beginning of humankind.

Mesozoic Era, from 245 to 66 million years ago. We're now living in the Cenozoic Era, which began 66 million years ago. The Paleozoic, Mesozoic, and Cenozoic eras are each divided into periods. Additionally, the Cenozoic Era is divided into even smaller units of time called epochs. The present epoch that we live in today, which spans the last ten thousand years (the time since the last significant Ice Age), is called the Holocene Epoch.

What new epoch has recently been proposed by Nobel Prize winner Paul Crutzen?

Most epochs last approximately more than 3 million years, enough time to deposit traces of subtle changes in the conditions on the planet, for scientists to discover in future years. Dutch atmospheric chemist and Nobel Prize winner Paul Crutzen coined the term "Anthropocene," or "New Man," epoch because changes to our planet, brought on by human activities like habitat destruction, environmental degradation, and the extinction of thousands of plant and animal species since the beginning of the Industrial Revolution two hundred years ago, will leave an indelible mark in the boundary layer of the surface of our planet. But some scientists disagree and assert that it really is not a new epoch, just a warm period during our current Holocene Epoch, or Ice Age, which should continue for another million years.

What is the AAG?

The Association of American Geographers (AAG) is a professional organization of academic geographers and geography students. The AAG was founded in 1904 and publishes two key academic journals in geography, the *Annals of the Association of American Geographers* and the *Professional Geographer*, as well as the *AAG Review of Books* and the online *AAG Newsletter*. The AAG also holds annual conferences and supports regional and specialty groups of geographers. Its membership spans more than sixty countries.

What is the NCGE?

The National Council for Geographic Education (NCGE) is an organization of educators that seeks to promote geographic education. The NCGE was founded by educator George Miller and was chartered in 1915 with a stated mission to enhance the status and quality of geography teaching from kindergarten through university. The NCGE publishes the *Journal of Geography* and holds conferences every year.

What is the National Geographic Society?

Founded in 1888, the National Geographic Society is one of the largest scientific and educational institutions in the world. It has supported exploration, cartography, and discovery, and it publishes the popular magazine *National Geographic*, in English and many other languages, and is the fifth-most-popular magazine in the United States. The society's publishing operations reach a global audience of over 600 million people each month.

What do modern geographers do?

While there are a few jobs with the title of "geographer," many geography students use their analytical ability and knowledge of the world to work in a variety of fields. Geography students often take jobs in fields such as city planning, cartography, marketing, real estate, environment, and teaching.

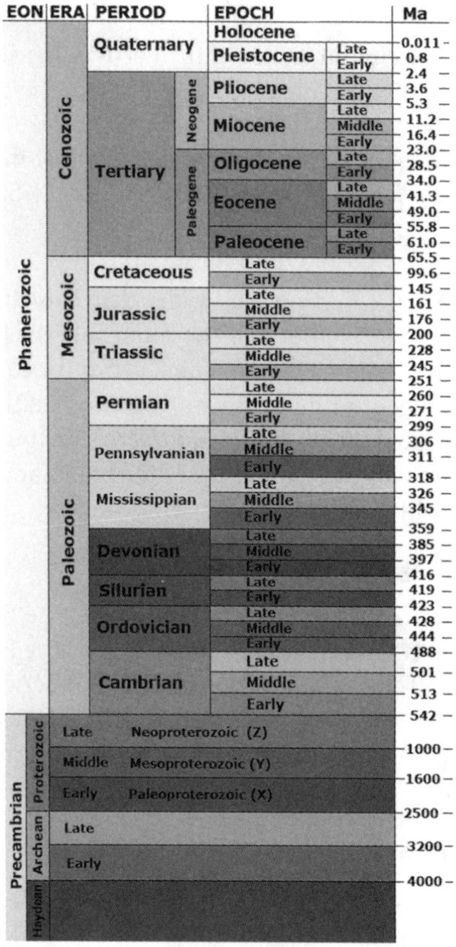

The geologic time scale

THE EARTH

How old is the Earth?

The Earth is approximately 4.54 billion years old (+/- 1%). We know this because scientists use radiometric dating techniques to analyze samples in order to determine its age. One of the oldest materials found in the Earth is a sample of crystals of the mineral zircon, found in Western Australia, that is at least 4.04 billion years old.

How was the Earth formed?

Scientists believe that the Earth was formed, along with the rest of the solar system, from a massive gas cloud. As the cloud solidified, it formed the solid masses such as the Earth and the other planets.

What is the circumference of the Earth?

The circumference of the Earth at the equator is 24,901.55 miles (40,066.59 km). Due to the irregular, ellipsoid shape of the Earth, a line of longitude wrapped around the Earth going through the north and south poles is 24,859.82 miles (40,000 km). Therefore, the Earth is a little bit (about 41 miles [66 km]) wider than it is high. The diameter of the Earth is 7,926.41 miles (12,753.59 km).

Is the Earth a perfect sphere?

No, the Earth is a bit wider than it is "high." The shape is often called a geoid (Earth-like) or an ellipsoid. The rotation of the Earth causes a slight bulge toward the equator. The circumference of the Earth at the equator is 24,901.55 miles (40,066.59 km), which is about 41 miles (66 km) greater than the circumference through the poles (24,859.82 miles [40,000 km]). If you were standing on the moon, looking back home, it would be virtually impossible to see the bulge, and the Earth would appear to be a perfect sphere (which it practically is).

What is a hemisphere?

A hemisphere is half of the Earth. The Earth can actually be divided into hemispheres in two ways: by the equator and by the Prime Meridian (through Greenwich, England) at 0 degrees longitude and another meridian at 180 degrees longitude (near the location of the International Date Line in the western Pacific Ocean. Zero and 180 degrees longitude divide the Earth into the Eastern (most of Europe, Africa, Australia, and Asia) and Western (the Americas) Hemispheres.

What is the equator?

The equator divides the Earth into Northern and Southern Hemispheres. There are seasonal differences between the Northern and Southern Hemispheres. When it is winter

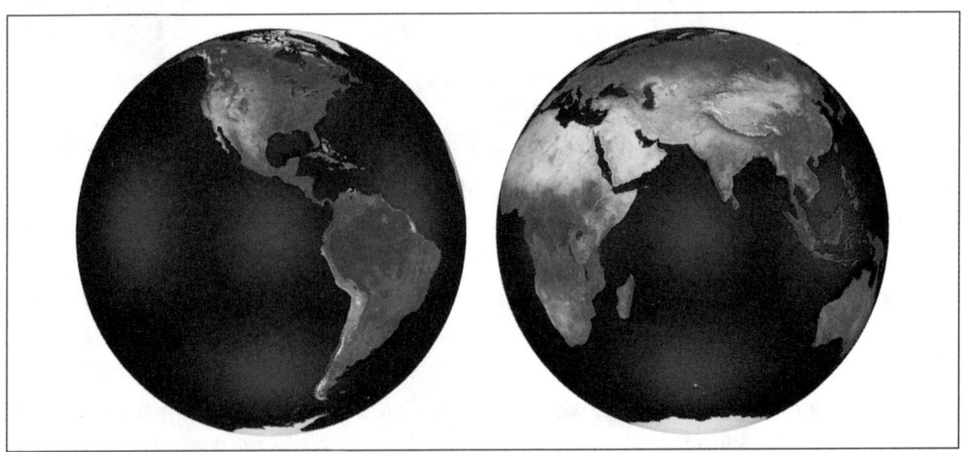

Western (left) and Eastern Hemispheres of our planet

How was the circumference of the Earth determined?

The Greek geographer and librarian at the Great Library of Alexandria, Eratosthenes (c. 273—c. 192 B.C.E.), was aware that the sun reached the bottom of a well in Egypt only once a year, on the first day of summer. The well that he studied was near the city of Aswan and the Tropic of Cancer (where the sun is directly overhead at noon on the summer solstice). Around the year 240 B.C.E., Eratosthenes estimated the distance between the well and the city of Alexandria based on the length of time it took camel caravans to travel between the two places. He measured the angle of the sun's shadow in Alexandria at the same time that the well was lit by the sun and then used a mathematic formula to determine that the circumference of the Earth was 25,000 miles (about 40,000 km)—amazingly close to the actual figure, which at the equator is 24,898 miles (40,070 km).

in the Northern Hemisphere, it is summer in the Southern Hemisphere. But there is no such difference between the Eastern and Western Hemispheres.

What are the Arctic and Antarctic Circles?

The circles are imaginary lines that surround the North and South Poles at 66.5 degrees latitude. The Arctic Circle is a line of latitude at 66.5 degrees north of the equator, and the Antarctic Circle is a line of latitude at 66.5 degrees south. Areas north of the Arctic Circle are dark for 24 hours near December 21 and areas south of the Antarctic Circle are dark for 24 hours near June 21. Almost all of the continent of Antarctica is located to the south of the Antarctic Circle.

If the Earth is so large, why did Christopher Columbus think that India was close enough to reach by sailing west from Europe?

The Greek geographer Posidonius did not believe Eratosthenes's earlier calculation, so he performed his own measurement of the Earth's circumference and arrived at the figure of 18,000 miles (28,962 km). Columbus used the circumference estimated by Posidonius when he argued his plan before the Spanish court. The 7,000-mile (11,263-km) difference between the actual circumference and the one Columbus used led him to believe he could reach India rather quickly by sailing west from Europe.

How fast does the Earth spin?

It depends on where you are on the planet. If you were standing on the North Pole or close to it, you would be moving at a very slow rate of speed—nearly zero miles per hour. On the other hand, those who live at the equator (and therefore have to move about 24,900 miles [40,000 km] in a 24-hour period) zoom at about 1,038 miles (1,670 km) per hour. Those in the mid-latitudes, as in the United States, breeze along from about 700 to 900 miles (1,126 to 1,448 km) per hour.

Why don't we feel the Earth moving?

Even though we constantly move at a high rate of speed, we don't feel it, just as we don't feel the speed at which we're flying in an airplane or driving in a car. It's only when there is a sudden change in speed that we notice, and if the Earth made such a change we would certainly feel it.

Does the Earth spin at a constant rate?

The rotation of the Earth actually has slight variations. Motion and activity within the Earth, such as friction due to tides, wind, and other forces, change the speed of the planet's rotation a little. These changes only amount to milliseconds over hundreds of years but do cause people who keep exact time to make corrections every few years.

What is the axis of the Earth?

The axis is the imaginary line that passes through the North and South Poles about which the Earth revolves. Because the Earth is tilted along this spin axis, as we make our way around the sun in an elliptical orbit for 365 days, we have our seasons (winter, spring, summer, and fall). As the Earth's axis is pointed toward the sun for people in the Northern Hemisphere, we experience summer. At the same time, for people in the Southern Hemisphere, as the axis points away from the sun, we experience winter.

What is inside the Earth?

The Earth is comprised of several layers: crust, mantle, and core. The outermost part of the Earth, the crust, is divided into huge plates that float atop the mantle, and are always in motion. The crust is comprised of the elements iron, oxygen, silicon, magnesium, sulphur, nickel, and trace amounts of many other elements. Beneath the crust is the mantle, which is about 1,800 miles (2,890 km) deep and is composed of silicate rocks, which are heated and cooled by the core. The mantle, which makes up the bulk of the interior of the Earth, is composed of three layers—two outer layers are solid and the inner layer (the asthenosphere) is a layer of rock that is easily moved and shaped. At the very center of the Earth is a dense and solid inner core of iron and other minerals that is about 1,520 miles (1,220 km) wide. Surrounding the inner core is a liquid (molten) outer core that is about 1,355 miles (2,180 km) thick.

If I dug through the Earth, would I end up in China?

If you are in North America and you were able to dig through the Earth (which is impossible due to such things as pressure, the molten outer core, and solid inner core), you would end up in the Indian Ocean, far from landmasses. If you were really lucky, you might end up on a tiny island, but you're surely not going to end up in China. The points at opposite sides of the Earth are called antipodes. Most antipodes of Europe fall into the Pacific Ocean.

When do earthquakes occur?

Earthquakes occur when the plates that comprise the Earth's crust rub against each other. We have mountains created when these gigantic plates collide and large trenches created when one plate slides underneath another.

What is the Mid-Atlantic Ridge?

We don't get to appreciate the beauty of this huge mountain range because it's located at the bottom of the Atlantic Ocean (with one exception: Iceland is a part of the ridge). It is part of one of the longest mountain ranges in the world. The ridge is a crack between tectonic plates where new ocean floor is being created as magma flows up from under the Earth. As more crust is created, it pushes the older crust farther away. The new crust at the ridge piles up to form mountains and then begins to move across the bottom of the ocean. Because the Earth can't get larger as more crust is created, the crust eventually has nowhere to go except back into the Earth. This is where subduction occurs.

What is subduction?

When two tectonic plates meet and collide, crust must either be lifted up, as in the case of the Himalayas, or it must be sent back into the Earth. When crust from one plate

In Iceland's National Park of Thingvellir one can visit the rift valley that is part of the Mid-Atlantic Ridge.

slides under the crust of another, it is called subduction, and the area around the subduction is called a subduction zone.

What is the North Magnetic Pole?

The North Magnetic Pole is where compass needles around the world point. Today it is located beyond Canada's Northwest Territories near 85.9 degrees north, 147 degrees west (latitude and longitude), about 900 miles (1,450 km) away from absolute North Pole. Since it moves continuously at about 34 miles per hour (55 km per hour), global research scientists are constantly trying to determine the current location of the pole. In order to determine true north, look at a recent topographic map for your local area. It should note the "magnetic declination," which means the degrees east or west that you'll need to rotate your compass to determine which way is actually north.

Who had the first idea that there was a magnetic North Pole?

As early as the year 1600, English physician and naturalist William Gilbert was the first person to define the North Magnetic Pole as the point on the Earth where the Earth's magnetic fields point vertically downward. It was not until 1831 that James Clark Ross found this point. Another famous explorer, Roald Amundsen, found the North Magnetic Pole in 1903, but in a different location. In a third major exploration in 1947, Paul Serson and Jack Clark found the pole on Prince of Wales Island.

CONTINENTS AND ISLANDS

What are continents?

Continents are the six or seven large landmasses on the planet. If you count seven continents, these include Europe, Asia, Africa, Australia, Antarctica, North America, and South America. Some geographers refer to six continents by combining Europe and Asia as Eurasia, due to the fact that it is one large tectonic plate and landmass. So whether you count Europe and Asia as one continent or two (divided at the Ural Mountains in western Russia) is up to the individual. Australia is the only continent that is its own country.

Why is the number of continents controversial?

We find that there is controversy about how many continents there truly are because of how experts classify and define the word "continent." It is generally accepted that a continent is a very large, continuous, and discrete landmass, likely separated by water. By using this definition, places like Greenland should be classified as a continent (since, like Australia, it is a landmass surrounded by water), but it is not. North and South America are typically classified as two continents, even though they are joined. Often, experts classify Europe and Asia as two continents, even though they are joined together, with

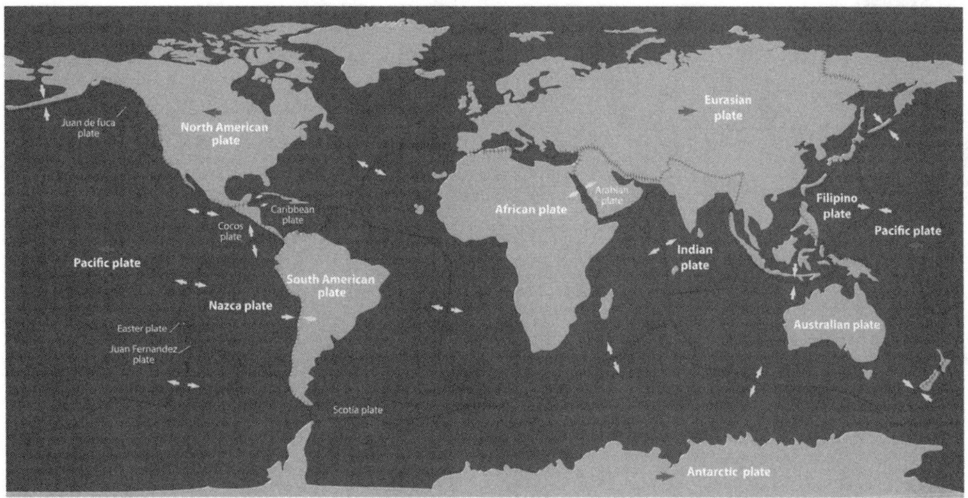

The crust of our planet is a network of several tectonic (or lithospheric) plates, ranging from just a few hundred miles across to thousands of miles, such as the Pacific and Antarctic plates.

no water between them. Depending on where you live on this planet, you may conclude that there are 6 or 7 continents. In most of the English- and Chinese-speaking world, there are seven continents (Asia, Africa, Europe, North America, South America, Australia, and Antarctica), while in Russia, Eastern Europe, and Japan, experts there consider Europe and Asia as one continent called Eurasia. And, in other parts of the world, experts consider North and South America as one landmass. So according to these two latter views of geography, we only have 6 continents, and not seven.

What is the largest continent?

If you were to combine the two landmasses of Europe and Asia into one, the largest continent is Eurasia at 21,100,000 square miles (54,649,000 square km). But even if you consider Europe and Asia to be two separate continents, Asia is still the largest, at 17,300,000 square miles (44,807,000 square km).

What is a subcontinent?

A subcontinent is a landmass that has its own continental shelf and its own continental plate. Currently, India and its neighbors form the only subcontinent, but in millions of years, Eastern Africa will break off from Africa and become its own subcontinent.

What was Pangea?

About 250 million years ago, during what is called the Paleozoic Period, all of the land on Earth was lumped together into one large continent known as Pangea. Faults and rifts broke the landmasses apart and pushed them away from each other. The continents slowly moved across the Earth to their present positions, and they continue to

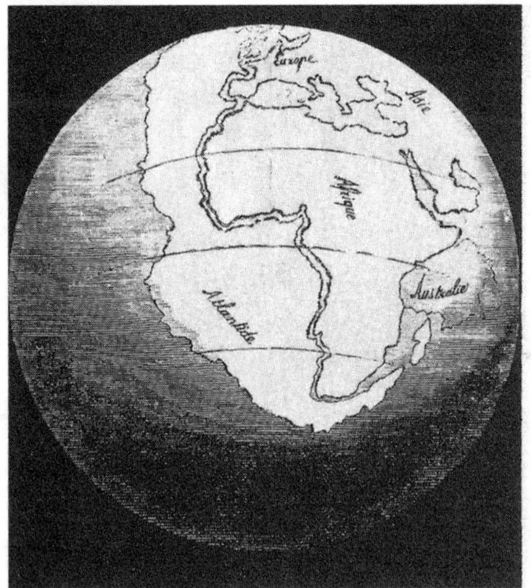

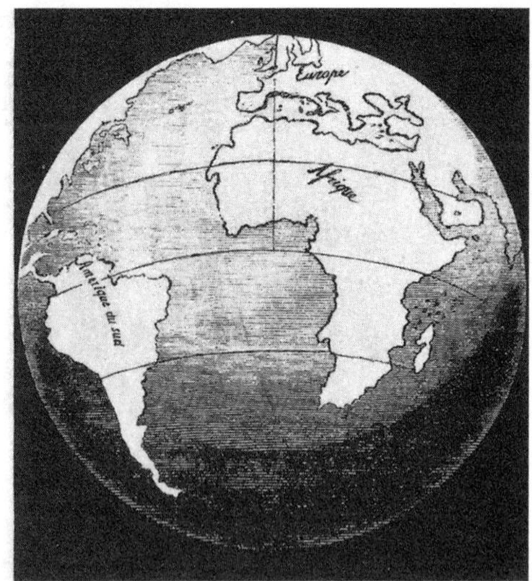

A rough drawing illustrating how continental drift caused the continents of Africa and South America to drift apart over millions of years.

move today. The Indian subcontinent (composed of India and its neighbors) continues to push into Asia and create the Himalayas. German scientist Alfred Wegener, who described his theory on continental drift, coined the term "Pangea" in 1912. More than ten years later, the word "Pangea" made its way into English scientific literature in 1926.

Where was Pangea located?

It was located near present-day Antarctica and has slowly drifted and split to form the continents as we know them today. The continents and their tectonic plates continue to move and will one day be in a much different arrangement than they are today.

How are mountains formed?

The process of orogeny, or mountain building, is related to continental drift. When two tectonic plates collide, they often form mountains. This is because the Earth's crust shortens and thickens, and the thicker crust gradually rises. Layers of rock far beneath the new mountain are superheated and become lighter, causing the rocks to gradually rise. The Himalayas are the result of the Indo-Australian Plate colliding with the Eurasian Plate. At these collision zones, volcanoes and earthquakes are common.

How did the Himalayas form?

About 30 to 50 million years ago, the landmass of India pressed into the landmass of Asia, pushing up land at the place of impact and creating the Himalayas. Even today, as the Indian subcontinent presses against Asia, the Himalayas continue to grow and change.

How much do the Himalayan Mountains grow?

Scientists estimate that the Himalayan Mountains grow .39 inches (1 cm) per year. In 1 million years, they will be approximately 6.2 miles (10 km) higher.

What is the highest point in the world?

At 29,029 feet (8,848 meters), the highest point above sea level in the world is Mt. Everest, which lies on the border of China and Nepal. This figure is based upon surveys of the height that were done in 1999 and 2005, without including the ice/snow cap to the figure. In one recent Chinese survey, the snow cap measured an additional 11 feet (3.35 meters).

Is Mt. Everest growing taller?

Because of the shifting plates underneath the surface of the Earth, Mt. Everest is actually changing its size by moving upward at an approximate rate of .16 inches (4.06 mm) and northeastward at a rate of .12–.24 inches (3.05–6.1 mm) per year.

What are the highest points on each continent?

In Asia, the highest peak is Mt. Everest at 29,029 feet (8,848 meters). The highest peak in South America, Aconcagua, lies in Argentina at 22,838 feet (6,961 meters). In North America, Alaska's Denali (also called Mt. McKinley from 1917 to 2015) is 20,322 feet (6,194 meters). Mt. Kilimanjaro at 19,341 feet (5,895 meters) is in Africa's Tanzania. Ice-covered Antarctica's high point is known as Vinson Massif, or Mt. Vinson, at 16,050 feet (4,892 meters). Europe's Mont Blanc is in the Alps between France and Italy at 15,771 feet (4,807 meters). A debate continues on whether or not to include in this list Mt. Elbrus, in the Caucasus Mountains, straddling the geographic border between what is Europe and Asia. At 18,510 feet (5642 meters), it is also considered by many to be the highest mountain in Europe. Australia's high point, Mt. Kosciuszko, is the lowest high point of all the continents at 7,310 feet (2,228 meters).

What else is interesting about Mt. Elbrus?

The mountain is actually a dormant volcanic cone and has two peaks, with the west peak approximately 69 feet (21 meters) higher than the east peak. Although the lower peak was first ascended in 1829, it was not until 1874 that the higher peak was scaled.

What is the highest mountain on Earth?

The highest mountain lies on the Big Island of Hawaii, with a height from the bottom of the sea floor, where it begins, rising 33,480 feet (10,205 meters) to the top of Mauna Kea, which is a volcano that rises 13,680 feet (4,170 meters) above sea level. The summit of Mt. Everest is actually considered to be the highest elevation above sea level.

What were the seven wonders of the ancient world?

While there was often disagreement by ancient and classical scholars as to which major works of art and architecture could be considered wonders, these seven were nearly al-

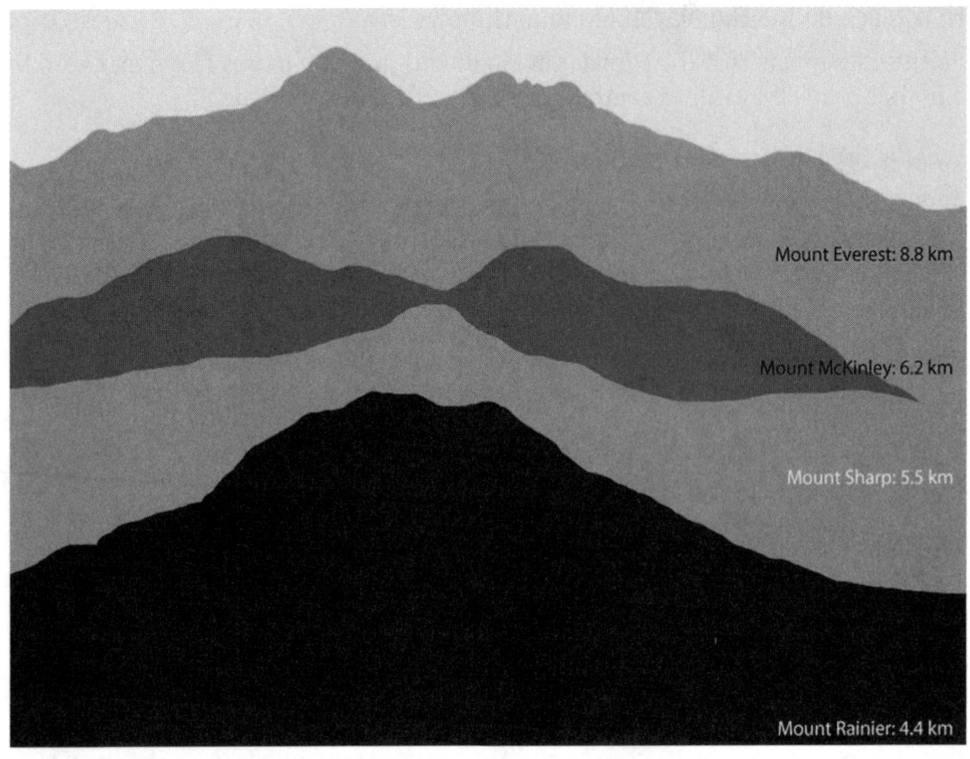

A comparison of Mt. Everest, Mt. McKinley, Mt. Sharp (on Mars), and Mt. Rainier.

ways on the list: the Pyramids of Egypt (the only remaining wonder), the Colossus of Rhodes (on the island of Rhodes in Greece), the Temple of Artemis at Ephesus (a marble temple in Turkey), the Mausoleum of Halicarnassus (Bodrum, Turkey), the Statue of Zeus at Olympia (an ivory and gold statue in southwestern Greece), the Hanging Gardens of Babylon (an enormous garden building, with plants of every kind, near Al Hillah, Iraq), and the Lighthouse of Alexandria (on the island of Pharos, near Alexandria, Egypt).

What are the seven wonders of the modern world?

According to the American Society of Civil Engineers, the seven wonders of the modern world include the Channel Tunnel between England and France; the CN Tower in Toronto, Ontario, Canada; the Empire State Building in New York City; the Golden Gate Bridge in San Francisco, California; the Itaipu Dam between Brazil and Paraguay; the Netherlands North Sea Protection Works; and the Panama Canal.

What are the seven natural wonders of the world?

These include the Aurora Borealis (northern lights), Mt. Everest (on the border of China and Nepal), Victoria Falls (near the border of Zambia and Zimbabwe in eastern Africa), the Grand Canyon (in Arizona in the United States), the Great Barrier Reef (in Australia), Parícutin (volcano in Mexico), and the harbor of Rio de Janeiro (in Brazil) with its stunning topography.

When did agricultural activity first begin?

Agricultural activity began about ten to twelve thousand years ago in a time period known as the first agricultural revolution. It was at this time that humans began to domesticate plants and animals for food. Before the agricultural revolution, people relied on hunting wild animals and gathering wild plants for nutrition. This revolution took place almost simultaneously in different areas of human settlement around the world.

HUMAN CIVILIZATION

Where did agriculture begin?

Agriculture simultaneously began in what is known now as the Middle East (Fertile Crescent), the Yangtze River Region of southern China, the Yellow River Region of northern China, Sub-Saharan Africa, South-Central Andes near modern-day Peru, Bolivia, and Chile, Central Mexico, and the eastern United States.

What is the difference between cultivation and domestication?

Cultivation is the deliberate attempt to sow and manage essentially wild plants and seed. Domestication is when people experiment and consciously select the right seeds to grow for various conditions.

When was the second agricultural revolution?

The second agricultural revolution occurred in the seventeenth century. During this time, production and distribution of agricultural products were improved through machinery, vehicles, and tools, which allowed more people to move away from the farm and into the cities. This mass migration from rural areas to urban areas coincided with the beginning of the Industrial Revolution.

What was the Industrial Revolution?

The Industrial Revolution began in the eighteenth century in England with the transformation from an agricultural-based economy to an industrial-based economy. It was a period of increased development in industry and mechanization that improved manufacturing and agricultural processes, thereby allowing more people to move to the cities. It included the development of the steam engine and the railroad.

How much of the world's population is devoted to agriculture?

In less-developed countries, such as Asia and Africa, a majority of the population is engaged in agricultural activity. In the more-developed countries of Western Europe and

What is the Green Revolution??

The Green Revolution began in the 1960s as an effort by international organizations (especially the United Nations) to help increase the agricultural production of less developed nations. Since that time, technology has helped improve crop output, which is reaching all-time highs throughout the world.

North America, less than one tenth of the population relies on agriculture for their livelihood.

How were animals first domesticated?

Dogs were probably some of the first animals to become domesticated. Wild dogs probably came close to human villages scavenging for food and were quickly trained as companions and protectors. Over time, early agriculturalists realized the value of domesticating other animals and proceeded to do so. Many different kinds of animals were domesticated in different areas of the world.

What are the El Castillo Cave Paintings?

The artwork that adorns the walls of this cave located in Cantabria, in the north of Spain, is currently thought to be the oldest works of human artwork found to date. The artwork has been dated to have been created around 39,000 B.C.E. and depicts three handprint stencils using a material called red ochre, created on a wall of the cave.

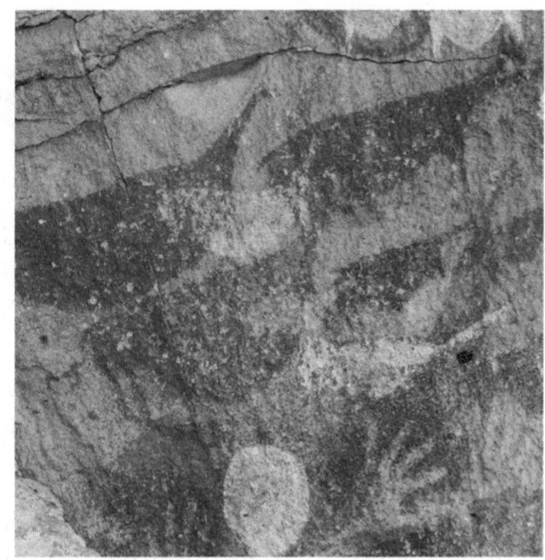

The Caves of Lascaux in France have clearly visible drawings of such things as human hands and animals, as well as star configurations.

What are the Caves of Lascaux?

There are many examples demonstrating that humanity has been interested in its surroundings since the beginning of recorded history. In fact, written on the walls of caves in Lascaux, France, are three dots indicating the brightest stars in the sky, which are estimated to have been drawn around 14000 B.C.E. Neolithic wall paintings found in Çatalhöyük, Turkey, show an early city plan from around 7500 B.C.E.

PEOPLE AND COUNTRIES

What is the largest country in the world, in terms of land?

Russia is by far the largest at about 6.592 million square miles (17.075 million square km) and occupies about one tenth of the Earth's habitable land. Russia is followed in size by Canada, China, the United States, Brazil, Australia, India, Argentina, Kazakhstan, and the Sudan.

Which country has the longest coastline?

The coastline of Canada and its associated islands is the longest in the world, about 164,663 miles (265,523 km) long. The United States has the second-longest coastline at about 82,836 miles (133,312 km).

Which countries have the fewest neighbors?

Approximately thirty-seven independent island nations (such as Australia, New Zealand, and Madagascar) have no neighbors. If you include territories, dependencies, and places managed by other countries, this number swells to seventy-five. Many nations, including Brunei, Cyprus, Haiti, the Dominican Republic, Papua New Guinea, Timor-Leste, Ireland, and the United Kingdom, are island nations that share an island.

Which non-island nations have the fewest neighbors?

There are ten non-island countries that share a land border with just one neighbor: Canada (neighboring the United States), Monaco (France), San Marino (Italy), Vatican City (Italy), Qatar (Saudi Arabia), Portugal (Spain), the Gambia (Senegal), Denmark (Germany), Lesotho (South Africa), and South Korea (North Korea).

Which countries are the most remote, in terms of how connected they are to the rest of the world?

According to the World Policy Institute, which publishes the *World Policy Journal*, several countries in the world are extremely remote in terms of how they are connected to the world. These countries include North Korea, Somalia, Myanmar, Democratic Re-

Which country has the most neighbors?

China is bordered by fourteen neighbors: Afghanistan, Bhutan, India, Kazakhstan, Kyrgyzstan, Laos, Mongolia, Myanmar, Nepal, North Korea, Pakistan, Russia, Tajikistan, and Vietnam.

public of the Congo, Niger, Timor-Leste, Madagascar, Burundi, Guinea-Bissau, and Papua New Guinea. These countries are very difficult places to visit.

Which countries in the world have the most neighbors?

There are quite a few countries in the world that share borders with other countries, including Austria (8), Brazil (10), China (14), France (8), Germany (9), Hungary (7), India (7), Iran (7), Mali (7), Niger (7), Poland (7), Russia (14), Saudi Arabia (7), Serbia (8), the Sudan (7), Tanzania (8), Turkey (8), Ukraine (8), and Zambia (8).

What percentage of the world's population can travel without obtaining a visa?

According to experts at the World Tourism Organization, 39% of the world's population does not need a visa in order to travel to their destination.

How does a city get chosen to host the Olympics?

The International Olympic Committee chooses a city as an Olympic site through a complex process lasting two years. Candidate cities (and their countries) must prepare answers to two rather detailed questionnaires, which describe how well the city is prepared for such criteria as environmental impact, climate, security, medical services, marketing, Olympic Village/housing, and transportation. Cities eagerly spend millions of dollars in construction and preparation for possible selection as a host city as an investment in the city's future.

What countries in the world do not use the metric system?

The metric system has been used for more than two hundred years. Yet, three countries still do not use the system. They are Liberia, Myanmar, and the United States.

How is a capital different from a capitol?

The capital is a city, and a capitol is a building. The capitol is located in the capital. To remember the difference, think about the "o" in the word capitol as being the dome of a capitol building. Capital cities are often the largest cities in a country or region.

What is geographic literacy?

In 1989, the National Geographic Society commissioned a survey to find out how much Americans and residents of several other countries knew about the world around them. The most recent study of 2006 assesses the geographic knowledge of young American adults between the ages of 18 and 24 and asks respondents how much they think they know about geography and other subjects, as well as their views on the importance of geographic, technological, and cultural knowledge in today's world. According to experts at the National Geographic Society, geographic literacy is achieved when someone understands the interactions and interconnections between people and our physical world and their implications.

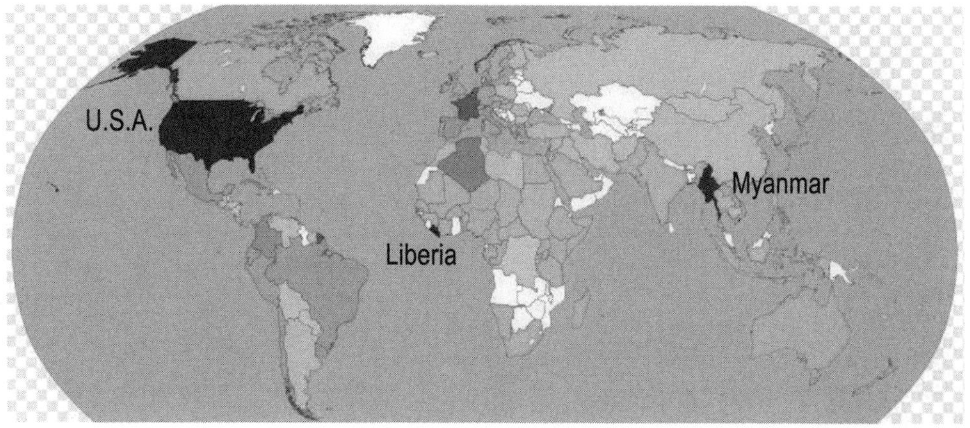

Almost all countries in the world use the metric system, with the exception of those marked in black on this map.

What were the results of the 2006 National Geographic–Roper Public Affairs Geographic Literacy Study?

The 2006 study is the latest in a series of surveys commissioned by the National Geographic Society, with the previous poll conducted in 2002. The countries that scored highest in the poll were Sweden, Italy, and France. The United States and Mexico scored the lowest.

The media subsequently reported the "geographic illiteracy" of the American population, the inability to use geographic understanding and reasoning in order to make decisions or draw conclusions. Due to the attention given to this problem, geographic education has since become a greater priority for educators.

As for the U.S. respondents, 60% cannot find the country of Iraq on a map of the Middle East, 40% do not know where Pakistan is located, and 33% could not find the state of Louisiana on a map. The study further found that less than 30% of all U.S. respondents said that it is absolutely necessary to know where countries in the news are located, and only 14% of U.S. respondents felt that fluency in another language is a necessary skill.

HISTORY AND INSTRUMENTS

How old is the oldest known map?

Researchers often point to early Mesopotamia and Egypt, around 5000 B.C.E. to 2000 B.C.E., when discussing the earliest of maps. Around 2700 B.C.E., the Sumerians drew sketch maps in clay tablets that represented their cities. These maps are some of the oldest-known maps. Debate continues whether or not examples of cave drawings, which appeared tens of thousands of years before, actually depict maps. Knowing and depicting where we are is as old as civilization itself.

What is the oldest map using the word America?

The map, acquired by the Library of Congress and dated 1507, incorporates many of the findings of the Italian explorer Amerigo Vespucci. The map was thought to be lost but was rediscovered in 1901 and was kept in a castle in Wolfegg, Germany, for more than 350 years.

What is the oldest known map drawn to scale?

In the sixth century B.C.E., the Greek geographer Anaximander created the first known map drawn to scale. His map was circular, included known parts of Europe and Asia, and placed Greece at its center.

What makes a piece of paper a map?

No matter what the medium, all maps must be a representation of an area of the Earth, celestial bodies, or space. Though maps are commonly printed on paper, they can come in a variety of forms, from being drawn in the sand to being viewed on handheld devices like phones/tablets and computers. A map should have a legend (a guide explaining the map's symbols), a notation of which way is north, and an indicator of scale. No map is perfect and every map is unique.

What is a good way to learn where places are?

The best way to learn about places that you have heard of is to look them up in an updated atlas, a collection of maps bound in a book. An atlas may include additional information, such as illustrations, statistical tables, topography, and other important information about a place or places. If you would rather use the Internet, you may try to enter the place in the search box at your favorite search engine, and then click on the results. You may also use online sites and apps for your mobile device, such as "Google Maps," Apple's "Maps," MapQuest, Rand McNally, or Yahoo Maps. You will be surprised by the amount of information you can find!

Why is a book of maps called an atlas?

The term "atlas" comes from the name of a mythological Greek figure, Atlas. As punishment for fighting with the Titans against the gods, Atlas was forced to hold up the planet Earth and the heavens on his shoulders. Because Atlas was often pictured on ancient books of maps, these became known as atlases.

What is interesting about Google's Street View?

Google's Street View is interesting because it allows users to begin their journey by seeing images of the Earth and allows users to drill down to a geographic location at the view of the place on the street. Google even has Street View images under the ocean at the Great Barrier Reef, off the northeast coast of Australia, and at many other ocean sites around the world. Google has transformed maps that we may use every day, by offering satellite imagery, road maps, and street-level images for use on our smartphones, mobile devices, and computers of all types.

What do cartographers do for a living?

Cartographers are often employed in publishing, government, military, land surveying, and conservation sectors. They contribute to the scientific, technological, and artistic components to the making of maps. Additionally, cartographers frequently provide complex information, using diagrams, charts, and spreadsheets.

How do cartographers shape our world?

Cartographers are mapmakers and cartography is the art of mapmaking. Cartographers map neighborhoods, cities, states, countries, the world, and even other planets. There are as many types of maps to make as there are cartographers to make them.

What country would purposely create false maps?

Within the former Soviet Union, incorrect maps were produced as a matter of course. Soviet maps purposely showed the locations of towns, rivers, and roads in incorrect places. Often, in different editions of the same map, towns would disappear from one version to the next. Street maps of Moscow were particularly incorrect and nonproportional. This cartographic mischief of the former U.S.S.R. was an effort to keep the geography of the country a secret, not only from foreigners but also from its own citizens. Even official government agencies were not allowed to have accurate maps.

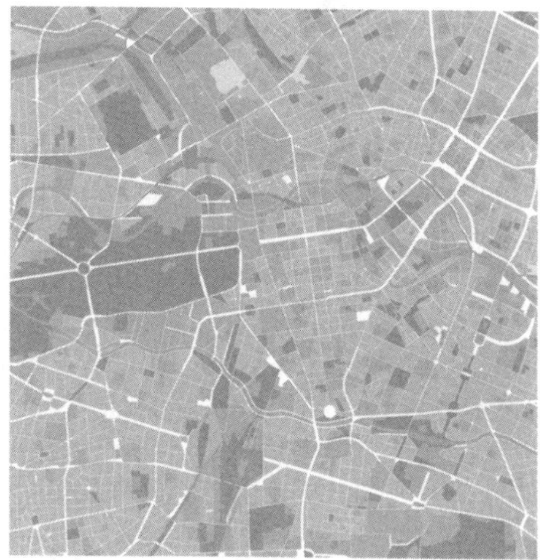

Cartographers make all kinds of maps, such as maps of the world, countries, states, or cities, such as this street map of Berlin.

Who decides which names go on maps?

In the United States, the U.S. Board on Geographic Names (BGN, comprising of representatives of many federal agencies) approves the official names and spellings of cities, rivers, lakes, and even foreign countries. If a town would like to change its name, it must petition the BGN for approval. Upon approval, the name is officially changed and updated in federal government gazetteers and records, which official and commercial mapmakers use to make their maps.

How can maps be used to start wars?

Throughout history, many countries have used maps as a means to justify aggression to a neighboring country by asserting ownership of some land, based upon their version of a map. For example, prior to its 1990 invasion of Kuwait, Iraq produced official maps that showed the independent country of Kuwait as Iraq's nineteenth province. Iraq used these maps as justification for its 1990 invasion and attempted annexation of Kuwait

How do I get to the refrigerator in the dark?

Not all maps are written on paper. When trying to reach the refrigerator at night, we do not smell our way to food, we use a map based upon our memory of the room. If we stumble on our way, it is usually over a misplaced toy or shoe that we did not remember leaving there. Everyone has this mental imagery in his or her mind. These mental maps help you find your way not only to the refrigerator in the dark but also to the grocery store, to work, and to every place that we go. People not only have mental maps of common trips they make but also of their city, country, and even the world. Every person's mental map is unique, based upon how wide an area that person travels and his or her knowledge of the world.

(Iraq was interested in Kuwait's oil reserves). Maps have been, and still are, used by a multitude of countries, provinces, and cities to prove or disprove the ownership of a certain piece of land.

What other countries have disputes over lines drawn on maps?

Japan, China, and Taiwan all have a dispute over the Diaoyu/Senkaku islands in the Pacific Ocean. India and Pakistan have an ongoing dispute over ownership of the northernmost region of Kashmir in India. China has disputed claims of sovereignty by Taiwan. North and South Korea are still in conflict over lines drawn by Korean and American forces that split the country decades ago.

How does a sextant help navigators?

In 1730, the sextant was invented independently by two men, John Hadley and Thomas Godfrey. Using a telescope, two mirrors, the horizon, and the sun (or another celestial body), the sextant measures the angle between the horizon and the celestial body. With this measurement, navigators could determine their latitude while at sea.

When was the compass invented?

Ancient south-pointing compasses have been found in China, dating back to as early as 200 B.C.E., during the Han Dynasty period. These early, metal compasses pointed south and helped Chinese mariners stay on course at sea. It was not until 1600 C.E. that the use and innovation of the compass had migrated to Europe. As early as the eleventh century, the Chinese were using a magnetic needle to determine direction and to foretell events. At approximately the same time, the Vikings may have also used a similar device. A compass is simply a metallic needle that points toward the Magnetic North Pole.

Have compasses always pointed north?

No, they have not. Though compasses always point to the magnetic pole, the magnetic pole has not always been in the north, as it always moves. Every 300,000 to 1 million years, the magnetic pole flips from north to south or from south to north. If compasses had been around before the last time the magnetic pole reversed, their arrows would have pointed south rather than north.

What is true north?

True north is the direction that one can map along the surface of the Earth to the geographic North Pole.

What is Magnetic North?

Magnetic North refers to the direction that corresponds to the magnetic field lines to which compasses point. The core of the Earth acts like a giant spherical magnet, with two poles: one North and one South. The strength of the magnetic field changes with time and location. The main field that lies far below the surface of the Earth generates approximately 90% of the total magnetic field. The other 10% of Earth's magnetic field is generated in our upper atmosphere and magnetosphere, where ions, electrons, and other particles help protect the planet from dangerous solar emissions.

What is a compass rose?

On old maps, the directions of the compass were represented by an elaborate symbol, known as a compass rose. Many of the older compass roses displayed 32 points, representing not only the four cardinal directions (north, south, east, and west) but also 28 subdivisions of the circle (southwest, south-southwest, etc.). This directional symbol resembled a rose, hence its name. Though compasses are now often drawn with only the four cardinal directions and the resemblance to the flower is minimal, the directional symbol is still called a compass rose.

What is an azimuth?

Azimuth is another method for stating compass direction. It is based on the compass as 360 degrees, with north at 0 degrees, east at 90, south at 180, and west at 270 degrees. You can refer to a direction as "head 90 degrees" instead of "head east."

How and why does Magnetic North move?

Scientists aren't sure why the Earth's magnetic pole moves, only that it does. The European Space Agency's SWARM mission launched three satellites in November 2013 to analyze the Earth's magnetic fields, contributing significant data for global researchers.

How much has the Magnetic North Pole moved?

Since it was documented in 1831 by James Ross to be slightly north of 70 degrees north latitude, it has moved to north of 80 degrees north latitude at an average rate of more than 24 miles (40 km) per year. The amount of movement from decade to decade varies, but it's never more than a few miles each year. During the twentieth century, the pole moved approximately 6.2 miles (10 km) per year. At the beginning

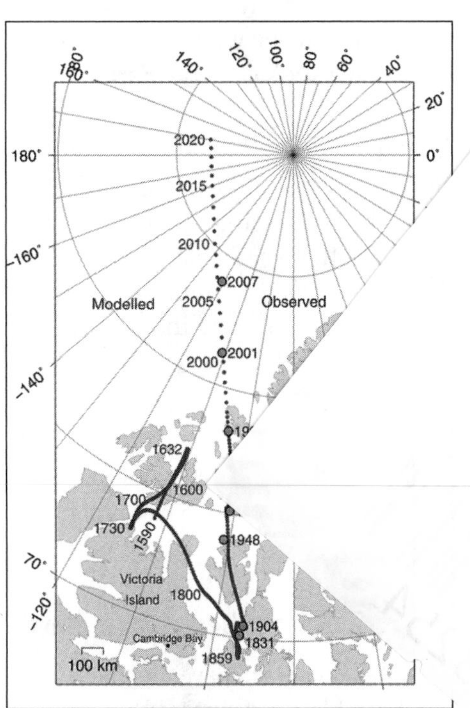

Unlike the geographic North Pole, the Magnetic North Pole does not stay in the same place but has shifted hundreds of miles over the last centuries.

of our current century, researchers have measured the movement to be as many as 24.85 (40 km) miles per year.

LATITUDE AND LONGITUDE

Where is the equator?

The equator is the line located equidistant between the North and South Poles. The equator evenly divides the Earth into the Northern and Southern Hemispheres and is zero degrees latitude.

What are lines of latitude and longitude?

Lines of latitude and longitude make up a grid system that was developed to help determine the location of points on the Earth. These lines run both north and south and east and west across the planet. Lines of latitude (those that run east and west) begin at the equator, which is zero degrees. They extend to the North Pole and the South Pole, which are 90 degrees north and 90 degrees south, respectively. Lines of longitude (those that run north and south) begin at the Prime Meridian, which is the imaginary line that runs through the Royal Observatory in Greenwich, England. The lines of longitude extend both east and west from the Prime Meridian, which is zero degrees, and converge on the opposite side of the Earth at 180 degrees.

Are lines of longitude and latitude all the same length?

No, they are not. Only the lines of longitude are of equal length. Each line of longitude equals half of the circumference of the Earth because each extends from the North Pole to the South Pole. The lines of latitude are not all equal in length. Since they are each complete circles that remain equidistant from each other, the lines of latitude vary in size from the longest at the equator to the smallest, which are just single points, at the North and South Poles.

How wide is a degree of longitude?

Though there are only a couple dozen lines of longitude shown on most globes and world maps, the Earth is actually divided into 360 lines of longitude. The distance be-

Do other planets have longitude and latitude lines?

Yes, scientists have divided the other planets and their moons into longitude and latitude systems like the Earth. They use these lines just as they do on Earth: to pinpoint exact locations on the planet or moon.

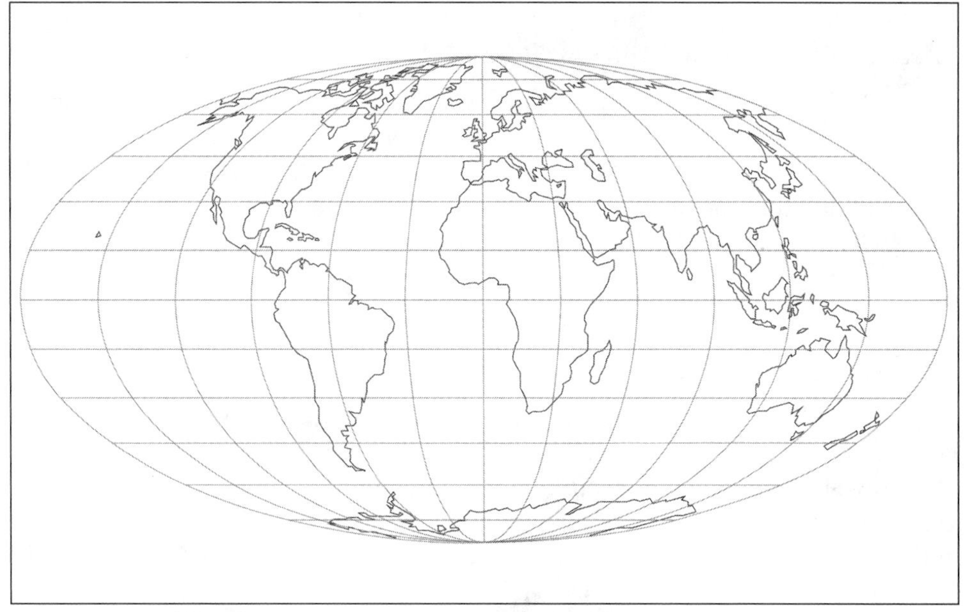

Lines of latitude go from side to side (east and west) on this map, while lines of longitude run top to bottom (north and south).

tween each line of longitude is called a degree. Because the lines of longitude are widest at the equator and converge at the poles, the width of a degree varies from 69 miles (111 km) wide to zero, respectively.

How wide is a degree of latitude?

Though there are only about a dozen lines of latitude shown on most globes and world maps, the Earth is actually divided into 180 lines of latitude. The distance between each line of latitude is called a degree. Each degree is an equal distance apart, at 69 miles (111 km).

What do minutes and seconds have to do with longitude and latitude?

Each degree of longitude and latitude is divided into 60 minutes. Each minute is divided into 60 seconds. An absolute location is written using degrees (°), minutes ('), and seconds (") of both longitude and latitude. Thus, the Statue of Liberty is located at 40°41'22" North, 74°2'40" West.

Which comes first, latitude or longitude?

Latitude is written before longitude. Latitude is written with a number, followed by either "north" or "south" depending on whether it is located north or south of the equator. Longitude is written with a number, followed by either "east" or "west" depending on whether it is located east or west of the Prime Meridian.

Why was the Prime Meridian established at Greenwich?

In 1675 the Royal Observatory in Greenwich, England, was established to study determination of longitude. In 1884, an international conference established the Prime Meridian as the longitudinal line that passes through the Royal Observatory. The United Kingdom and United States had been using Greenwich as the Prime Meridian for several decades before the conference.

Why was computing longitude so difficult?

It wasn't until the sixteenth century that clocks were fabricated in such a way that they could accurately tell time both on land and at sea. The only way of determining how far east or west one could go is by plotting the stars in two locations and recording the exact time in both locations simultaneously, and then recording the time and position at the destination. As clocks became more accurate, the ability to measure speed and distance became possible.

How can I remember which way latitude and longitude run?

You can remember that the lines of latitude run east and west by thinking of lines of latitude as rungs on a ladder ("ladder-tude"). Lines of longitude are quite "long" because they run from the North Pole to the South Pole.

How can a gazetteer help me find latitude and longitude?

A gazetteer is an index that lists the latitude and longitude of places within a specific region or across the entire world. Many atlases include a gazetteer, and some are published separately.

How can I find the latitude and longitude of a particular place?

To find latitude and longitude of a particular location, you will need to consult either a gazetteer, database, or website that includes longitude and latitude data. Though gazetteers are readily accessible, they don't include as many places as online databases. There are a number of sites on the Internet that have extensive databases of latitude and longitude and even include such specific places as public buildings.

READING AND USING MAPS

What is the difference between a physical map and a political map?

A physical map shows natural features of the land such as mountains, rivers, lakes, streams, and deserts. A political map shows human-made features and boundaries such as cities, highways, and countries. The maps we use in atlases and see on the walls of classrooms are typically a combination of the two.

What is a topographic map?

A topographic map shows human and physical features of the Earth and can be distinguished from other maps by its great detail and by its contour lines indicating elevation. Topographic maps are excellent sources of detailed information about a very small area of the Earth. The United States Geological Survey (USGS) produces a set of topographic maps for the United States that are at a scale of 1:24,000 (one inch equaling 2,000 feet [or 1 centimeter equaling about 240 km]). You can purchase these maps online, at sporting goods stores, or through the USGS itself.

Why are road maps so difficult to fold?

The problem lies with the multitude of folds required to return the map to its original, folded shape. The easiest way to fold a road map is to study the creases and to fold the map in the order that the creases will allow. But once you've made a mistake, the folds have lost their telltale instructions. To fold a road map, begin by folding it accordion style, making sure that the "front" and "back" of the folded design appear on top. Then, once the entire map is folded accordion style, fold the remaining slim, long, folded paper into three sections. And, voilà, your road map is folded!

Why is color important on a relief map?

A relief map portrays various elevations in different colors. But, a common color scheme found on relief maps causes a problem. On these maps, mountains are displayed as red or brown, while lowlands are shown in shades of green. This is confusing because the green areas on the map are often misconstrued as fertile land, while brown areas are mistaken for deserts. For example, an area such as California's Death Valley, which is shown in green on relief maps because it lies below sea level, seems fertile, when actually it is an inhospitable desert.

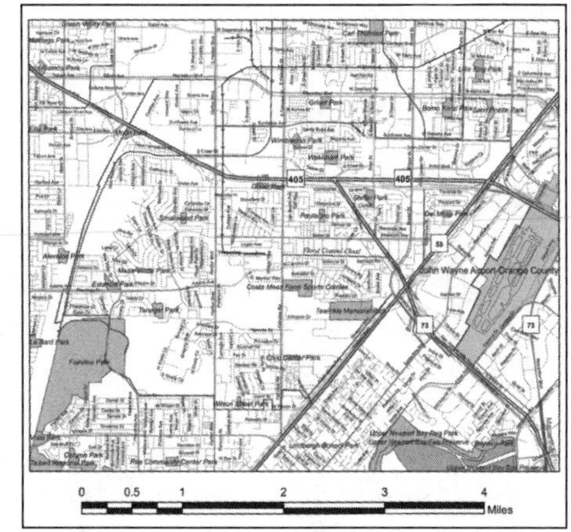

A map scale will show you how a distance on a map translates to mileage in the real world.

What does the scale of a map tell me?

A scale indicates the level of detail and defines the distances between objects on a map. On a map, scales can be written as a fraction, a verbal description, or as a bar scale. A fraction, or ratio, using the example of 1/100,000 or 1:100,000, indicates that one unit of any form of measurement on the map is equivalent to 100,000 units of the same measurement in the area being represented. For instance, if you use inches as the unit of measurement, then one inch on the map would equal 100,000 inches in the area represented by the map.

A verbal description describes the relationship as if it were a verbal instruction, such as “one inch equals one mile.” This allows the versatility of having different units of measurement.

What is a bar scale?

A bar scale uses a graphic to show the relationship between distance on the map to distance in the area represented. The bar scale is the only type of scale that allows a reduction or enlargement of the map without distorting the scale. This is because when you increase the size of the map, the bar scale is increased proportionally. For a fraction or verbal scale, the proportion (1:1,000) is only true for the map at that size. For example, when enlarging a map, the map might become twice as large, but the numbers in a ratio do not change as they would need to in order to stay accurate.

How can I determine the distance between two places by using a scale?

By using a ruler, compare the distance between two points on a map with the information on the scale to calculate the actual distances between the two points. For example, if you measure the distance between two towns as being five inches and the ratio says 1:100,000, then the actual distance between the towns is 500,000 inches or 7.9 miles (12.7 km).

What is the difference between small- and large-scale maps?

A small-scale map shows a small amount of detail over a wide area, such as the world. A large-scale map shows a large amount of detail while representing a limited area, such as neighborhoods or towns.

Why is every map distorted?

No map is completely accurate because it is impossible to accurately represent the curved surface of the Earth on a flat piece of paper. A map of a small area usually has less

How is the Earth’s surface like an orange peel?

Any attempt to represent a sphere, like the Earth, in a flat representation results in distortions. The surface of the Earth is like the peel of an orange. When we try to take the “peel” of the surface of the Earth and lay that onto a flat piece of paper like a map, large, open areas are created. Mapmakers correct for these distortions, so there are as few distortions as possible. The strategies for correcting for these problems are referred to as “projections.”

distortion because there is only a slight curve of the Earth to contend with. A map of a large area, such as maps of continents or the world, are significantly distorted because the curvature of the Earth over such a large area is extreme. This means that many places around the world may appear smaller or larger than they actually are, depending on where on the map the place is located.

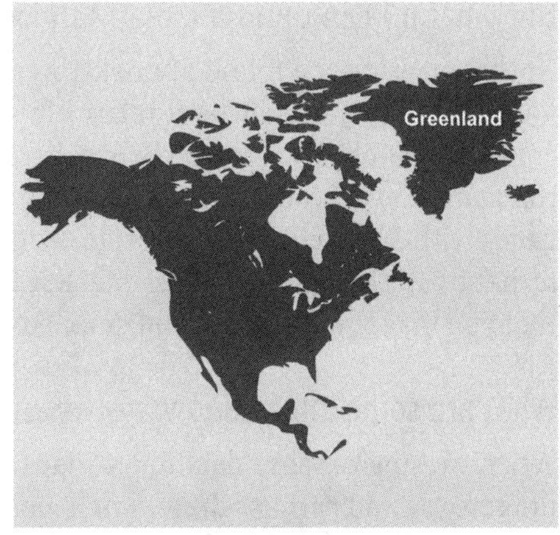

In many types of projection maps, Greenland appears much larger than it actually is because of distortion caused by keeping the lines of latitude and longitude straight.

Why does Greenland appear larger on most maps than it actually is?

Because of the distortions that must appear on all maps, many maps place the distortion in the northern and southern extremes of the Earth. In one of the common projections, known as Mercator, Greenland appears to be similar in size to South America, despite the fact that South America is actually eight times larger than Greenland. The advantage of the Mercator projection is that the lines of latitude and longitude remain perpendicular; thus, the map is useful for navigation.

How can a legend help me read a map?

The legend, usually found in a box on the map, is information that explains the symbols used on a map. Though some symbols seem standard, like a railroad line, even those can be represented differently on different maps. Since there really are no standard symbols, each map's legend should be consulted when reading a map.

Why is there often a cross next to the east direction on maps?

On old maps, a cross often sits next to the east direction on a compass rose. This cross represents the direction to Paradise and the Holy Land.

Where can I buy maps?

There are many places you can buy maps. Most brick-and-mortar and online bookstores offer an extensive collection of local and foreign travel maps, wall maps, and atlases. Also, many cities have specialty travel and map stores that offer a larger and more varied collection of maps, as well as maps of more exotic locales. Maps are also available to view or download at many websites. Just put in the name of the place, region, city, or country that you wish to find, and you will discover both maps that are either free or available for purchase. Look for the most up-to-date maps when selecting a map from more than one available vendor.

But what if I can't find the map I'm looking for?

Not all maps can be found at bookstores or even in specialty stores. If you are looking for an extremely specific and relatively uncommon map, visit a local university's map collection to obtain a copy. Their collections are often far greater in size and breadth than any store. If you need help locating a map, you should be able to discuss your map needs with a friendly map librarian. You can also try constructing a search on the Internet that is as specific as possible. Use advanced search options to filter out results so that you may find a precise and detailed map.

What are some interesting ways we can use maps?

When we superimpose data upon maps of all kinds, we are able to visualize data in a unique way and perhaps draw conclusions that we cannot see just by looking at tables, graphs, and raw data in columns. According to experts at the *Washington Post* in a blog post called "40 Maps That Explain the World," some interesting maps include: the world's major writing systems, the best/worst places to live, maps of world religions, language, psychological comparisons of people, early maps showing how countries were formed over time, racial tolerance/intolerance, and ethnic diversity.

What is the difference between relative and absolute location?

There are two different ways to describe where a place is located: relative location and absolute location. Relative location is a description of location using the relation of one place to another. For instance, using relative location to describe where the local grocery store is, you might say that it's on Main Street, just past the high school. Absolute location describes the location of a place by using grid coordinates, most commonly latitude and longitude. For instance, the local grocery store would be described as being located at 23°23'57" North and 118°55'2" West.

MODERN MAPPING

Besides supporting telecommunication, how do satellites help us to understand our world?

Satellites capture images of the Earth's weather patterns, the growth of cities, the health of plants, and even individual buildings and roads. Satellites circle the Earth, or remain geostationary (in the same place with respect to the Earth), and send data back to the Earth via radio signals.

How have satellites changed mapmaking?

Satellite images, which are accurate photographs of the Earth's surface, allow cartographers to precisely determine the location of roads, cities, rivers, and other features on

Satellites have made much more detailed maps possible and are also actually easy to produce.

the Earth. These images help cartographers create maps that are more accurate than ever before. Since the Earth is a dynamic and ever-changing place, satellite images are great tools that allow cartographers to stay up-to-date.

How old is Google Maps?

Google Maps first started in 2004, when Google acquired an Australian mapping-related company, Where 2 Technologies, which began the year before. Originally conceived as a downloadable piece of software, Google management transformed the software to purely Web based and began releasing developer versions (The Maps API) of the software in 2005 to further spawn development and use of maps to more sites and users. Google acquired several other firms in order to realize its dream of bringing maps of all kinds to all people. In 2008, Street View was introduced to give users a digital, street-level view of places of interest in all parts of the world, and it continues today.

How has GIS revolutionized cartography?

Geographic Information Systems (GIS) began in the 1960s with the popularity of computers. Though very simplistic in its beginning, new technology and inventions have expanded and enhanced the functions of GIS. GIS has revolutionized cartography by using computers to store, analyze, and retrieve geographic data, thus allowing infinite numbers of comparisons to be made quickly. The program formulates information into various "layers," such as the location of utility lines, sewers, property boundaries, and streets. These layers can be placed together in a multitude of combinations to create a plethora of maps, unique and suitable to each individual query. The versatility of GIS makes it indispensable to local governments and public agencies.

How can GIS help my town?

Your community can use GIS on a day-to-day basis and in emergency situations. GIS allows public works departments, planning offices, and parks departments to monitor the status of the community's utilities, roads, and properties. In an emergency, GIS can give emergency teams the information they need to evacuate endangered areas and respond to the crisis.

What is GPS?

GPS, or the Global Positioning System, is a network of satellites in orbit around the Earth that may give us a precise location of either ourselves or a point of interest. It began as early as 1973 with military defense in mind but was later expanded for general use in the early 1990s.

How does GPS work?

We often take GPS for granted because it is ubiquitous, found on phones, mobile devices, stand-alone units, and embedded in the navigation systems of our cars and many other products. It works by analyzing how long it takes for a time signal to travel from one of the three to four satellites to your receiver. This process is called trilateration. When satellites are above the horizon, the results are much more accurate.

How many GPS satellites are there?

Although we need only 24 functioning satellites in order to provide an accurate measurement of a location, there are currently 69 satellites orbiting the Earth. Some are not functional and had to be replaced by a working unit, while others are on reserve in case they are needed.

How does a GPS unit know where I am?

Individual Global Positioning System (GPS) units on the Earth receive information from a U.S. military-run system of 24 satellites that circle the Earth and provide precise time and location data. The individual GPS unit receives data from three or more satellites that triangulate its absolute location on the Earth's surface. If you are carrying such a device, your absolute location is the same as that of the device.

How can GPS keep me from getting lost?

A GPS unit provides precise latitude and longitude for the location of the device. By using a handheld GPS unit or a GPS-enabled device, along with a map that provides latitude and longitude (such as a topographic map), you can determine your precise location on the Earth's surface. This is a valuable tool for those who hike or travel in remote regions and for ships at sea. GPS is now widely available in cars; as stand-alone, portable, pocket-sized devices; on phones; and even on the boxes that ship products that you buy. In short, GPS is used in all aspects of our lives.

THE EARTH'S MATERIALS AND INTERNAL PROCESSES

What are the three main layers of the interior of the Earth?

Scientists do not know very much about what is below the surface of our planet. But we do know that it is divided into three layers: crust, mantle (upper and lower), and core (outer and inner).

How thick is the Earth's crust?

The thickness of the Earth's crust varies at different points around the planet. Under continents, the crust is approximately 27–31 miles (45–50 km) thick, but under the oceans, it is a mere 6.2 miles (10 km) thick, usually surrounded by an 18-mile- (30-km-) thick contour.

What did Alfred Wegener discover?

Although Wegener was not the first to discover that the continents seem to appear to fit together like pieces of a jigsaw puzzle, he was the first to create a theory that the continents seemed to have drifted apart, in 1912. His original paper was entitled "The Formation of the Major Features of the Earth's Crust (Continents and Oceans)." It was not until long after his death in 1930 that modern technology and scientific analyses proved that he was correct.

What is continental drift?

The Earth is divided into massive pieces of crust that are called tectonic plates. These plates lie wedged together like a puzzle. The plates slowly move, crashing into each

This contour map indicates the differing levels of thickness in the Earth's crust as marked in ten-kilometer (six-mile) intervals.

other to form mountain ranges, volcanoes, and earthquakes. The plates are like rafts floating on water; this is called continental drift. By 1968, the theory of continental drift was gradually replaced by the theory of plate tectonics.

How do we know that the continents were once joined together?

There is extensive evidence, dating back many millions, perhaps billions, of years, that the continents were joined together. Fossil evidence of plant and animal life that are similar have been found on different continents, as well as similar glacial sediment.

How many tectonic plates are there?

There are seven significant plates on the planet. Some of the largest include the Eurasian Plate, North American Plate, South American Plate, African Plate, Indo-Australian Plate, Pacific Plate, and Antarctic Plate. Some smaller plates are located between the major plates. The smaller plates include the Arabian Plate (containing the Arabian Peninsula), the Nazca Plate (located to the west of South America), the Philippine Sea Plate (lo-

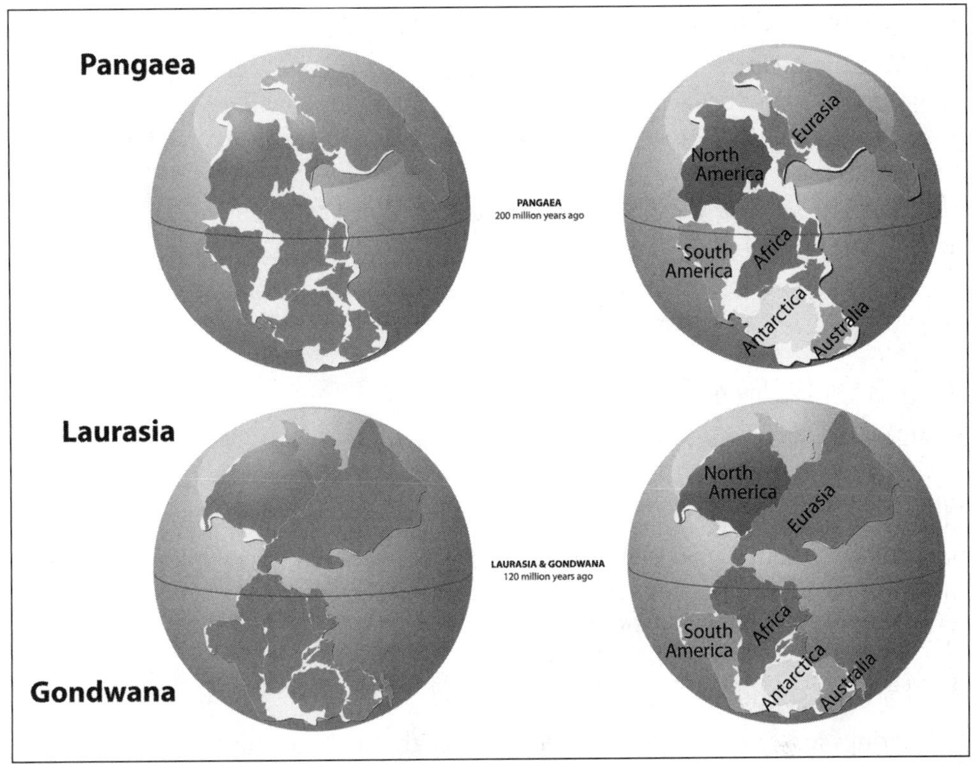

Pangea was a supercontinent that existed about 200 million years ago. It gradually separated into two large continents—Laurasia and Gondwana—about 120 million years ago, and then those continents divided further over time.

cated southeast of Japan, containing the northern Philippine islands), the Indian Plate (containing Southeast Asia, India, and South China to Eastern Indonesia), the Cocos Plate (located southwest of Central America), the Juan de Fuca Plate (just off the coast of Oregon, Washington, and Northern California), and the Caribbean Plate (beneath Central America and the Caribbean Sea).

What type of rocks are formed by lava?

Igneous rocks are formed when liquid magma under the surface of the Earth, or lava on the surface of the Earth, cools and hardens into rock.

What type of rocks are formed from particles?

Sedimentary rocks are formed by the accumulation and squeezing together of layers of sediment (particles of rock or remains of plant and animal life) at the bottom of rivers, lakes, and oceans or even on land. The continual accumulation of more and more layers of sediment places a great amount of pressure on the lowest layers of sediment and, over time, compresses them into rock.

What are recycled rocks?

Metamorphic rocks are rocks that had a prior existence as sedimentary, igneous, or other metamorphic rocks. Underground heat and pressure metamorphose or change one type of rock into another, creating a metamorphic, or recycled, rock.

What type of sand do we often see on beaches near volcanoes?

Because the content of the sediment near volcanoes is comprised of dark black igneous rock, the resulting beach sand in such places as Hawaii and Indonesia is actually a dark brown or even black in color.

Which is larger, clay or sand?

A single grain of sand is 1,500 times larger than a grain of clay.

What is a dike?

A dike is magma that has risen up through a crack between layers of rock. When this magma solidifies, it becomes very solid rock. If the rock around it is eroded, a dike can form great rock monoliths above the ground.

What are hot springs?

Hot springs are created by underground water that is heated and percolates to the Earth's surface. Aside from being natural baths, the steam from hot springs can be used to drive turbines, which create electricity. This type of energy production is called geothermal energy.

Sinkholes have become a growing problem in the United States, destroying roads and sometimes even swallowing houses.

Where are hot springs used by people around the world?

Hot spring baths have been used in cultures throughout the world since ancient times. They are found in such places as Japan, Taiwan, Australia, the United States, Iceland, and Sicily.

Why does the ground sink?

In many places around the world, seemingly solid land lies over vast oil deposits or water aquifers. Without the liquid supporting it, the ground sinks into the space left behind. In some parts of California's Imperial Valley, the land has dropped more than 25 feet (7.6 meters) due to underground water being removed from the area. Unless the pumping of underground water and oil is stopped, the land will continue to sink.

Why do houses fall into sinkholes?

Houses that sit upon limestone rock have the proclivity to fall into sinkholes. As underground water wears away the limestone rock, it creates underground caverns. If the water wears away too much limestone, the cavern may collapse, taking anything on the surface with it. A sinkhole is just one of the many reasons to have your home inspected by a geologist.

NATURAL RESOURCES

What is a renewable resource?

A renewable resource is one that can be replenished within a generation. Forests, as long as they are replanted, are a renewable resource. Materials such as oil, coal, and natural gas are known as nonrenewable resources because they require millions of years to be created. So, once the world's supply of oil is gone, it's gone for a long, long time.

When will we run out of oil on Earth?

Scientists estimate that there are approximately 1.3 trillion barrels of oil on reserve (not yet pumped) in the world's oil fields, deep beneath the Earth. Debate continues whether or not we have reached the point of maximum production of this oil, which is called

What are perpetual resources?

Perpetual resources are natural resources, such as solar energy, wind, and tidal energy, that have no chance of being used in excess of their availability. They can be used for power generation and conversion to electric energy indefinitely.

"peak oil." Based upon today's rate of demand and consumption, many experts believe that we have approximately 40 years before the world's oil supplies will be nearing depletion. Many other scientists believe that new technologies and findings will enable the world to extend this period past this date. By 2040, many forecast that the production of oil will be 20% of what it is today. Because of technology, like horizontal drilling and fracking, used to extract oil from relatively inaccessible places, we may delay the time when the world runs out of oil.

What are fossil fuels?

Underground fuels such as natural gas, oil, and coal are all known as fossil fuels because they are encased in rocks, just like fossils. It takes millions of years of compression and the building up of dead plants, animals, and organic material to create these fuels.

What is a fossil?

The outline of the remains of a plant or animal embedded in rock is called a fossil. Fossils are formed when a plant or animal dies and becomes covered up by sediments. Over time, the layers compress the remains, which are then embedded into the rock.

LANDSCAPES AND ECOSYSTEMS

What are basins and ranges?

Basins and ranges are sets of valleys and mountains that are spaced close together. Most of Nevada and the western part of Utah are composed of sets of basins and ranges.

What is permafrost?

Permafrost describes ground (including soil, rock, organic material, and ice) that is permanently at or below 32 degrees Fahrenheit (0 degrees Celsius) for a minimum of two consecutive years. Its thickness depends on location but can extend from 9.94 to 4,921 feet (3 meters to 1,500 meters) beneath the surface. Permafrost occurs in higher latitudinal regions, which have cold climates.

Is the permafrost thawing?

Yes, the permafrost is thawing. In places with great areas of permafrost, such as Alaska, the permafrost has warmed to the highest level in 10,000 years. During the last 50 years, the Arctic regions have been warming to record-high temperatures. During this time, Alaska's average temperature has warmed an average of 3.3 degrees above normal.

Why is permafrost so important?

Aside from the mediating effects on the Earth's temperature, permafrost provides a cover for the habitat that supports wild flora and fauna, which provide sustenance for the entire ecosystem of the Arctic regions, including the many indigenous people who live there. Permafrost also covers organic rich sediment that contains methane hydrates, which may be released into the air as "greenhouse gases." As the permafrost thaws, these greenhouse gases are released into the atmosphere and may contribute to global climate change. Permafrost makes up 24% of all land in the Northern Hemisphere.

Evidence is piling up that greenhouse gases from human industry are contributing to global warming of the planet.

What are greenhouse gases?

Greenhouse gases are chemical compounds that absorb and emit thermal radiation, are present in our atmosphere, and allow direct sunlight to reach the Earth's surface. But as the heat bounces back from the Earth to the atmosphere, the gases absorb this energy, and less heat from the surface of the Earth is able to escape back into the atmosphere and is trapped in the lower atmosphere. Many greenhouse gases are found in nature, like carbon dioxide, water vapor, methane, nitrous oxide, and ozone, and are regulated by a process called the carbon cycle. The trouble is that man-made greenhouse gases from chlorofluorocarbons (CFCs), hydrofluorocarbons (HFCs), perfluorocarbons (PFCs), and carbon dioxide created by man-made activities such as the burning of forests and the use of fossil fuels (coal, oil, and natural gas) cause these gaseous emissions to unnaturally rise. For example, human activity causes carbon, which exists in nature in a solid form (as coal), to move to a gaseous form, increasing the concentrations of these gases in our atmosphere.

When did man-made greenhouse gases begin to be an issue for mankind?

Scientists believe that man-made greenhouse gases began when people began to burn coal at the beginning of the Industrial Revolution in 1750. Measurements of carbon dioxide occurring in our atmosphere in the present day, compared to 1750, show a 40% increase. Experts believe this is a major contributor to the effects of a phenomenon called global warming. The effects of the sudden warming impacts many aspects of our global ecosystem, including our marine ecology, the lives of billions of people, the air that we breathe, and the increase in surface temperatures felt throughout the world.

What percentage of animal species have become extinct over the last 40 years?

According to a 2014 report issued by experts at the World Wildlife Fund, 52% of all animal species on our planet have become extinct in the past 40 years, due to the degradation of our environment.

What is the cryosphere?

The cryosphere is the area of the Earth where water has solidified and includes ice, floating ice, glaciers, permafrost, and snow.

How much of the Earth's surface is frozen?

About one-fifth of the planet is permafrost, or frozen for all or most of the year.

What do deserts and polar regions have in common?

Deserts and polar regions both have average rainfall of less than four inches (10 cm) per year. Both regions contain some of the driest places on Earth.

What is a jungle?

A jungle is a forest that is composed of very dense vegetation. The term tropical rain forest can often be used interchangeably with jungle. Jungles occur most often in tropical areas such as the Amazon and Congo river basins.

What is a rain forest?

A rain forest is any densely vegetated area that receives over 40 inches (100 centimeters) of rain a year.

What is a tropical rain forest?

A tropical rain forest is a rain forest that lies between the Tropic of Cancer in the north (northernmost latitude around the Earth, where the sun appears directly overhead during the Northern Solstice) and the Tropic of Capricorn in the south (the southernmost latitude around the Earth, where the sun appears directly overhead during the Southern Solstice) or within the "tropics." Tropical rain forests are known for their very diverse species of plant and animal life. Tropical rain forests exist throughout Central America, northern Brazil, the Congo River Basin, and Indonesia.

Where is the northernmost rain forest?

Juneau, the capital of Alaska, is in the middle of one of the largest and most northerly rain forests in the world, which is located within the Chugach National Forest and Chugach State Park. Other rain forests extend to the northern regions of Japan and Siberia, Russia.

What is a desert?

A desert is an area of light rainfall. Deserts usually have little plant or animal life due to the dry conditions. Contrary to the popular image, deserts are not just warm, sand-swept areas like the Sahara Desert; they can also be frigid areas like Antarctica, one of the driest places on Earth.

How can an area be lower than sea level?

Land in the midst of a continent can be lower than sea level because the land is not close enough to the sea to be flooded with water. Movement of the tectonic plates pushes areas like the Dead Sea in Israel and Death Valley in California to elevations lower than sea level.

What was the highest temperature ever recorded?

World temperature records have been sources of controversy for many years, with many temperature records decertified by such groups as the World Meteorological Organization. For example, the world's highest temperature, thought to have been recorded in error in the desert in Libya in 1922 as 136 degrees Fahrenheit (57.78 degrees Celsius) was decertified in 2012. California's Death Valley, in the United States, holds the record set in 1913 for the highest temperature in the world of 134 degrees Fahrenheit (56.7 degrees Celsius).

Do oases really exist?

Oases do exist and they are quite prevalent throughout the eastern Sahara Desert in Africa. An oasis has a source of water, often an underground spring, that allows vegetation to grow. Small towns are located at some larger oases in the desert. Oases have been traditional stopping places for nomads traveling across deserts.

How do sand dunes move?

Sand dunes are created and transported by wind. Wind blows sand from the windward side of the dune to the opposite side, slowly transporting it across the landscape.

What causes erosion?

Wind, ice, and water are the most common agents of erosion. They wear down and carry away pieces of rock and soil. The process is accelerated when trees that help hold the soil in place have been destroyed by fire or have been chopped down. With fewer trees, the soil is easily eroded and washed away, leaving a barren surface where plants can no longer grow.

What does a glacier leave behind?

When a glacier moves across the land, it acts like a giant bulldozer, pushing and collecting rock, dirt, and debris. A moraine is a deposit of rock and dirt carried by a glacier and left behind once the glacier melts and recedes.

What is a tree line?

A tree line is the point of elevation at which trees can no longer grow. The tree line is caused by low temperatures and frozen ground (permafrost).

Glaciers retreating as the last ice age ended left behind the spectacular valley and mountains of Yosemite National Park.

How high is a tree line?

Tree lines vary, depending on where they are in the world and how close they are to a geographic pole. Trees stop growing in mountain regions from around 2,600 feet (800 meters) in places such as Sweden to over 17,000 feet (5,200 meters) in the Andes Mountains in Bolivia.

How do forest fires help forests?

An occasional fire is often necessary for a forest. Forest fires clear undergrowth, giving more room for trees to grow, thus rejuvenating the forest. Since forest fires are usually extinguished by firefighters as rapidly as possible, the amount of undergrowth in forests has increased. This extra undergrowth can become extremely flammable, making fires even more dangerous to people. It is thought by many experts that a policy of allowing the forest to burn naturally, while protecting human structures, produces a more natural environment.

What is tundra?

A tundra is a dry, barren plain that has significant areas of frozen soil or permafrost. Tundra is common in the northernmost parts of North America, Greenland, Europe, and Asia. Although rather inhospitable, there is plant life on the tundra. This life consists of low, dense plants such as shrubs, herbs, and grasses. There are even some species of insects and birds that can survive the harsh conditions of tundra.

ASTEROIDS AND NEAR EARTH OBJECTS

Did an asteroid cause a mass extinction of the dinosaurs?

Known as the Cretaceous-Paleogene extinction event approximately 66 million years ago, a six-mile wide (about 10 km) asteroid struck the Earth. Most scientists believe that this impact might have started a chain of events that led to the extinction of between 71% and 81% of the Earth's species of animals and plants, including nonavian dinosaurs. Today's birds, having evolved from surviving dinosaurs over millions of years of evolution, continue living, even after such an event. Many scientists believe that an asteroid that large would have created a layer of dust that would have surrounded the Earth, lowering temperatures and causing deadly, highly acidic, rain.

Will another large asteroid strike the Earth?

Yes, asteroids have struck the Earth in the past and are likely to strike again in the future. Small asteroids strike the planet about every 1,000 to 200,000 years. Large asteroids are much less likely to threaten the planet; they strike only about three times every million years. Huge asteroids, such as the one that may have killed dinosaur species, are even less frequent.

How many places on Earth have evidence of asteroid impacts?

There are about 160 places on Earth that display evidence of asteroid impacts. Depending on the size of an asteroid, it can have significant effects on the planet. If an asteroid strikes the ocean, it can create huge, destructive tsunamis or tidal waves. If an asteroid strikes land, it can create a huge crater, cause earthquakes, and propel debris into the atmosphere, creating major climatic changes.

What is a Potentially Hazardous Asteroid (PHA)?

A PHA is an object that is 3.28 feet (1 meter) in length, which is large enough to be detected by astronomers on Earth. It has an orbital path that could bring it in close proximity to Earth. It is very hard to predict the orbital path of a PHA, since its trajectory is always influenced by its proximity to a nearby planet's gravitational pull.

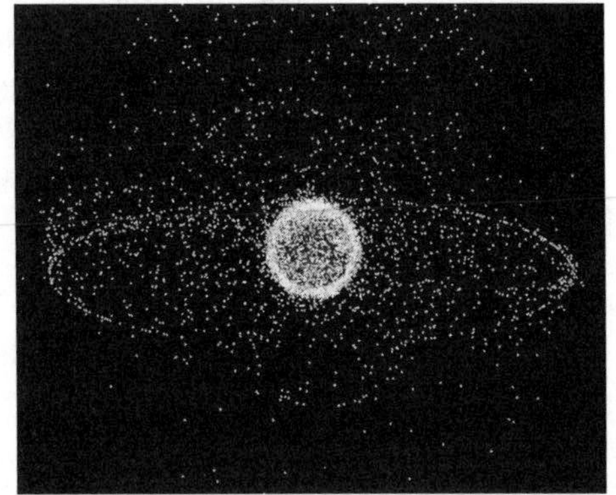

There are essentially two levels of space junk around our planet: objects in geosynchronous orbit and those in low Earth orbit.

How much junk is orbiting our planet?

According to scientists at NASA who track orbital debris, there are approximately 500,000 pieces traveling around the Earth, some at speeds of up to 17,500 miles per hour, from tiny screws to booster rockets. Of the 500,000 pieces, more than 20,000 pieces are larger than a softball. Some are as small as flecks of paint, which may damage orbiting spacecraft like the International Space Station, because of their speed. NASA, the U.S. Department of Defense, and other space agencies are able to track about 21,000 pieces of debris.

How many Potentially Hazardous Asteroids is NASA tracking that could impact the Earth in the twenty-first century?

There are approximately 1,572 PHAs that are on NASA's list of most likely objects whose orbital paths may bring them close to the Earth's orbit sometime in the next 100 to 200 years. Scientists have not fully calculated the risks because they still need to figure out the effects of gravity of various planetary bodies on the exact course of the asteroids. At best, scientists may calculate the probability of a collision with Earth, during a certain period of time.

What is a Near Earth Object (NEO)?

A Near Earth Object is any object in space that is relatively close to Earth and is of any size. The difference between an NEO and a Potentially Hazardous Asteroid (PHA) is that PHAs are anything larger than 500 feet (150 meters) in diameter and have orbits that bring them closer than 4,650,000 miles (7,480,000 km) to Earth. All PHAs are also NEOs, but not all NEOs are PHAs.

How many Near Earth Objects is NASA tracking?

Since the first Near Earth Object was discovered in 1973, approximately 12,000 objects are currently being tracked by NASA.

When and where was the last great impact on the Earth from a Near Earth Object?

The last big impact was in June of 1908 in Tunguska, Russia, in Siberia. Scientists do not know for sure what hit the Earth, as it could have been a comet or asteroid. But whatever it was, it had an estimated explosive energy of fifteen megatons. The impact event is significant because it was the first time during the modern era that we had eyewitness accounts by locals living nearby. The object is estimated to have been 120 feet (36.58 meters) wide and traveled at a speed of 33,500 miles per hour (53,913 km/hr).

What is the Space Guard Survey?

The Space Guard Survey refers to many initiatives by organizations such as NASA to catalogue all Near Earth Objects that are larger than 0.6 miles (1 kilometer) in diameter. The survey assesses their potential intersection with Earth's orbital path to determine if they are a threat.

WATER AND ICE

How much of the Earth is covered by water?

About 71% of the surface area of the Earth is covered by water. Oceans hold 96.5% of the Earth's water. The other 29% of the Earth is land, located primarily in the Northern Hemisphere. If you look at a globe, you'll notice that the Southern Hemisphere has a much larger surface area covered with water.

How much is the sea level rising?

Scientists believe that sea levels are rising by approximately .12 inches (3.05 mm) per year. The rising sea levels have been more extreme during the past century because of the increase in temperature of the world's oceans as well as the gradual melting of ice in the Earth's polar regions. The recent increase in sea levels are more extreme than over the past one thousand years.

What countries are threatened the most from rising sea levels and may cease to exist in the twenty-first century?

Many of the world's small island nations are concerned about the rising sea levels. Low-lying island nations of the Pacific and Indian oceans are most at risk, most notably Tuvalu, the Maldives, Palau, and Micronesia. Other candidates for severe flooding and reclamation of coastal land by the impending sea include Bangladesh, India, Thailand, Vietnam, Indonesia, and China, affecting the lives of hundreds of millions of people.

How much land disappears when the sea level rises?

Scientists believe that as the seawater rises by 0.04 inches (1 millimeter), the shoreline disappears by 4.9 feet (1.5 meters). This means that if the sea level rises by 3.28 feet (1 meter), the shorelines will extend another 1 mile (1.6 km) inland.

What is a nautical mile?

Used for measuring ocean-based distances, a nautical mile is equivalent to approximately 6,076 feet (1,852 meters) or 1.15 miles (1.85 km). The speed of ships is measured in knots. One knot is equivalent to one nautical mile per hour.

What is the average depth of the Earth's oceans?

The average depth of the Earth's oceans is approximately 14,000 feet (4,267.2 meters) deep.

What percentage of the Earth's surface is covered by water?

Approximately 70.8% of the Earth's surface is covered by saltwater. Although we have but one giant, interconnected ocean covering the planet, we divide it into large sections called oceans and smaller bodies of water called seas.

How is water distributed on the Earth?

The oceans contain approximately 97% of all of the world's water. Of this ocean water, 96% is saltwater or saline. The other 4% is comprised of freshwater flows, ice, snow, and precipitation. Of the 3% of nonocean water, approximately 69% of that is frozen in the form of glaciers and ice caps; 90% of this water is in Antarctica, and about 9% is in Greenland. What remains is the source of most of the water that we use each day. Of this useful water, 30% is merely groundwater, and .3% of this water is actually found in rivers and lakes. So less than 1% of all available fresh water on the planet can be found in rivers and lakes. Only .1% of water is found in the atmosphere.

What are the names of the oceans and seas?

The five largest bodies of water are named the Atlantic Ocean, the Pacific Ocean, the Indian Ocean, the Arctic Ocean, and the Southern Ocean. Each of the world's oceans have numerous seas, including the Baltic Sea, the Mediterranean Sea, the Greenland Sea, the Ross Sea, the Andaman Sea, and the Coral Sea.

How many seas are landlocked and not connected to the world's oceans?

There are four seas that are completely surrounded by land: the Aral Sea, the Caspian Sea, the Dead Sea, and the Salton Sea.

What is so dead about the Dead Sea?

The Dead Sea was considered to be dead because of its high salinity (at 34.2%, which is approximately 9.6 times more salty than the world's oceans). So aquatic life cannot survive in such conditions. But there are several other places that have more salinity than the Dead Sea, including Lake Vanda (Antarctica), Lake Assal (Djibouti), Lagoon Garabogazköl (Caspian Sea), and many lakes in the McMurdo Dry Valleys (Antarctica), where the salinity in each of them has been measured at more than 40%.

What are trenches?

Trenches are deep, "v"-shaped depressions on the Earth's surface, lying in the deepest parts of the Earth's oceans, mostly in the Pacific Ocean. They are caused by subduction,

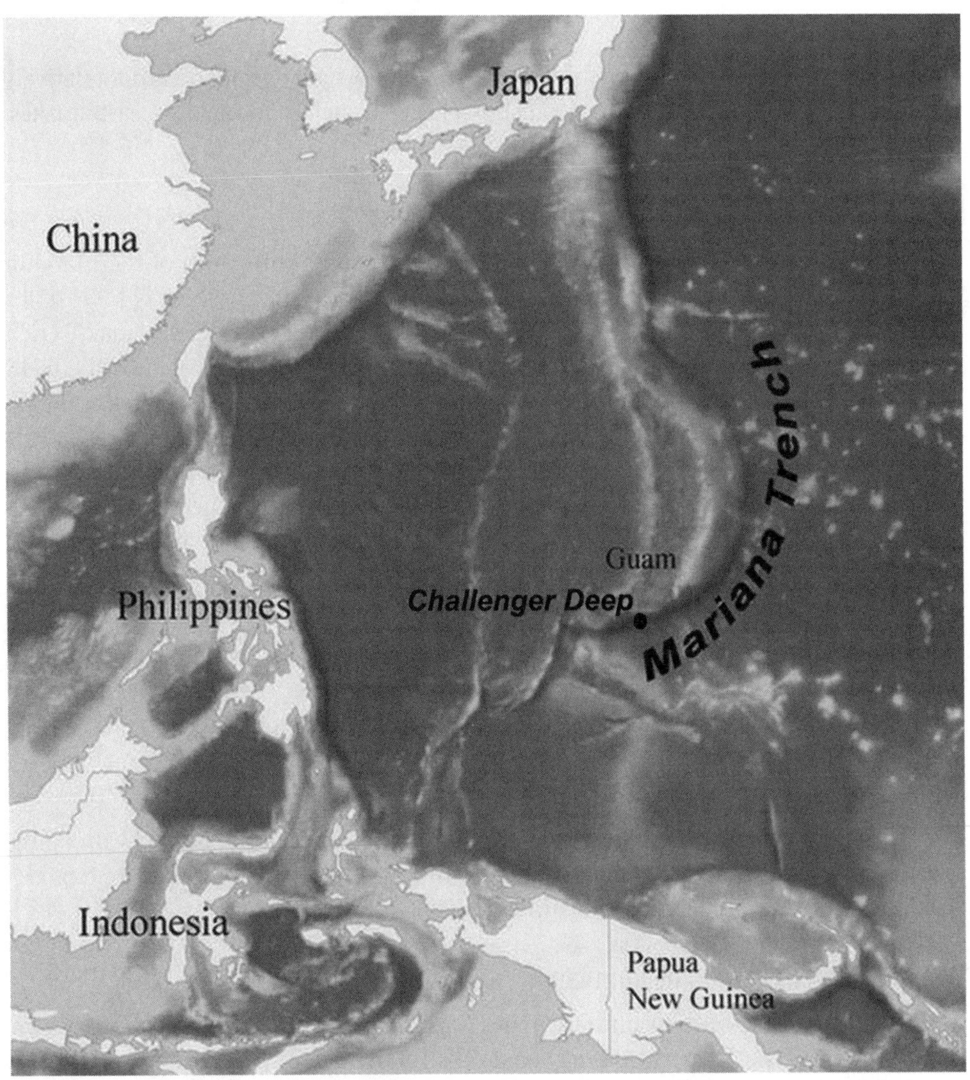

The deepest point in the oceans of the planet is the Mariana Trench in the South Pacific Ocean. The above map also shows where the Challenger Deep is located, with a depth of 35,814 feet (10,916 meters).

What island did Robinson Crusoe shipwreck on?

Daniel Defoe based his novel *Robinson Crusoe* on the real-life story of Alexander Selkirk. Selkirk was an English sailor who had an argument with the captain of his ship and asked to go ashore on the island of Más a Tierra (also known as Robinson Crusoe Island or Isla Robinson Crusoe), about 400 miles (644 kilometers) west of Santiago, Chile, in the South Pacific Ocean. Selkirk was stranded on the island from 1704 to 1709, when he was rescued by another English ship.

where tectonic plates collide, pushing one heavier, older plate underneath a relatively less dense, newer plate, causing the formation of deep trenches, mountains, earthquakes, and volcanic islands.

What are some of the deepest points in the oceans?

Lying deep below the Pacific Ocean, about 200 miles (322 km) south of the island of Guam, is the Marianas Trench (also known as the Mariana Trench), which is 1,554 miles (2,550 km) long and 44 miles (71 km) wide. The deepest point of the Marianas Trench is named Challenger Deep at 35,814 feet (10,916 meters), first discovered by the HMS *Challenger* in 1875. In the Atlantic Ocean, the Puerto Rico Trench is 28,373 feet (8,648 meters) below the surface. In the Arctic Ocean, the Eurasia Basin at the Litke Deep is 17,881 feet (5,449 meters) deep. The Java Trench or Sunda Trench in the Indian Ocean is 25,344 feet (7,725 meters) deep. Another deep point of note in the Pacific Ocean is Monterey Canyon off the coast of northern California. It is about 95 miles (153 km) long and 11,800 feet (3,600 meters) deep. The cold waters generated in the trench create a perfect environment rich in foods that support a diverse range of wildlife.

In comparison to all of these ocean canyons, the most famous land canyon—the Grand Canyon in Arizona—is 277 miles (446 km) long and 6,000 feet (1,829 meters) deep. The world's ocean canyons are much more impressive, but most people will never see them.

What is a sounding?

A sounding is a method once used for determining the depth of the ocean by dropping a weighted line into the water and measuring the length of that line when it stops descending at certain predetermined points.

Where is the farthest point from land?

In the middle of the South Pacific Ocean lies the Pacific pole of inaccessibility, a spot on the Earth that is the farthest from any land. It is 1,670 miles (2,688 km) from any land. Located at 48°52.6' South and 123°23.6' West, this spot is approximately equidistant from Antarctica, Australia, and Pitcairn Island.

Where is the farthest point from an ocean?

The Eurasian pole of inaccessibility is located in Xinjiang Province, in northern China, and is over 1,600 miles (2,574 km) from any ocean. Located at 46°17' North, 86°40' East, the land is approximately equidistant from the Arctic Ocean, Indian Ocean, and Pacific Ocean.

What is the world's largest island?

The world's largest island is Greenland (in Greenlandic, Kalaallit Nunaat). Greenland is located in the North Atlantic Ocean near Canada, between the latitudes of 59' and 83' North, and longitudes of 11' and 74' West. It is an autonomous country within the kingdom of Denmark, with the queen of Denmark as the head of state, but with a local referendum in 2008, it was granted self-rule and self-government. It is approximately 840,000 square miles (2,175,600 square km). Australia, while it also meets the usual definition of an island (surrounded by water) and is larger than Greenland, is not considered an island but a continent.

Why is Greenland considered an island while Australia is a continent?

Australia is three and a half times larger than Greenland and comprises most of the land on the Indo-Australian Plate, while Greenland is distinctly part of the North American Plate.

What is an archipelago?

An archipelago is a chain (or group) of islands that are close to one another. The Aleutian Islands of Alaska and the Hawaiian Islands are both archipelagos. They are usually formed by plates pushing into one another or by volcanic activity.

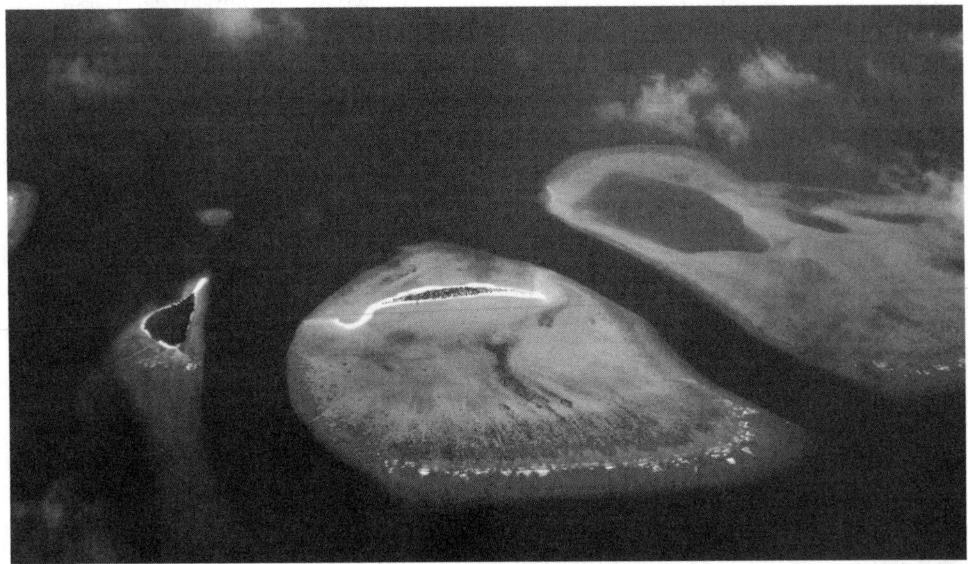

Atolls like these in the Maldives are formed of coral reefs surrounding what was once a volcano.

What is a coral reef?

Coral reefs are formed by the accumulation of calcium carbonate that comes from the external skeletons of tiny animals called coral polyps. The polyps live in shallow, warm water and thus congregate around islands in the tropics, where coral reefs are abundant.

What is an atoll?

In addition to reefs, coral can also form atolls. Atolls are formed when a volcano, around which coral often grows, erodes away, leaving a circular wall of coral with a lagoon at the center.

What is a strait?

A strait is a narrow body of water between islands or continents that connects two larger bodies of water. Two of the most famous straits are the Strait of Gibraltar, which connects the Mediterranean Sea and the Atlantic Ocean, and the Strait of Hormuz, which connects the Persian Gulf to the Gulf of Oman.

HIGH, LOW, BIG, SMALL, AND WONDROUS

Where is the lowest point on Earth?

To reach the lowest point on Earth, you would have to go far below the Pacific Ocean, south of Guam to the Marianas Trench, at a place known as Challenger Deep. Its maximum known depth is 6.831 miles or 36,070 feet (10,994 meters).

Is the lowest point on Earth actually closer to the center of the Earth?

No, because of the Earth's irregular shape—it is not a perfect sphere—places in the Arctic Ocean are technically closer to the Earth's center than the Mariana Trench, although not as deep.

Where is the world's lowest point on dry land?

The world's lowest point is at the shore of the Dead Sea on the border of Palestine, Israel, and Jordan. It is 1,378 feet (420 meters) below sea level.

What are the lowest points on each continent?

In Africa, the lowest point is Lake Assal in Djibouti, 509 feet (155 meters) below sea level. In North America, California's Death Valley lies at 282 feet (86 meters) below sea level. Argentina's Laguna del Carbón is the lowest point in South America at 345 feet (105 meters) below sea level. The Caspian Sea in Europe lies at 92 feet (28 meters) below sea level. In Asia the shoreline of the Dead Sea lies at 1,378 feet (420 meters) below sea level. And Australia's lowest point is a mere 52 feet (16 meters) below sea level at Lake Eyre.

OCEANS AND SEAS

How does the hydrologic cycle work?

The movement of water from the atmosphere to the land, rivers, oceans, and plants and then back into the atmosphere is known as the hydrologic cycle. We can pick an arbitrary point in the cycle to begin our examination. Water in the atmosphere forms clouds or fog and falls (precipitates) to the ground. Water then flows into the ground to nourish plants or into streams that lead to rivers and then to oceans, or it can flow into the groundwater (underground sources of water). Over time, water sitting in puddles, rivers, and oceans evaporates into the atmosphere. Water in plants is transpired into the atmosphere.

What is evapotranspiration?

The process of water moving into the atmosphere is collectively known as evapotranspiration. It is the combination of water vapor being evaporated from the surface of the Earth (such as from lakes, rivers, or puddles) into the atmosphere, and transpiration, which is the movement of water from plants to the air.

Where is all the water?

Over 97% of the world's water is contained within oceans and is too salty to drink or to irrigate crops with (except when the water is cleaned through a desalination plant, which is not done very often). About 2.8% of the world's water supply is fresh water. Of that 2.8%, about 2% is frozen in glaciers and ice sheets. This leaves only about 0.8% of the world's water that is accessible through aquifers, streams, lakes, and in the atmosphere. The water that we use primarily comes from this 0.8%.

How many people in the world have limited access to clean water?

According to experts at MIT, approximately 700 million people have limited access to clean water. This number will swell to approximately 1.8 billion people over the next ten years as the global population grows and demand for industrial water used for manufacturing, power generation, agriculture, and energy production continues to accelerate.

What is desalination?

Desalination is the process of removing some salt and minerals from ocean water in order to produce consumable fresh water for a population. According to experts at the International Desalination Association, there are approximately 17,000 desalination plants worldwide in 150 countries, producing 21.1 billion gallons (66.5 cubic meters) per

Desalination plants remove salt from seawater to create fresh drinking water. The process has improved over the years, but there is still the problem of what to do with all the salt and other minerals that are generated from the process.

day. Three hundred million people in the world rely on desalinated water for their daily drink water needs. Twenty percent of all water used in Israel is desalinated water that comes from the Mediterranean Sea. It is estimated that by 2016, 50% of Israel's water supply will be produced in the country's desalination plants.

What is a desalination plant?

A desalination plant is a facility that pumps ordinary seawater through a myriad of expensive equipment and processes, transforming the salty water into fresh water. This process has been used with some success in many parts of the world, including the United States, the Caribbean, and the Middle East. It is much more efficient and less expensive, however, to clean wastewater (water that has been used for bathing, cooking, cleaning, etc.) than it is to clean and desalinate seawater.

How many desalination plants are there?

There are approximately 17,000 plants located throughout the world. Saudi Arabia alone accounts for 24% of the world output of fresh water from desalination. More than half of the world's desalination plants are located in the Middle East due to the need for fresh water for the petroleum industry. The United States accounts for another 15% of all plants, Europe and Asia 10% each, and Africa approximately 6%.

Where is the biggest desalination plant in the United States?

The biggest desalination plant is in Carlsbad, California, where it opened in December 2015 at a cost of approximately $1 billion. It is capable of producing 50 million gallons (188 million liters) of drinking water to arid San Diego County per day. This is a fraction of the largest plant in the world, the Jebel Ali Plant in the United Arab Emirates, which produces over 140 million gallons (530 million liters) per day.

What is an aquifer?

An aquifer is an underground collection of water that is surrounded by rock. The creation of an aquifer is a very slow process, as it relies upon precipitated water (rainwater) to percolate through the soil and rock layers and into the aquifer. An aquifer lies above a lower layer of rock that holds the water in place and keeps it from moving further underground.

What is the Ogallala Aquifer?

The Ogallala Aquifer is a huge aquifer that spans an area from western Texas to South Dakota, including parts of Colorado, Kansas, Nebraska, Oklahoma, and New Mexico. The oldest water deposited in the aquifer is over one million years old, and only a very small amount of water is added each year. The Ogallala Aquifer is being pumped rapidly by the farms in the region, causing a reduction in the amount of water in the aquifer. Consequently, wells have to be continually deepened so that they can continue to pump water. The aquifer supports 20% of the wheat, corn, cotton, and cattle produced in the United States.

Why are we losing groundwater?

Water is pumped from aquifers around the world for irrigation, industrial, and household needs. Aquifers do not refill as rapidly as water is being pumped out, so in many areas, there is a danger that some aquifers may disappear altogether.

What are ice ages?

Ice ages began during the Precambrian Era approximately 600 million years ago. Throughout the life of the planet, the climate has warmed and cooled many times. During the cooling periods, ice ages have occurred. Large sheets of ice cover large portions of land. In the most recent ice age, which began approximately 3 million years ago and ended about 10,000 years ago, large parts of northern Europe and North America were covered by ice sheets. Some experts argue that there is evidence to suggest that the most recent ice age continues to this very day.

Will there be another ice age?

Yes, eventually the Earth will again cool and ice will cover land at higher latitudes and elevations. It may be a hundred years from now or it may be thousands of years away, but the Earth's climate is always slowly changing.

If I keep walking in a straight line, will the Coriolis effect cause me to veer off in one direction or another?

If your body were completely symmetrical (it is not) and neither leg were longer and you were walking on perfectly flat land, then, yes, you might start veering due to the Coriolis effect.

What is the Coriolis effect?

Due to the rotation of the Earth, any object on or near the Earth's surface will veer to the right in the Northern Hemisphere and to the left in the Southern Hemisphere. This applies especially to phenomena such as ocean currents and wind. Imagine an airplane flying to New York from Los Angeles. As the plane flies over the United States, the Earth continues to rotate under the plane, and it lands in New Jersey instead. Pilots need to factor the spinning of the Earth into their trajectories in order to end up in the right place. North of the equator, ocean currents and winds rotate clockwise, but south of the equator, the opposite is true.

Does the Coriolis effect make the water in my toilet, sink, and bathtub swirl clockwise?

No, the Coriolis effect has very little effect on such small bodies of water. The flow down the drain is mostly a function of the shape of the container that holds the water.

What is the difference between a bay and a gulf?

Both a bay and a gulf are bodies of water partially surrounded by land, but a bay is a smaller version of a gulf. Famous bays include the San Francisco Bay (California), the Bay of Pigs (Cuba), Chesapeake Bay (Maryland/Virginia area), Hudson Bay (Canada), the Bay of Bengal (near India and Southeast Asia), and the Bay of Biscay (France). Famous gulfs include the Gulf of Mexico (southern United States), the Persian Gulf (between Saudi Arabia and Iran), and the Gulf of Aden (between the Red Sea and the Arabian Sea).

Where does the Loch Ness Monster live?

Loch Ness lies along a natural fault line in Scotland. The fabled monster is supposed to live in Loch Ness. The term "loch" is Gaelic and is used in Scotland to refer to a lake or narrow inlet of the sea. Loch Ness is fully surrounded by land and is therefore a lake. The first recorded sighting of this alleged monster was in the sixth century C.E. by St. Columba. In 1933, the first photograph of the supposed monster, known as Nessie, was taken.

How are waves created?

Waves are created by wind blowing across the surface of water. Though waves appear to move along the surface of the water, they are simply the movement (oscillation) of water up and down due to the friction of the air. When waves occur near the shore, they may become steeper and "break."

How does the Old Faithful geyser shoot water into the air?

A geyser, such as the famous Old Faithful, located in Wyoming's Yellowstone National Park, is the result of an underground aquifer that is warmed by heated rocks and magma. There is a small fissure or crack in this aquifer's surface that allows the steam and heated water to jet from the ground (about every hour).

Possibly the most famous geyser on the planet is Old Faithful in Yellowstone National Park.

How does water wash away the land?

Drops of rain hit soil and rock and displace grains of material. When water flows over the surface, it loosens and carries away pieces of rock or soil. There is a tremendous amount of energy in a raindrop. Over days, weeks, months, years, centuries, and millennia, the erosive power of water can cut through even the strongest rocks. The material that the flowing water picks up is eventually deposited when the flow of the stream slows down, a process known as deposition.

How much does a gallon of water weigh?

Water is quite a heavy substance. One gallon of water at room temperature weighs about 8.33 pounds (3.78 kilograms).

How is water used in the home?

In the United States, we use approximately 80 to 100 gallons of water per day. A typical family of four people might use up to 400 gallons of water per day. About 26.7% of household water is used for flushing toilets; 21.7% is used for washing clothes; 20% may be used for washing dishes. The remaining water that we use is for showers/baths, drinking, and drips/leaks.

What waterfall has the largest flow of water?

Boyoma Falls (formerly known as Stanley Falls), on the Congo River in the central African nation of the Democratic Republic of the Congo, has the greatest flow of water in the world, estimated to be 4.49 million gallons per second (17 million liters per second), over the course of a drop of 196.85 feet (60 meters).

What is the most visited waterfall in the world?

Niagara Falls, on the border of Canada and the United States between Lake Ontario and Lake Erie, has 212,000 cubic feet (6,000 cubic meters) of water flowing over its 173- and 182-foot-high (52.7- and 55.5-meter-high) waterfalls.

The spectacular Yosemite Falls in California's Sierra Nevada Mountains is 2,425 feet (739 meters) high.

What is the highest waterfall in the world?

The highest waterfall in the world is Angel Falls, located in Canaima National Park in southeast Venezuela. It is 3,212 feet (979 meters) high.

How does the boiling point of water help determine altitude?

The boiling point of water at sea level is 212 degrees Fahrenheit (100 degrees Celsius). The boiling point drops about one degree for every 500-foot (152-meter) increase in altitude. Therefore, in Denver, Colorado, in the United States, at 5,280 feet (1,609 meters) above sea level, water boils at about 202 degrees Fahrenheit (94.4 degrees Celsius). The change in the boiling point is why cooking instructions are nearly always modified for higher altitudes, when they involve the use of boiling water.

What is the difference between a sea and an ocean?

While a sea can be any body of saltwater, it usually refers to a body of saltwater partially or completely enclosed by land. Oceans, though they can also be referred to as seas, are large areas of saltwater, unobstructed by continents.

How salty is seawater?

About 3.5% of the weight of seawater is salt (not just sodium chloride or table salt, but also potassium chloride, calcium chloride, and other types of salts). This would equal just over three and one-third cups of salt in one gallon of seawater.

What are ocean currents?

The oceans don't remain still; their water is constantly moving in giant circles known as currents. In the Northern Hemisphere, currents move clockwise, while in the Southern Hemisphere, they move counterclockwise. Currents help to moderate temperatures

> **How tall is the tallest waterfall in the United States?**
>
> Yosemite Falls in California, which is comprised of three separate falls, has a height of 2,425 feet (739 meters).

on land in such northern places as the British Isles, which are farther north than the United States/Canadian border, by sending warm water from the Caribbean northeast across the Atlantic Ocean to northern Europe. A current known as the Antarctic Circumpolar Current circles the southern continent. The North Atlantic and North Pacific oceans each have a large clockwise current, while the South Atlantic and South Pacific oceans each have a large counterclockwise current.

What are the largest seas?

Parts of oceans that are surrounded by islands or otherwise partially enclosed are often known as seas. The five largest seas in order are: the South China Sea, the Caribbean Sea, the Mediterranean, the Bering Sea, and the Gulf of Mexico.

Why is the Mediterranean Sea so salty?

Due to the high temperatures in the Mediterranean region, evaporation of the Mediterranean Sea occurs more rapidly than in other bodies of water; therefore, more salt is left behind. The warm, dense, salty water in the Mediterranean is replaced by less salty and dense Atlantic water in the western part of the sea at the Strait of Gibraltar. Water that flows into the Mediterranean from the Atlantic Ocean usually remains in the sea for anywhere from 80 to 100 years before returning to the Atlantic Ocean.

Has the Mediterranean Sea always been there?

Salt and sediment found at the bottom of the Mediterranean Sea prove that on several occasions the Mediterranean Sea has dried up, leaving a large layer of salt behind. Scientists speculate that the Strait of Gibraltar has, on occasion, closed up, keeping water from being able to flow back and forth between the Atlantic Ocean and the Mediterranean Sea.

Where are the four colored seas: Black, Yellow, Red, and White?

The four colored seas are not geographically associated with one another. The Black Sea is located near the Balkan Peninsula and is bordered by Turkey, Russia, and Ukraine (it is also the home of the port city of Odessa). The Red Sea is located to the south of the Black Sea, between the Arabian Peninsula (Saudi Arabia) and Africa. The Red Sea has been a major trade route for hundreds of years and has been especially useful since the completion of the Suez Canal. The White Sea is in northern Europe. It is part of the

Arctic Ocean and is a Russian sea (it lies to the east of Finland). The Yellow Sea is far to the east, between China and the Korean Peninsula.

What are the seven seas?

The "seven seas" spoken of by mariners from long ago are oceans or parts of oceans. The Atlantic and Pacific oceans are so large that they were each divided into two "seas." The Antarctic, Indian, and Arctic were also considered seas, thus totaling seven. If a sailor had sailed upon all seven seas, he had sailed around the world. There are not just seven but dozens of seas in the world.

Is the Black Sea really black?

No, it is not. This sea, located to the north of Turkey, is quite deep and has darker-looking water than most water bodies, but receives its name from the severity of the waters when sailing.

Where is the Putrid Sea?

The Putrid Sea, also known as the Syvash Sea or Rotten Sea, lies to the east of the Isthmus of Perekop, between Crimea and Ukraine. It is a swampy area of salty lagoons that lie along the west coast of the Sea of Azov and cover an area of approximately 990 square miles (2,560 square km).

RIVERS AND LAKES

What is the longest river in the world?

Egypt's famous Nile River is the longest in the world. It is more than 4,100 miles (6,597 km) long from its sources in the Ethiopian Highlands (the source of the portion of the Nile called the Blue Nile) and Lake Victoria (the source of the White Nile). The Nile Valley is the center of contemporary and ancient Egyptian civilization. Following the Nile in length are the Amazon (in Brazil), the Missouri-Mississippi (United States), the Chang or Yangtze (China), the Huang He or Yellow (China), and the Ob (Russia).

Do rivers always flow from north to south?

No, they do not! Rivers always flow from higher ground to lower ground. Though we are familiar with rivers like the Mississippi River in the United States, which flows from north to south, rivers always flow the way gravity takes them. There are many major rivers in Europe, Asia, and North America that flow from south to north, such as the Ob in Russia, the Nile in Africa, and the Mackenzie in Canada.

The Nile River meanders through much of the eastern part of Africa and is a source of water for seven countries.

Why are the Blue Nile and White Nile rivers both called Niles?

The Nile River begins as two separate tributaries—the White Nile and the Blue Nile. The White Nile begins its flow from Lake Victoria in Eastern Africa, and the Blue Nile originates in the Ethiopian Highlands. The Blue Nile and the White Nile converge in Khartoum, the capital of the Sudan, and form the Nile River, which continues on to the Mediterranean. It is approximately 4,258 miles (6,853 km) long.

Was the flooding of the Nile predictable before dams were built?

The summer floods of the Nile River were so predictable that the Egyptian calendar was based on their rise and fall. Flooding on the Nile occurred from late June until late October. The floods brought nutrients and sediments beneficial to the nearby agricultural lands, making farming productive throughout the remainder of the year. Measuring scales called "nilometers" were placed along the river and not only measured the river height but also served as calendars. When Egypt's Aswan Dam was completed in 1970, flooding on the lower stretches of the Nile ceased, but it still occurs in parts of the Sudan.

What is the longest river in the United States?

The Missouri-Mississippi River is the longest in the United States, approximately 3,860 miles (6,211 km) in length.

Where are the highest rivers in the world?

The highest rivers are in Tibet, where there are two that stand out from all the rest: the Ating River at 20,013 feet (6,100 meters) and the Brahmaputra River at 19,751 feet (6,020 meters).

Why are the Missouri and Mississippi rivers lumped together?

Actually, the Missouri River was incorrectly named. The Missouri River is the main feeder river of what is now known as the Mississippi River. Usually, the main feeder bears the same name as the rest of the river. Therefore, the full length of the Mississippi River, including the Missouri River, is known as the Missouri-Mississippi River.

Which river carries the most water?

By far, Brazil's Amazon River carries more water to the sea than any other river in the world. The discharge at the mouth of the river is about seven million cubic feet (170,000 cubic meters) per second, which is about four times the flow of the Congo in Africa, the river ranked second in terms of discharge. It would take the Amazon only about 28 days to fill up Lake Erie. The Yangtze, Brahmaputra, Ganges, Yenisey, and Mississippi are other rivers with very high discharges.

What is a delta?

A delta is a low-lying area where a river meets the sea. Often, the river divides into many tributary streams, forming a triangular-shaped area. The river deposits a large amount of sediment at its mouth, creating excellent soil for farming once the channel of the stream moves. One of the most famous deltas is where the Nile River meets the Mediterranean Sea. Other major deltas include the Mississippi River Delta in Louisiana, the Ganges River Delta in India, and the Yangtze Delta in China. The word delta comes from the Greek letter delta, referring to its triangular shape when written.

Where are some of the world's largest freshwater marshes?

The Okavango Delta, in the southern African country of Botswana, forms a marsh that is 5,800 square miles (15,000 square km) in area. The Okavango River empties into the arid Kalahari Desert, forming the largest freshwater marsh in the world. The Everglades in Florida is another one of the world's largest marshes, consisting of 2,185 square miles (5,659 square km). The water across this southern Florida marsh averages six inches (15 centimeters) in depth. The Everglades is an endangered ecosystem, threatened by excess drainage and the introduction of exotic plants.

What is a drainage basin?

The area that includes all of the tributaries for an individual stream or river is its drainage basin. For example, the drainage basin for the St. Lawrence River includes the area surrounding the Great Lakes. Rivers such as the Platte (which has its own drainage basin) flow into the Missouri, and the Missouri flows into the Mississippi. The combined area drained by the Platte, the Missouri, the Mississippi, and all other Mississippi River tributaries combined create the third-largest drainage basin in the world. The Amazon has the largest drainage basin of any river, while the Congo has the second largest.

What is a tributary?

Any stream that flows into another stream is a tributary. Most major rivers have hundreds of tributaries, which on a map look like branches of a tree. One classification system of rivers is based upon the number of tributaries a river has.

What is a watershed?

A watershed is the boundary between drainage basins. It is usually the crest of a mountain where water flows on either side into two different drainage basins.

What is a wadi?

Wadi is the Arabic word for a gully or other stream bed that is dry for most of the year. A wadi is a channel for streams that develop during the short rainy season. The channels of wadis were probably initially carved when the desert regions of today had more rainfall.

What is a meander?

Streams and rivers that have carved a flat floodplain commonly flow in curves known as meanders. These S-shaped curves vary by the size and flow of the river. The river flows faster on the outside curve of the meander and therefore continues to cut and create a larger curve.

What is an oxbow lake?

An oxbow lake is a crescent-shaped lake that is formed when the meander, or curve of a river, is cut off from the rest of the river during a flood, or when the curve of the meander becomes so large that the river begins flowing along a new path. The curve that remains becomes its own lake. These can commonly be seen along the Mississippi River system.

What is the world's largest lake?

The Caspian Sea (which is really a lake) is the largest lake in the world. Until 11 million years ago, it was contiguous with the world ocean, so it is a saltwater lake. It is surrounded by Russia, Kazakhstan, Turkmenistan, Iran, and Azerbaijan and is over 143,200 square miles (370,888 square km) in area. The second-largest lake in the world, Lake Superior in North America, is a mere 31,820 square miles (82,414 square km) in area and has more water than the other four neighboring Great Lakes combined. It is considered by many to be the largest freshwater lake in the world by surface area.

What is Africa's largest lake?

Located in eastern Africa, Lake Victoria is Africa's largest lake by area, with a surface area of approximately 26,828 square miles (69,485 square km). It is also the world's second-largest freshwater lake, after Lake Superior. Lake Victoria is bordered by Uganda, Kenya, and Tanzania. The lake was named by British explorer John Hanning Speke, the first European to see the lake (in 1858), in honor of the reigning British queen. A map, drawn by Arab explorers in the late twelfth century C.E., describes the lake in great detail.

The Caspian Sea is actually a very large lake located in southern Russia, which also touches Turkmenistan and Kazakhstan.

Where is the highest lake in the world?

A yet-to-be-named small crater lake found atop Ojos del Salado (Spanish for "eyes of salty water"), a volcanic mountain on the border of Argentina and Chile, has an elevation of 20,965 feet (6,390 meters). Lake Titicaca, a lake sandwiched between Peru and Bolivia, is at 12,507 feet (3,812 meters) above sea level and is the highest navigable lake in the world, which means that you can go for a boat ride on this lake.

PRECIPITATION

How do we know how much rain actually falls?

Agencies like the National Weather Service use very accurate devices that measure rainfall to the nearest one-hundredth of an inch. The devices, known as rain gauges or tipping-bucket gauges, collect rainwater, usually at a point unaffected by local buildings or trees that may interfere with the rain.

Where in the world does it rain the most each year?

Cherrapunji and the nearby town of Mawsynram, in India, near the border of Bangladesh, have the highest average rainfall per year in the world, averaging 467 inches (11,872 mm). They are in a subtropical climate zone and are impacted by monsoon rains that are heavy during the months of April through October. Mt. Waialeale, on the island of Kauai in Hawaii, receives an average of 452 inches (11,500 mm) of rain a year and often records some of the highest average precipitation in the world.

Where does it rain the least?

The Dry Valleys of Antarctica are considered to be among the driest places on Earth, with annual precipitation of less than .08 inches (2 mm). Because of its proximity to the Atacama Desert, Arica, Chile, is also considered by many to be one of the world's driest places with an average annual rainfall of .03 inches (.761 mm). The northern Sudan's Wadi Halfa (which is in the Sahara Desert) receives an average of less than .10 inches (2.54 mm) of rain per year. That's hardly a drop at the bottom of a bucket.

How much water is in snow?

When about 10 inches (25 centimeters) of snow melts, it turns into about one inch (2.54 cm) of water. Snow has pockets of air between snowflakes when they are on the ground, so it takes ten times the amount of snow to make an equivalent amount of water.

How can I measure the amount of rain that falls where I live?

Any container with a flat bottom and flat sides can measure rainfall. The width of the top of the container must be the same as at the bottom of the container, but the diameter does not matter. It could be a device purchased for measuring precipitation or something as simple as a coffee can.

Do oceans get more rain than land?

The oceans receive about 77 percent of the world's precipitation. The remaining 23 percent of precipitation falls on the continents. Some areas of the world receive far more precipitation than others. Some parts of equatorial South America, Africa, Southeast Asia, and nearby islands receive over 200 inches (500 centimeters) of rain a year, while some desert areas receive only a fraction of an inch of rain per year.

Where was the most snowfall ever recorded?

Washington State's Mt. Baker, at an elevation of 4,200 feet (1,280 meters), recorded the most snowfall in a single season (July 1998–June 1999): 1,140 inches (2,896 centimeters), which is 95 feet (28.96 meters) of snow.

What is the difference between snow and hail?

Snow is water vapor that freezes in clouds before falling to the Earth. Hail is water droplets (raindrops) that have turned to ice inside of clouds.

How is hail formed?

Hail is ice that is formed in large thunderstorm clouds. Hail begins as droplets of water, normally destined to become raindrops, that are blown upward and subsequently freeze. They then fall lower within the cloud, where they collect more water, are blown upward again, and refreeze. The hailstone grows larger as it collects more and more ice and eventually falls to the ground.

How big was the largest hailstone?

In 2010, a hailstone that fell near Vivian, South Dakota, measured 8 inches (20.32 cm) in diameter with a circumference of 18.62 inches (47.3 cm), even after six hours of thawing. It weighed 1 pound, 15 ounces (878.84 grams). In 2003, a hailstone was recovered near Aurora, Nebraska, with a diameter of 18.75 inches (47.63 centimeters). The previous record was in 1970, when people recovered a hailstone with a 17.5 inches (44.45 centimeters) diameter in Kansas.

GLACIERS AND FJORDS

What is a glacier?

A glacier is a mass of ice that stays frozen throughout the year and flows downhill. Glaciers are capable of carving rock with their weight and slow, steady movement. They are responsible for the stunning landscape of Yosemite National Park in California. Large glaciers that cover the land are also known as ice sheets.

Are there still glaciers in the United States?

Yes, small glaciers exist throughout Alaska, within the Cascade Range of Washington State, sporadically across the Rocky Mountains, and also in the Sierra Nevada Mountains of California.

Glaciers around the world, including Serrano Glacier in O'Higgins National Park, Chile, have been receding due to global warming.

How old are glaciers?

Glaciers present today were created during the last stage of glaciation, the Pleistocene epoch, which lasted from 1.6 million years ago to about 10,000 years ago.

Did glaciers create the Great Lakes?

Yes, the Great Lakes are the world's largest lakes formed by glaciers. During the Pleistocene epoch, glaciers inched over the Great Lakes area, moving weak rock out of their way and leaving behind huge, carved basins. As the glaciers began to melt, the basins filled with water and created the Great Lakes.

Are glaciers only found in cold, northern places?

No, glaciers are found in all six continents.

What is a tropical glacier?

Tropical glaciers are those found high in the mountains of tropical regions in the world. The Andes Mountains in South America contain 70% of the world's tropical glaciers.

Is global warming causing the world's glaciers to melt?

Many scientists believe that greenhouse gases from human activities, and the effect of this on global temperatures, are directly causing glaciers in all parts of the world to melt and recede at an unprecedented rate. It is thought that by 2030, there will be no glaciers in Glacier National Park in Montana. In East Africa, Mt. Kenya's Lewis Glacier in Kenya has lost 40% of its size in just the last twenty-five years.

What are the consequences when glaciers melt?

Glaciers that have melted in the Himalayas, home to the world's largest mountains, have filled up and burst the banks of nearby glacial lakes, filling rivers and causing wide-

From where does the word "fjord" originate?

The word fjord comes from the Norse language and means "where you travel across." It is significant to early Norwegians as a place to travel across to get to the sea when there were no bridges available.

spread flooding and death to nearby populations downstream. Similar consequences will likely befall those now living near other glaciers around the world.

What is a fjord?

During the ice ages, glaciers, which were prevalent at higher latitudes and elevations, became so large that gravity drove them to lower elevations, eventually all the way to the sea. On their way, glaciers would carve deep canyons in the surface of the Earth. At the end of the ice age, as the ice melted and the ocean level rose, these glacial troughs filled with seawater. These very dramatic-looking canyons with high cliffs hanging over a thin bay of water are known as fjords. Fjords are very common in Norway and Alaska.

What is the highest fjord in Norway?

The highest fjord is Sognefjord, which begins at a depth of 4,291 feet (1,308 meters) in the ocean and rises to more than 3,280 feet (1,000 meters).

Are all fjords found in Norway?

No. In fact, fjords are found throughout the world, wherever glaciers retreated and have cut into the earth, filling in and creating a huge valley of seawater. Notable fjords are found in Alaska and on New Zealand's South Island.

Where is the longest fjord?

The longest fjord is in Greenland, at Scoresby Sund. It stretches more than 217 miles (350 km). English explorer William Scoresby mapped the fjord in 1822.

CONTROLLING WATER

What is a dam?

By blocking the flow of a river, a dam allows a reservoir of water to build up. Dams are built in order to minimize floods, to provide water for agriculture, and to provide water for recreational uses. Dams in the United States are somewhat controversial, as the U.S. Bureau of Reclamation (established in 1902) and the Army's Corps of Engineers battle to build more dams and control more water in the western United States. Many outdoor enthusiasts and environmentalists feel that dams are not always necessary.

What is the tallest dam in the world?

China is home to three of the top four highest dams in the world. The highest in the world, the Jinping-I Dam on the Yalong River in Sichuan Province in China, is 1,001 feet (305 meters) high. Tajikistan is home to the second-highest dam in the world, the Nurek, which stands 980 feet (380 meters) high. The United States' tallest dam, Oroville Dam (in Northern California), is currently ranked twenty-fourth on the world list at 750 feet (228.6 meters) high.

How do farmers water their crops?

The process of artificially watering crops is called irrigation. In some areas of the world, agriculture can rely on rainfall for all of its water needs. In drier areas, usually those receiving less than 20 inches (51 centimeters) of rainfall per year, irrigation is required. Water is pumped from aquifers or delivered via an aqueduct to the fields, where it flows through small channels between plants or is sprayed through sprinklers. In very water-conservative regions such as Israel, water is scientifically dripped onto plants, thereby providing the exact amount of water necessary.

Although they were constructed many centuries ago, Roman aqueducts such as Pont del Diable in Tarragona, Spain, still stand today, a legacy of amazing engineering from ancient times.

How did the ancient Romans and Mesopotamians get water to their cities?

The ancient Romans and Mesopotamians built aqueducts to transport water between a source and areas where it was needed for agriculture or civilization. The Roman system was very extensive and was constructed throughout its empire. Some portions of these ancient aqueducts are still in use. Today, modern, concrete-lined channels transport water hundreds of miles. The most extensive aqueduct systems in the world today are those that bring water to southern California from the Colorado River in the east and from the Sacramento River in the north.

Were the Romans the only civilization to develop water resources in an advanced way?

No, recent excavations in Henan Province in China uncovered a network of clay pipes built during the Eastern Zhou Dynasty (1122 to 256 B.C.E.). The pipes were connected to many reservoirs around the cities, using technology that may actually predate Roman water works.

Where is the biggest hydroelectric dam in the world?

The Three Gorges Dam, which spans the Yangtze River in China's Hubei Province, is the biggest hydroelectric dam in the world. Completed in 2012, it generates more power than any power station in the world, producing 98.8 terawatt hours of power. It is capable of generating 22,500 megawatts of power using 32 turbines. Over 1,300,000 people were displaced as result of living in the valleys above the site of the dam. This fact, along with the potential loss of thousands of archaeological sites, has been a source of controversy since the project was conceived in the 1990s.

CLIMATE

DEFINITIONS

What is the difference between climate and weather?

Climate is the long-term (usually 30-year) average weather for a particular place. The weather is the current condition of the atmosphere. So, on a particular day, the weather in Barrow, Alaska, might be a hot 70° Fahrenheit, but its tundra climate is generally polar-like and cold.

How are different types of climates classified?

In 1884, the German climatologist Wladimir Köppen developed a climate classification system that is still used today, albeit with some modifications. He classified climates into five categories: tropical/megathermal, dry, temperate/mesothermal, continental/microthermal, polar, and alpine. He also created subcategories for these classifications. His climate map is often found in geography texts and atlases. In the 1960s, the climate classifications were updated to take into account vegetation that is native to certain climatic zones, providing more accuracy to the classification system.

What is a Mediterranean climate?

A Mediterranean climate is a climate similar to the one found along the Mediterranean Sea: warm, hot, and dry in the summer and mild, cool, and wet in the winter. Areas that are renowned for having a Mediterranean climate but are not near the Mediterranean Sea are California, southwestern Oregon, southwestern South Africa, and Chile.

What is global warming?

Global warming is the gradual increase of the Earth's average temperature—which has been rising since the Industrial Revolution (late eighteenth century–early nineteenth

How did the Inca civilization experiment with climate?

In the Urubamba Valley in Peru, in a city called Moray, are the remains of a great amphitheater-like terraced system. Archaeologists and scientists now believe that this was a great agricultural laboratory, where each area of the terrace exhibited completely different climates, allowing the Incas to experiment on cultivated vegetation with different climates and growing techniques.

century C.E.). If temperatures continue to increase, some scientists predict major climatic changes, including the rise of ocean levels due to ice melting at the poles. According to many scientists, global warming is primarily due to the greenhouse effect.

What is the greenhouse effect?

The greenhouse effect is a natural process of the atmosphere that traps some of the sun's heat near the Earth. The problem with the greenhouse effect, however, is that it has been unnaturally increased, causing more heat to be trapped and the temperature on the planet to rise. The gasses that have caused the greenhouse effect were added to the atmosphere as a byproduct of human activities, especially combustion from automobiles, output from factories, and the burning of forests.

What is the effect of global warming and climate change on Earth?

By the year 2100, relative to 1990, world temperatures could rise from 2 to 11.5° Fahrenheit (1.1 to 6.4° Celsius), and sea levels may rise 7.2 to 23.6 inches (18 to 59 centimeters). According to scientists at NASA, for 650,000 years, atmospheric carbon dioxide, a main contributor to warming, has never been above 300 parts per million. Since 1950, this number has steadily increased to nearly 400 parts per million in 2014.

What is air pollution?

Air pollution is caused by many sources. There are natural pollutants that have been around as long as the Earth, such as dust, smoke, volcanic ash, and pollens. Humans have added to air pollution with chemicals and particulates due to combustion and industrial activity.

What are the most polluted cities in the world in terms of air quality?

Scientists measure pollutants that are present in the air that we breathe by measuring the size of the particulate, including cancer-causing ammonia, carbon, nitrates, and sulfate. These pollutants easily pass into our bloodstream when we breathe. The places in the world where we may find severe concentrations of these chemicals in the air include New Delhi, India; Patna, India; Gwalior, India; Raipur, India; and Karachi, Pakistan. Of the top twenty most polluted cities in the world, according to the World Bank, thirteen are in India. The rest include Doha, Qatar; Igdir, Turkey; and Khorramabad, Iran.

What are the sources of air pollution?

Air pollution has two main sources: anthropogenic (man-made) and natural. Man-made sources of pollution include factories, cars, motorcycles, ships, incinerators, wood burning, oil refining, chemicals, consumer product emissions like aerosol sprays and fumes from paint, methane from garbage in landfills, and pollution from nuclear and biological weapons production and testing. Natural sources of pollution may include dust, methane from human and animal waste, radon gas, smoke from wildfires, and volcanic activity.

THE ATMOSPHERE

How much pressure does the atmosphere exert upon us?

Average air pressure is 14.7 pounds per square inch (1.0335 kilograms per square centimeter) at sea level.

Why is the sky blue?

This is one of the world's most frequently pondered questions, and, contrary to what some people believe, the sky's blue color is not due to the reflection of water. Light from the sun is composed of the spectrum of colors. When sunlight strikes the Earth's atmosphere, ultraviolet and blue waves of light are the most easily scattered by particles in the atmosphere. So, other colors of light continue to the Earth while blue and ultraviolet waves remain in the sky. Our eyes can't see ultraviolet light, so the sky appears the only color remaining that we can see: blue.

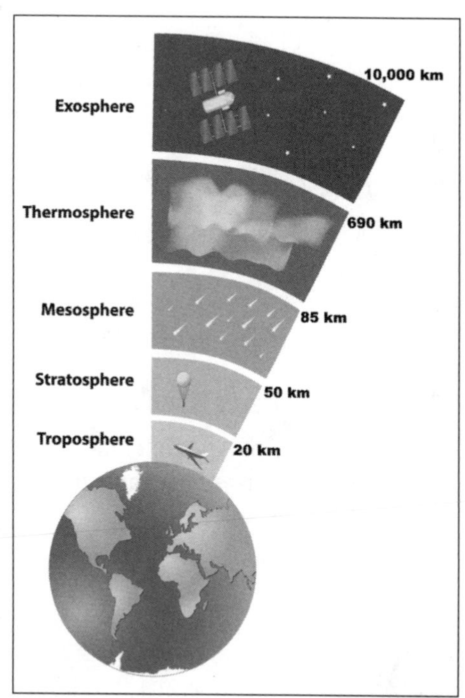

Earth's atmosphere is much thicker than most people think. Many just consider it the part that is the troposphere, but it actually extends about 6,200 miles (10,000 kilometers) into space.

How many layers are in the atmosphere?

There are five layers that make up the Earth's atmosphere. They extend from just above the surface of the Earth to outer space. The layer of the atmosphere that we breathe and exist in is called the troposphere and extends from the ground to about 10 miles (16 km) above the surface. From about 10 miles to 30 miles (16 to 48 km) up lies the stratosphere. The mesosphere lies from 30 to 50 miles (48 to 80 km) above the surface. A very thick layer, the thermosphere, lies from 50 miles all the way to 125 miles up (80 to 200 km). Above the 125-mile (200-km) mark lies the exosphere and space.

What is the air made of?

The air near the Earth's surface is primarily nitrogen and oxygen—nitrogen comprises 78.09% and oxygen 20.95%. The remaining 1% is mostly argon (0.93%), a little carbon dioxide (0.039%), and other gasses (0.06%).

Why can I hear an AM radio station from hundreds of miles away at night but not during the day?

At night, AM radio waves bounce off of a layer of the ionosphere, the "F" layer, and can travel hundreds, if not thousands, of miles from their source. During the day, the same reflection of radio waves cannot occur because the "D" layer of the ionosphere is present and it absorbs radio waves.

Why don't FM radio waves travel very far?

FM radio waves are "line of site," which means they can only travel as far as their power and the height of their radio antenna will allow. The taller the antenna, the farther the waves can travel along the horizon (as long as they have enough power).

Does air pressure change with elevation?

Yes, it does. The higher you go, the less air (or atmospheric) pressure there is. Air pressure is also involved in weather systems. A low-pressure system is more likely to bring rain and bad weather versus a high-pressure system, which is usually drier and brings clear skies. At about 15,000 feet (4,572 meters), air pressure is half of what it is at sea level.

What are the different kinds of clouds?

There are dozens of types of clouds, but they can all be classified into three main categories: cirriform, stratiform, and cumuliform. Cirriform clouds are feathery and wispy;

What is albedo?

Albedo is the amount of the sun's energy that is reflected back from the surface of the Earth. Overall, about 33 percent of the sun's energy bounces off the Earth and its atmosphere and travels back into space. Albedo is usually expressed as a percentage.

they are made of ice crystals and occur at high elevations. Stratiform clouds are sheet-like and spread out across the sky. Cumuliform clouds are the ubiquitous cloud that we often see—puffy and individual. These clouds can be harmless, or they can be the source of torrential storms and tornadoes.

How much of the Earth is usually covered by clouds?

At any given time, about one-half of the planet is covered by clouds.

How do airplanes create clouds?

When the air conditions are right and it's sufficiently moist, the exhaust from jet airplane engines often creates condensation trails, known as contrails. Contrails are narrow lines of clouds that evaporate rather quickly. Contrails can turn into cirrus clouds if the air is close to being saturated with water vapor.

OZONE

What is the ozone layer?

The ozone layer is part of the stratosphere, a layer of the Earth's atmosphere that lies about 10 to 30 miles (16 to 48 km) above the surface of the Earth. Ozone is very important to life on the planet because it shields us from most of the damaging ultraviolet radiation from the sun.

Is the ozone layer being depleted?

Scientists have recognized that a hole has developed in the ozone layer that has been growing since 1979. The hole is located over Antarctica and has been responsible for increased ultraviolet radiation levels in Antarctica, Australia, and New Zealand. As the ozone hole grows, it will increase the amount of harmful ultraviolet light reaching the Earth, causing cancer and eye damage and killing crops and microorganisms in the ocean.

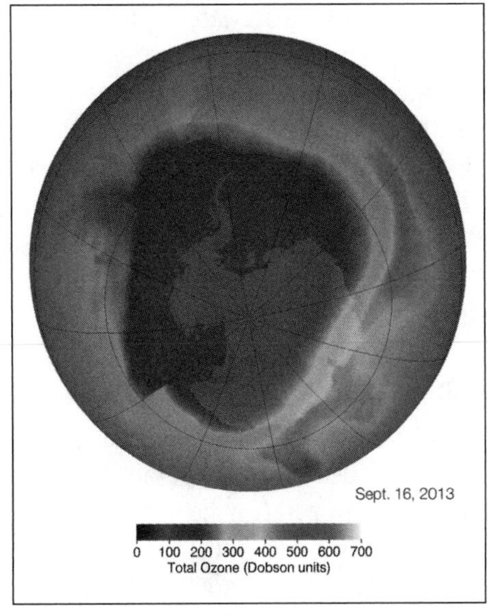

This 2013 graphic shows the extent of ozone depletion over Antarctica.

How much of the ozone layer is being depleted?

Since 1975, scientists believe that more than 33% of the ozone layer has disappeared. There is a seasonal factor to the reduction in ozone at any given time during the year, too. At different times, the ozone layer naturally declines or rises. But scientists also know that chlorofluorocarbons (CFCs), which are used for air conditioning, aerosol sprays, halon in fire extinguishers, and the interaction of man-made chemicals with nitrogen in our atmosphere directly cause ozone depletion. It is a man-made problem that requires a man-made solution. By 1989, 193 countries signed an agreement, called the Montreal Protocol, limiting the use of CFCs. And by 2011, all countries in the world signed this agreement, which ultimately has had a mitigating effect on the ozone hole, which was being observed as growing over Antarctica. Scientists at NASA who monitor the ozone in our atmosphere believe that the ozone depletion issue for the inhabitants on Earth is no longer a great problem today because concentrations of man-made ozone-depleting chemicals in use have stopped increasing and are actually declining.

How do CFCs destroy ozone?

When CFCs rise up in the atmosphere to the ozone layer, ultraviolet rays break them down into bromine and chlorine, which destroy ozone molecules.

CLIMATIC TRENDS

What is El Niño?

El Niño (also known as the El Niño Southern Oscillation [ENSO]), is a large patch of warm water that moves between the eastern and western Pacific Ocean near the equator. When the warm water of El Niño, about 1.8 degrees Fahrenheit (1 degree Celsius) warmer than normal, is near South America, the warm water affects the weather in the southwestern United States by increasing rainfall and is responsible for changes in the weather throughout the world. El Niño lasts for about four years in the eastern Pacific Ocean and then returns to the western Pacific near Indonesia for another four years. When the warm water is in the western Pacific, it is known as La Niña, the opposite of El Niño. When La Niña is in action, we have "normal" climatic conditions. Scientists at the National Oceanic and Atmospheric Administration (NOAA) monitor the effects of these types of changes to determine how they may influence our weather each year.

Can people live in a torrid zone?

The ancient Greeks divided the world into climatic zones that were not accurate. The three zones included frigid, temperate, and torrid. They believed that civilized people could only live in the temperate zone (which, of course, was centered around Greece). From Europe northward was part of the inhospitable frigid zone, while most of Africa was torrid. Unfortunately, this three-zone classification system stuck and was later expanded to five zones once the Southern Hemisphere was explored. People identify everything north of the Arctic Circle (near northern Russia) and south of the Antarctic Circle (near the coast of

Antarctica) as frigid, everything between the tropics and the Arctic and Antarctic circles as temperate, and the zone between the Tropics of Cancer and Capricorn as torrid.

Where does the name El Niño come from?

The phenomenon of El Niño was discovered by Peruvian fishermen who noticed an abundance of exotic species that arrived with the warmer water. Since this usually occurred around the Christmas season, they called the phenomenon El Niño, which means "the baby boy" in Spanish, in honor of the birth of Jesus Christ. La Niña, the opposite cycle of El Niño, means "the baby girl."

What are ice core samples and why are they important?

An ice core sample is a thick column of ice, sometimes hundreds of feet long, that is produced by drilling a circular, pipelike device into thick ice and then pulling out the cylindrical piece. Ice core samples from places like Greenland and Antarctica provide scientists with important clues about past climates. Air trapped in the ice remains there for thousands of years, so when scientists collect ice cores, they can analyze the air to determine the chemical composition of the atmosphere at the time the ice was formed. Sediments and tiny bugs are also found in the ice and provide additional clues to the state of the natural world at the time the ice was first deposited.

What is continentality?

Areas of a continent that are distant from an ocean (such as the central United States) experience greater extremes in temperature than do places that are closer to an ocean. These inland areas experience continentality. It might be very hot during the summer, but it can also get very cold in winter. Areas close to oceans experience moderating effects from the ocean that reduce the range in temperatures.

What are the horse latitudes?

Horse latitudes are high-pressure regions, more formally known as subtropic highs, which are warm and don't have much wind. Legend has it that the lack of wind sometimes caused sailors of the sixteenth and seventeenth centuries to throw their horses overboard in an effort to conserve water on board. That's how the region, centered around 30° latitude, got its name.

WEATHER

How does land turn into desert?

The process known as desertification is complicated and results from such activities as overgrazing, inefficient irrigation systems, and deforestation. It is most widespread in the Sahel region of Africa, a strip of land along the southern margin of the Sahara

As water use rises due to agriculture and increasing populations, lakes and even small seas are disappearing. The Aral Sea in Uzbekistan, for example, has almost disappeared over the last fifteen years.

Desert. The Sahara grows larger because of desertification. Desertification can be reversed by changing agricultural practices and by replanting forests.

What are Fahrenheit and Celsius?

Fahrenheit and Celsius are two common temperature scales used throughout the world. Temperature in Fahrenheit can be converted to Celsius by subtracting 32 and multiplying by five; divide that number by nine, and you have Celsius. Conversely, you can convert Celsius to Fahrenheit by adding 32, multiplying by nine, and finally dividing by five. Kelvin, a system used by scientists, is based on the same scale as Celsius. All you have to do is add 273 to your Celsius temperature to obtain Kelvin. Zero degrees Kelvin is negative 273° Celsius.

What is a low high temperature and a high low temperature?

When meteorologists look at daily temperature, there is always a low and a high temperature for each day. If the high temperature is the coldest high temperature for that day or for the month, you have a new record—a new low high. Conversely, if the low temperature for a day is quite warm and breaks records, that's a new high low!

Why is it hotter in the city than in the countryside?

Cities have higher temperatures due to an effect known as the urban heat island. The extensive pavement, buildings, machinery, pollution from automobiles, and other things urban cause an increase in warmth in the city. Cities such as Los Angeles can be up to five degrees hotter than surrounding areas due to the urban heat island effect. The term comes from temperature maps of cities where the hotter, urban areas look like islands when isotherms (lines of equal temperature) are drawn.

What are some world weather records?

The following are some amazing weather records. The wettest: Cherrapunji, and the town of Mawsynram nearby in India, near the border of Bangladesh, have the highest average rainfall per year in the world, averaging 467 inches (11,872 mm); the coldest: the East Antarctic Plateau, Antarctica, with a measurement of –136° Fahrenheit (–93.2° Celsius); the driest: Dry Valleys of Antarctica which receives 0 inches (0 centimeters) of rainfall per year; and the hottest: Lut Desert, Iran, which has sizzled at 159.3° Fahrenheit (70.7° Celsius) in five of the seven years from 2004 to 2009.

Why is it more likely to rain in a city during the week than on the weekend?

Urban areas have an increased likelihood of precipitation during the workweek because intense activity from factories and vehicles produce particles that allow moisture in the atmosphere to form raindrops. These same culprits also produce warm air that rises to create precipitation. A study of the city of Paris found that precipitation increased throughout the week and dropped sharply on Saturday and Sunday.

What does a 40% chance of rain really mean?

When the morning weather report speaks of a 40% chance of rain, it means that throughout the area (usually the metropolitan area), there is a four in ten chance that at least 0.001 of an inch of rain (0.0025 centimeters) will fall on any given point in the area.

Why is it more wet on one side of a mountain than the other?

It's much more wet on one side of a mountain than the other because of a process known as orographic precipitation. Orographic precipitation causes air to rise up the side of a mountain range and cool off, creating storms. The storms deposit a great deal of precipitation on that side of the mountain and create a rain shadow effect on the opposite side

Why are there so many discrepancies in the world records of weather?

The discrepancies in the data reflect the length of time that we use to measure weather phenomena. Some records were set by observing the weather over decades; others only occurred during the span of a few years, months, or even hours or minutes.

of the range. The Sierra Nevada Mountains are an excellent example of orographic precipitation because the mountains of the western Sierras receive considerable rainfall (far more than California's Central Valley), while the eastern Sierras are quite dry.

What is a rain shadow?

When the moisture in the air is squeezed out by orographic precipitation, there's not much left for the other side of the mountains. The dry side of the mountain experiences a rain shadow effect because they are in the shadow of the rain.

What is a thunderstorm?

Thunderstorms are localized atmospheric phenomena that produce heavy rain, thunder and lightning, and sometimes hail. They are formed in cumulonimbus clouds (clouds that are big and bulbous) that rise many miles into the sky. Most of the southeastern United States has over forty days of thunderstorm activity each year, and there are about 100,000 thunderstorms across the country annually. Thunderstorms are different from typical rainstorms because of their lightning, thunder, and occasional hail.

What are monsoons?

Occurring in southern Asia, monsoons are winds that flow from the ocean to the continent during the summer and from the continent to the ocean in the winter. The winds come from the southwest from April to October and from the northeast (the opposite direction) from October to April. The summer monsoons bring a great deal of moisture to the land. They cause deadly floods in low-lying river valleys, but they also provide the water southern Asia relies upon for agriculture.

What is the origin of the word "monsoon"?

The word "monsoon" comes from several source languages, including from the Portuguese word *moncau*, the Arabic word *mawsim*, and the early Dutch word *monsun*.

WIND

What are dust devils?

These columns of brown, dust-filled air, which can rise dozens of feet, are not as evil as the name suggests. They are caused by warm air rising on dry, clear days. Winds associated with dust devils can reach up to 60 miles (96.5 km) per hour and cause some damage, but they are not as destructive as tornadoes and usually die out pretty quickly.

What causes the wind to blow?

The Earth's atmospheric pressure varies at different places and times. Wind is simply caused by the movement of air from areas of higher pressure to areas of lower pressure.

The greater the difference in pressure, the faster the wind blows. Some detailed weather maps show wind speed along with isobars (areas of equal air pressure) indicating the level of air pressure.

Dust devils are like very weak tornadoes that never cause any damage.

In which direction does the west wind blow?

It blows from the west to the east. Wind direction is always named after the direction from where it originates.

What is the jet stream?

The jet stream is a band of swiftly moving air located high in the atmosphere, meandering across the troposphere and stratosphere, up to 30 miles high (48 km). The jet stream affects the movement of storms and air masses closer to the ground.

What are the westerlies?

These westerly winds flow at mid-latitudes (30 to 60 degrees north and south of the equator) from west to east around the Earth. The high-altitude winds known as the jet stream are also westerlies.

What is katabatic wind?

Katabatic wind is high-density air that moves from a higher elevation down a slope because of the force of gravity. These winds are sometimes known as "fall winds."

What is the windiest place on Earth?

Because of katabatic winds, Antarctica frequently wins the top honors for being the windiest place on Earth. Winds near Commonwealth Bay, which was discovered in 1912, are frequently recorded to be 150 miles per hour (240 km per hour), with an average wind speed over the course of a year of 50 miles per hour (80 km per hour).

What is a willy-willy?

Willy-willy is a word that traces its roots to indigenous Australian language. It is a term used to describe a dust devil, when air is heated up on the ground, drawing forcefully nearby cooler air, causing the air to spiral upward.

Is Chicago really the "Windy City"?

Chicago is not the windiest big city in the lower 48 states of the United States. Chicago's average wind speed of 10.3 miles (16.58 km) per hour is beat by Boston (12.3 mph/19.79 kph), Dallas (10.7 mph/17.22 kph), Oklahoma City (12.2 mph/19.63 kph), Buffalo (11.8 mph/18.99 kph), and Milwaukee (11.5 mph/18.51 kph).

What is the origin of the name "Windy City"?

Although Chicago is not really that windy compared with other American cities, the name has been used since the nineteenth century. It refers not necessarily to the weather but to the observations by many when describing Chicago politicians, metaphorically, as "talkative," "boastful," and "self-promoting."

What world weather record does the United States hold?

The United States claims the world's highest surface wind, 318 miles per hour (511.77 km per hour), during a tornado in Oklahoma in 2009.

HAZARDS AND DISASTERS

What is the difference between a watch and a warning?

The U.S. National Weather Service issues watches and warnings for a variety of hazards when they may be imminent. A watch (such as a tornado watch or a flood watch) means that such a hazardous event is likely to occur or is predicted to occur. A warning is more serious. It means that a hazard is already occurring or is imminent. Warnings are usually broadcast on television and radio stations via the Emergency Alert System (formerly known as the Emergency Broadcast System).

How should we prepare for disaster?

Disasters can and do happen everywhere. You should prepare for disaster by having a disaster supply kit with supplies for you and everyone in your family available at home and work, as well as a minikit in your automobile. It should include food, water, first-aid equipment, sturdy shoes, an AM/FM radio (with batteries kept outside of the radio), a flashlight (with batteries kept outside of the flashlight), vital medication (especially prescription medication), blankets, cash (if the power and computers are down, credit and ATM cards won't work), games and toys for children, and any other essentials. Contact your local chapter of the Red Cross for more information about disaster preparedness.

Should we use candles after a disaster or power outage?

Many deaths and a great deal of property damage have been caused by fires resulting from people using candles following a disaster. People leave candles burning as a source of light, but these can fall over and start fires. It is strongly advised that people not use candles when the power goes out. There are many flashlights and battery-operated lanterns that are available commercially and should be part of your disaster supply kit. Most cell phones have flashlights as part of the rear-camera light, too.

What's the difference between the old Emergency Broadcast System and the Emergency Alert System?

The Emergency Broadcast System (EBS), created in 1964 to warn the country of a national emergency such as nuclear attack, became the Emergency Alert Service (EAS) in 1997. The old EBS system relied on one primary radio station in each region to receive an emergency message and then broadcast it to the public and other media outlets. The new system, which also includes cable television, operates via computer and can be automatically and immediately broadcast to the public. It also allows additional local governmental agencies the opportunity to broadcast emergency messages. Future plans for the EAS include radios and televisions that will automatically turn on when an alert is announced.

What is the leading cause of disaster-related deaths in the United States?

Lightning is the leading cause of disaster-related deaths in America. From 1940 to 1981, about 7,700 people died from lightning strikes, 5,300 from tornadoes, 4,500 from floods, and 2,000 from hurricanes. So, it's best to avoid open spaces, elevated groundwater, tall, metal objects, and metal fences during an electrical storm. In an analysis of eight years of data from 2006 to 2013 by experts at the National Weather Service of the NOAA, fishermen accounted for three times the number of deaths from lightning strikes than golfers. About 261 people were killed by lightning during this period, roughly three per year.

What is the best way to help after a disaster?

Disaster relief agencies such as the Red Cross are in vital need of money after a disaster to purchase necessary items for victims or provide financial support to them. Go online or call your local chapter of the Red Cross to find out how to help. Donating food or clothing is burdensome on the agencies in the immediate aftermath of a disaster, as personnel are not available to sort, clean, or distribute donated goods.

How did a map help stop the spread of cholera?

During an 1854 cholera outbreak in London, a physician named John Snow mapped the distribution of cholera deaths. His map showed that there was a high concentration of deaths in an area surrounding one specific water pump (water had to be hand-pumped and carried in buckets at the time). When the handle was taken off of the water pump, the number of cholera deaths plummeted. When it was determined that cholera could be spread through water, future epidemics were curbed. This was the beginning of medical geography.

What are incidence maps?

Researchers at such institutions as the Centers for Disease Control and Prevention (CDC) use incidence maps, which plot where and how people have been infected or exposed to such potentially harmful viruses as influenza, Ebola virus, West Nile virus, and HIV in order to understand the rate of transmission as related to geography. An incidence map may help scientists figure out the origin of a disease and where and how quickly it is spreading. Global incidence maps are of increasing importance in the fight against potentially harmful biological disasters.

How does medical geography help control the spread of diseases?

Medical geographers and epidemiologists (scientists who study diseases and epidemics) use mapping to monitor the spread of diseases and locate the source of a disease. For example, by mapping a group of inordinately high numbers of cancer patients in a city, we may find that all live close to a factory that has been releasing cancer-causing toxins into the groundwater. By identifying the source and spread of a disease, the disease can often be combated. Organizations like the Centers for Disease Control and Prevention use maps to describe the outbreak of such diseases as Ebola and to determine quarantine areas and where to focus treatment efforts.

Which natural disasters doesn't southern California experience?

Urban southern California is plagued by many natural disasters, including earthquakes, wildfires, floods, landslides, and tornadoes. Thankfully, they rarely receive snowstorms or hurricanes.

What causes wildfires to occur?

Approximately 10% of all wildfires are caused by lightning that ignites material and forms a wildfire. Experts at the U.S. Park Service assert that the remaining 90% of all wildfires are started by human activity, like campfires, the burning of trash, discarded cigarettes, and arson.

How can I learn more about disasters in my town?

Each community should have its own disaster plan that includes a history of past disasters (those that have happened in the past are likely to occur in the future) along with plans for dealing with future disasters. You should be able to consult this plan to learn how your community would cope with disaster and to find out the locations of evacuation routes and shelters. Many communities place important disaster-planning information online for easy reference.

VOLCANOES

How are volcanoes formed?

Volcanoes are the result of magma rising or being pushed to the surface of the Earth. Hot liquid magma, which is located under the surface of the Earth, rises through cracks and weak sections of rock. The mountain surrounding a volcano is formed by lava (called magma until it arrives at the Earth's surface) that cools and hardens, making the volcano taller, wider, or both.

What is the difference between magma and lava?

Magma is hot, liquefied rock that lies underneath the surface of the Earth. When magma erupts or flows from a volcano onto the Earth's surface, it becomes lava. There is no difference in substance; only the name changes.

What is the Ring of Fire?

If you were to look at a map of the world's major earthquakes and volcanoes, you would notice a pattern circling the Pacific Ocean. This dense accumulation of earthquakes and volcanoes is known as the Ring of Fire. The ring is due to plate tectonics and the merger of the Pacific Plate with other surrounding plates, which creates faults and seismic activity (especially Alaska, Japan, Oceania, and coastal North and South America), along with volcanic mountain ranges, such as the Cascades of the U.S. Pacific Northwest and the Andes of South America.

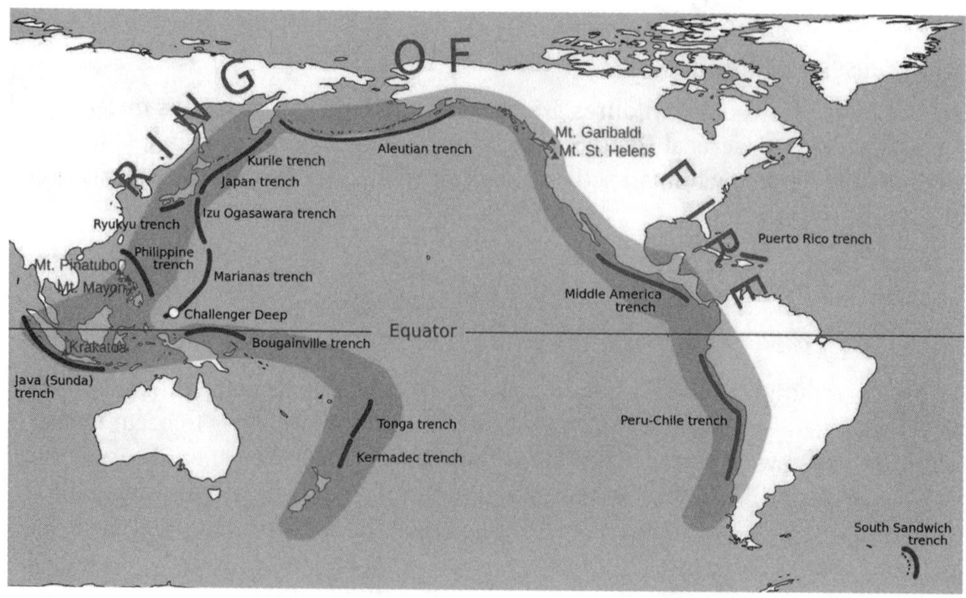

The Ring of Fire encircles the Pacific Ocean with volcanic and earthquake activity.

How many active volcanoes are there in the world?

There are about 1,500 potentially active volcanoes around the world, but the number often depends on how we define active. Experts at the National Geographic Society consider approximately 1,900 volcanoes on Earth to be active, with evidence of activity and a strong likelihood to explode again. Most are located in the Ring of Fire surrounding the Pacific Ocean. About one-tenth of the world's active volcanoes are located in the United States. A volcano is considered active if it has erupted in the last 10,000 years. If you were to consider how many volcanoes may be at the bottom of the sea floor, there could be many thousands more.

What are some of the world's most active volcanoes, in terms of numbers of years of eruptions?

The volcanoes that have been active the most number of years include Mt. Etna in Italy (3,500 years), Mt. Stromboli in Italy (2,000 years), and Mt. Yasur in Vanuatu (800 years).

How many active volcanoes are located in Europe?

There are more than sixty active volcanoes in Europe and more than forty dormant ones, many of which are located in Italy, Sicily, and Greece.

Which volcano poses the most risk to people in Europe?

Mt. Vesuvius, on the southwestern coast of Italy, lies very near a city of more than one million people: Naples. Although it last erupted in 79 C.E., it is still active, and geologists predict there is a very good chance it will erupt again, potentially putting the city of Naples and the surrounding area at risk.

How many active volcanoes does Iceland have?

Iceland, formed by volcanoes, lies between two tectonic plates along the Mid-Atlantic Ridge and is home to approximately 130 volcanoes. Of these, 30 are considered to be active.

Where are the active volcanoes in the United States?

Alaska, Washington, Oregon, and California have many potentially active volcanoes. The most recent large-scale eruption in the United States was that of Mt. St. Helens in southern Washington State in 1980. Other volcanoes in the region, such as Mt. Shasta, Mt. Lassen, Mt. Rainier, and Mt. Hood, could erupt with little warning.

What is so interesting about volcanoes in Alaska?

Alaska contains approximately 130 volcanoes, with 90 that have been active in the last 10,000 years. Volcanoes in the state of Alaska account for over 75% of all volcanoes that have erupted in the United States in the last two hundred years.

The ruins of Pompeii attract tourists today, but it is difficult to imagine the horror of an entire city wiped out by a volcano.

How was Pompeii destroyed?

In the year 79 C.E., the volcano Mt. Vesuvius erupted and buried the ancient Roman town of Pompeii under 20 feet (6 meters) of lava and ash. Pompeii is famous because excavations of the city, which began in 1748 and continue to this day, provide an excellent look at Roman life at the beginning of the millennium. The covering of the city by debris preserved not only the places where people last stood but also paintings, art, and many other artifacts. The nearby city of Herculaneum was also buried and perfectly preserved. Although a much smaller version of Pompeii, it contains some of the best art, architecture, and examples of daily life in Roman times and is only twenty minutes away from Pompeii. Even loaves of bread that were baking on the day of the eruption are preserved in a nearly 2,000-year-old bakery.

EARTHQUAKES

What creates earthquakes?

The tectonic plates of the Earth are always in motion. Plates that lie side by side may not move very easily with respect to one another; they "stick" together, and occasionally they slip. These slips (from a few inches to many feet) create earthquakes and can often be very destructive to human lives and structures.

What is an epicenter?

An epicenter is the point on the Earth's surface that is directly above the hypocenter, or the point where earthquakes actually occur. Earthquakes do not usually occur at the surface of the Earth but at some depth below the surface.

What is a fault?

A fault is a fracture or a collection of fractures in the Earth's surface where movement has occurred. Most faults are inactive, but some, like California's San Andreas Fault, are quite active. Geologists have not discovered all of the Earth's faults, and sometimes earthquakes occur that take the world by surprise, like the one that occurred in 1994 in Northridge, California. When earthquakes occur on faults that were previously unknown, they are called blind faults.

What is the significance of the infamous San Andreas Fault?

The infamous San Andreas Fault lies at the border between the North American and the Pacific tectonic plates. This fault is situated in California and is responsible for some of the major earthquakes that occur there. Los Angeles is on the Pacific Plate, but San Francisco is on the North American Plate. The Pacific Plate is sliding northward with respect to the North American Plate, and, as a result, Los Angeles gets about half an inch closer to San Francisco every year. In a few million years, the two cities will be neighbors.

Was San Francisco destroyed by earthquake or by fire in 1906?

In 1906, a very powerful earthquake struck San Francisco, California, which sparked a fire that destroyed much of the city. In an effort to preserve San Francisco's image with residents and would-be visitors, official policy regarding the disaster stated that it was not the earthquake but mostly the fire that destroyed the city. Official books and publications produced after the earthquake referred to both the fire and the earthquake as having caused the damage. In fact, the earthquake did considerable damage to the city and killed hundreds.

Will California eventually fall into the ocean?

No, it will not. The famous San Andreas Fault, which runs along the western edge of California from the San Francisco Bay area to southern California, is known as a transverse fault. This means that the western side of the fault, which includes places like Monterey, Santa Barbara, and Los Angeles, is sliding northward with respect to the rest of the state. In a few million years, the state's two largest urban areas, San Francisco and Los Angeles, will be right next to each other. The fault is moving at about two centimeters (just under an inch) a year.

A view of San Francisco taken from the Union Ferry Building and looking toward Market street shortly after the 1906 earthquake and fire.

Which states are earthquake-free?

While a twenty-year period isn't an excellent indicator, there are four states that had no earthquakes between 1975 and 1995: Florida, Iowa, North Dakota, and Wisconsin.

Is there a high risk of earthquakes in the Midwestern United States?

Great earthquakes struck the New Madrid, Missouri, area in 1811 and 1812. They caused considerable damage (some areas experienced shaking at the level of XI on the Mercalli scale) and were felt as far away as the East Coast. The potential exists for future earthquakes in the region, since earthquakes have occurred there before. Planning and preparedness continues throughout the region, centered at the junction of Missouri, Arkansas, Illinois, Kentucky, Tennessee, and Mississippi.

What should I do in the event of an earthquake?

Duck, cover, and hold! Duck under a table, counter, or any area that can provide protection from falling objects. Cover the back of your head with your hands to help protect against flying debris. Hold on to the leg of the table or anything solid to ride out the shaking.

Is it safe to stand in a doorway during an earthquake?

While a doorway is a nice, structurally sound place to be during an earthquake, officials have found that many people are injured when a door swings open and closed during an earthquake, so you may want to avoid standing in a place where your fingers can become crushed.

What is the Richter scale?

The Richter scale measures the energy released by an earthquake. It was developed in 1935 by California seismologist Charles F. Richter. With each increase in Richter magnitude, there is an increase of thirty times the energy released by an earthquake. For example, a 7.0 earthquake has thirty times the power of a 6.0 earthquake. Each earthquake only has one Richter magnitude. The strongest earthquakes are in the 8.0–9.0 range—8.6 for Alaska's 1964 earthquake and 8.0 for China's 1976 earthquake in Tangshan.

What is the Mercalli scale?

The Mercalli scale measures the power of an earthquake as felt by humans and structures. Italian geologist Giuseppe Mercalli developed it in 1902. The Mercalli scale is written in Roman numerals, and it ranges from I (barely felt) to XII (catastrophic). The Mercalli scale can be mapped surrounding an epicenter and will vary based on the geology of an area.

The Mercalli Scale of Earthquake Intensity

Scale	Effects
I	Barely felt
II	Felt by a few people, some suspended objects may swing
III	Slightly felt indoors as though a large truck were passing
IV	Felt indoors by many people, most suspended objects swing, windows and dishes rattle, standing autos rock
V	Felt by almost everyone, sleeping people are awakened, dishes and windows break
VI	Felt by everyone, some are frightened and run outside, some chimneys break, some furniture moves, causes slight damage
VII	Considerable damage in poorly built structures, felt by people driving, most are frightened and run outside
VIII	Slight damage to well-built structures, poorly built structures are heavily damaged, walls, chimneys, monuments fall
IX	Underground pipes break, foundations of buildings are damaged and buildings shift off foundations, considerable damage to well-built structures
X	Few structures survive, most foundations destroyed, water moved out of banks of rivers and lakes, avalanches and rockslides, railroads are bent
XI	Few structures remain standing, total panic, large cracks in the ground
XII	Total destruction, objects thrown into the air, the land appears to be liquid and is visibly rolling like waves

What does an earthquake feel like?

Smaller earthquakes or tremors feel disorienting at first. You feel a sense that the room is spinning, as if you are becoming dizzy. Usually preceding an earthquake, when the initial tremors hit, you can hear the sounds of things rattling that you have never heard before, like glasses rubbing against each other and windows vibrating. With larger earthquakes, as the earth nearby tears or opens, you can hear a very loud rumbling sound that is similar to a passing train.

How many mini-earthquakes happen each year on our planet?

Experts at the U.S. Geological Survey believe that if we consider earthquakes of low magnitudes of between 2 and 2.9, there are an estimated 1.3 million mini-earthquakes each year, somewhere on Earth.

How many really big earthquakes occur each year?

On average, there are about 134 earthquakes of a magnitude 6.0–6.9, about fifteen of a magnitude 7.0–7.9, and one huge magnitude 8.0–8.9 earthquake each year. Many of these really big earthquakes occur in the ocean, so we don't hear much about them.

Is a magnitude ten the top of the Richter scale?

While the media often refers to the Richter scale as being on a scale of one to ten, there is no upper limit, even though the strongest quakes are not as high as ten.

TSUNAMIS

What causes a tsunami?

A tsunami, also known as a seismic sea wave, is usually caused by an earthquake that occurs under the ocean or near the coast of a landmass. The seismic energy creates a large sea wave that can cause heavy damage hundreds or even thousands of miles from its source. The state of Hawaii is frequently struck by tsunamis.

How does Hawaii protect itself from tsunamis?

There is a sophisticated global monitoring network that provides warnings about possible tsunamis, allowing the islands of Hawaii and other coastal areas to prepare for impending disaster. Hawaii also has a thorough evacuation system to protect lives in the face of tsunami danger.

What caused the great Indian Ocean tsunami of December 2004?

The great Indian Ocean tsunami of December 2004 was caused by a magnitude 9.0 earthquake in the ocean off the coast of Sumatra, Indonesia, which then caused a reverberating swell of water to move toward the countries of Indonesia, Thailand, Sri Lanka, India, and the Maldives. Its effects were felt as far away as Africa before its energy finally dissipated.

What was the effect of the 2011 Tohoku earthquake and tsunami that occurred in Japan?

The magnitude 9.0 earthquake that occurred off the coast of Japan in March 2011 caused a tsunami, which melted down three nuclear reactors at the Fukushima Daiichi Nuclear Power Plant. Approximately 16,000 people lost their lives, and many millions of people were without electricity and water for several weeks after the incident.

What is the Pacific Tsunami Warning Center?

The United States has two tsunami warning centers that are administered by the National Oceanic and Atmospheric Administration: the Pacific Tsunami Warning Center based in Hawaii and the West Coast and Alaska Tsunami Warning Center. The Pacific Tsunami Warning Center is the command center for monitoring and warning all nations that may be affected by a tsunami. With data from a network of thirty-nine detection buoys called the DART array, the center can issue alerts of real-time earthquake activity in the Pacific basin and the tsunamis that may result, giving residents affected time to head to safe ground, away from low-lying coastal areas.

What is the DART array?

DART stands for the Deep-Ocean Assessment and Reporting of Tsunamis, which consists of an array of thirty-nine buoys that float in critical spots in the Pacific. Each DART system consists of an anchored seafloor bottom-pressure recorder (BPR) and a companion moored surface buoy for real-time communications. An acoustic link transmits data from the BPR on the seafloor to the surface buoy. The BPR collects temperature and pressure at fifteen-second intervals. In normal mode, it transmits the data every fifteen minutes. If there is an event, the system reports back data collected in fifteen-second intervals every minute.

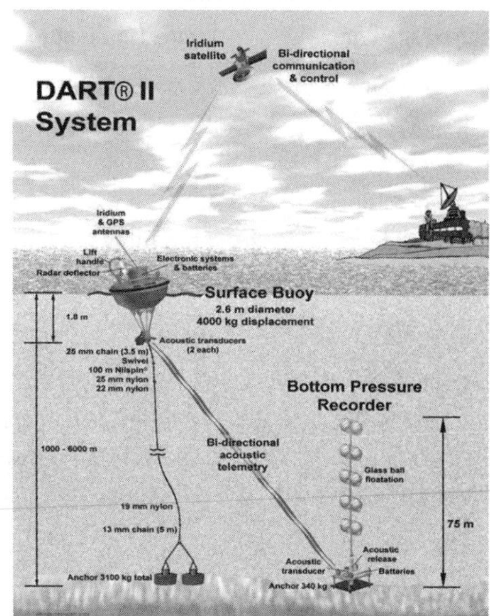

The Deep-Ocean Assessment and Reporting of Tsunami (DART) array helps to warn people early of an approaching tsunami.

HURRICANES

Why are hurricanes so destructive?

It is the floods caused by hurricanes that are the most destructive element. The low-pressure center of a hurricane causes a mound of water to rise above the surrounding water. This hill of water is pushed by the hurricane's fierce winds and low pressure onto the land, where it floods coastal communities, causing significant damage. Hurricanes also spark tornadoes that contribute to the devastation.

How fast do hurricane winds blow?

The strongest hurricanes, or category 5 hurricanes, often have winds that reach speeds well over 150 miles (240 km) per hour.

How are hurricanes ranked?

We use the Saffir-Simpson Hurricane Wind Scale to rank the intensity and destructive capacity of hurricanes. Hurricanes are ranked on a scale of one to five, with category one hurricanes being the weakest and category five being the strongest and most destructive. The rankings for damage caused by each category of hurricane are: 1, minimal; 2, moderate; 3, extensive; 4, extreme (such as Hurricane Andrew in 1992); and 5, catastrophic.

Hurricane Type	Winds (mph/kph)	Surge Levels
Category One	74–95 mph/119–153 kph	4–5 feet/1.2–1.5 meters
Category Two	96–110 mph/154–177 kph	6–8 feet/1.8–2.4 meters
Category Three	111–130 mph/178–209 kph	9–12 feet/2.75–2.4 meters
Category Four	131–155 mph/210–249 kph	13–18 feet/4–5.5 meters
Category Five	>155 mph/249 kph	>18 feet/5.5 meters

What was Hurricane Katrina?

Hurricane Katrina was the name given to the hurricane that developed in the Gulf of Mexico and struck New Orleans and many other cities along the southern coast of the United States in late August 2005. Winds from Katrina were initially only a category two hurricane, with a tidal surge ranked as a category three.

FLOODS

How many people died as a result of the subsequent failure of the levees and flooding after Hurricane Katrina struck?

Approximately 1,460 people lost their lives following the landfall of Hurricane Katrina.

Was the 2005 New Orleans disaster caused by a flood or a hurricane?

The initial cause of the disaster was Hurricane Katrina, which whipped up tides and forced rain and seawater against a very fragile levee system that protected New Orleans. Since the city is 49% below sea level, when the man-made levees broke, flood waters moved in and inundated much of the city.

How much precipitation may cause flooding?

The amount varies widely for different areas. In some U.S. western deserts, or in some large urban areas, just a few minutes of strong rain will cause a flash flood in canyons and low-lying areas, both urban and rural. In areas prone to greater rainfall amounts, it often takes quite a bit more rain (sometimes a few days' or weeks' worth) to cause rivers to overflow and dams to fill up, raising concerns of those who live downstream. Areas that normally receive more rainfall have better natural drainage systems and are usually home to plants that readily absorb the extra water.

What have been some of the most destructive floods in history?

Some of the world's most catastrophic flooding takes place in China. A flood on the Huang He River in 1931 killed up to 3.7 million people. By comparison, a flood in the United States in 1889 that caused the failure of a dam upstream from the community of Johnstown, Pennsylvania, killed only 2,200 people.

What is a floodplain?

A floodplain is the area surrounding a river that, when unmodified by human structures, would normally be flooded during a river flood. A floodplain can be a few feet or many miles wide, depending on the river flow as well as the local terrain. Some floodplains are as large as an entire country, like Vietnam's Mekong River Delta, which covers 7,450 square miles (12,000 square feet).

The plain is usually a flat area with areas of higher elevation on either side. Even though levees and floodwalls can be built (with homes and businesses built just behind them), the floodplain does not vanish. If the structures break or are damaged, the water from a flood can fill a floodplain, just as it did before humans occupied it.

What is a 100-year flood?

A 100-year flood refers not only to the size of a flood but also to the odds of it occurring. A 100-year flood has a one percent (or 1 in 100) chance of occurring in any given year. It has no relationship to the frequency of occurrence. The magnitude of such a flood is relative to the frequency of occurrence, so a 100-year flood is much larger than any run-of-the-mill annual flood. A 500-year flood only has a one in 500 (0.2%) chance of oc-

Why do people live in floodplains?

People have lived in floodplains for thousands of years. Fertile land for agriculture lines the floodplain, and the nearby water source makes life easier. Unfortunately, when the river does flood, these communities are severely damaged and people suffer. Hazard mitigation, such as levees, dams, dikes, and other structures, attempt to limit damage during floods. Sometimes, when the structures fail (such as a levee breaking), large areas are inundated with water. Inhabitants of floodplains must balance the risks with the rewards of living in such an unpredictable environment.

curring in any given year and would be much larger and more devastating than a 100-year flood.

What is the National Flood Insurance Program?

The National Flood Insurance Program (NFIP) was established by the U.S. federal government in 1956 as a subsidized insurance program for home and business owners, seeking to reduce the socioeconomic impact of floods on private and public structures. The government began the program by creating Flood Insurance Rate Maps (FIRM) that showed the boundaries of 100-year and 500-year flood zones. The cost of the insurance is based on the flood risk. The Federal Emergency Management Agency (FEMA) oversees the program and requires the purchase of flood insurance by any owners potentially affected by a disaster before they can be provided with disaster assistance. This way, the next time a flood occurs, they will be insured.

How can I obtain a flood map of my community?

The best way to see a Flood Insurance Rate Map for your area would be to contact your local government. Their planning or emergency management agency should have the FIRM maps available. Users may search FEMA's Flood Map Service Center by entering an address and viewing a map on their site. Obtaining maps from FEMA may not be recommended because the maps change often and are best interpreted by a planning or emergency expert.

What should I do in the event of a flood?

If a flood is expected, turn on a battery-powered radio and listen for information about when and where to evacuate. If a flood or flash flood is coming toward you, move quickly to a higher elevation—but don't ever try to outrun a flood. Also, don't drive through standing water, as it can quickly rise and stall your vehicle, possibly trapping you among swiftly moving water.

TORNADOES

What are tornadoes?

Tornadoes are very powerful, yet tiny, storms that have destructive winds capable of leveling buildings and other structures. Winds in a tornado form a dark gray column of air, with the center of the tornado acting like a vacuum, picking up objects and moving them along the storm's path. Tornadoes can last from a few minutes to an hour.

What should I do when a tornado approaches?

Try to get to the lowest level of the building (unless you are in a mobile home or outdoors, in which case you should seek a sturdy and safe shelter). Go to the center of the room and hide under a sturdy piece of furniture. Stay away from windows, hold on to the leg of a table or something else stable, and protect your head and neck with your arms.

What is the Fujita scale of tornado intensity?

The Fujita scale measures the strength of a tornado based on observed damage and effects. The scale ranges from F0 (a weak tornado) through F6 (an almost inconceivable tornado, having close to no chance of actually occurring). About 75% of all tornadoes are weak (F0–F1), while only 1% are violent (F4–F5). It was developed by T. Theodore Fujita of the University of Chicago in 1971.

Where is Tornado Alley?

Tornadoes occur more frequently in the central United States than anywhere else in the world. Tornado Alley is an area stretching from northwest Texas, across Oklahoma (the tornado capital of the world), and through northeast Kansas. On average, over 200 tornadoes occur across Tornado Alley each year.

The awesome power of tornadoes is still not fully understood by meteorologists.

What is the most dangerous state to live in due to tornadoes?

Massachusetts is considered the most dangerous state to live in due to tornadoes. While Oklahoma receives far more tornadoes than Massachusetts does, the population density and risk of death or severe injury is greater in the New England state.

How many people on average are killed by tornadoes each year in the United States?

Approximately eighty people are killed each year due to the destructive power of tornadoes. Many more people are injured and displaced as a result.

What states have the most tornadoes each year?

Using data analyzing tornado frequency each year over a long period of time (1953–2004), researchers at the National Climatic Data Center found the following average annual occurrence in the United States. Some of the highest states include Texas (139), Oklahoma (57), Florida and Kansas (55), and Nebraska (45).

What were some of the most destructive tornadoes in U.S. history?

Some of the worst tornadoes in U.S. history include: the Tri-State tornado, which struck Missouri, Indiana, and Illinois in 1925, killing 695 people and injuring 2,027; the 1840 Natchez tornado, which struck Mississippi, killing 317 people and causing injuries to another 109; and the St. Louis/East St. Louis tornado of 1896, which killed 296 people and injured 1,000.

Are there tornadoes in Europe?

While 90% of tornadoes occur in the United States, there are tornadoes in Europe, especially in western France. Other tornado regions of the world include northern South America (especially Brazil and Argentina), southern Mexico, Australia, Bangladesh, South Africa, and Japan.

LIGHTNING

How many bolts of lightning strike our planet each year?

Each year lightning flashes forty to fifty times per second, or approximately 1.4 billion times per year. Most of these flashes of light only occur for a few microseconds and are detectable by scientists. Seventy percent of all lightning occurs in the tropics. But only 25% of all lightning actually moves from cloud to ground.

What different types of lightning are there?

There are four types of lightning: cloud-to-cloud, within a cloud, cloud-to-ground, and cloud-to-air. Of course, cloud-to-ground is the most dangerous form of lightning, especially in the spring and summer months, when more people are more likely to be outside.

Lightning strikes occur someplace on the planet 24/7. The safest thing to do during a lightning storm is to stay indoors.

How much energy does one bolt of lightning contain?

A bolt of lightning contains enough energy to light a 100-watt light bulb for three months.

How many people are killed in the United States by lightning?

In the year 2013, only twenty-three people died from being killed by lightning, which is the fewest fatalities since 1940, when records of fatalities were first recorded. During the 1940s, because of the large rural population in the United States at the time, more people were exposed to the hazards of lightning. For example, 432 people were killed by lightning in 1943.

Looking at data from 1959 to 2003, approximately seventy-three people die each year after being struck by lightning, and about 300 are injured annually. The highest death rates are in Florida, with 425 fatalities.

Does lightning ever strike twice in the same place?

Lightning can and often does strike in the same place twice. Since lightning bolts head for the highest and most conductive point, that point often receives multiple strikes of lightning in the course of a storm. So to be safe, stay away from something that has already been struck by lightning! Tall buildings (such as the Empire State Building) often receive numerous lightning strikes during a storm.

OTHER HAZARDS AND DISASTERS

What is acid rain?

Motor vehicles and industrial activity release tons of wet and dry pollutants into the air. When mixed together, the pollutants form sulfuric and nitric acids that later fall to the ground in rain or snow. This precipitation is known as acid rain. Acid rain is responsible for damaging lakes by killing plant and animal life and for killing trees around the world. Canada has been especially hard-hit by acid rain caused by industrial activities in the United States. Two-thirds of all sulphuric acid and one-quarter of all nitric acid found in acid rain can be traced to U.S. electric power generation, which relies upon fossil fuel, such as coal, in order to operate.

Does radiation from a nuclear plant stop at the 10-mile (16-km) zone?

American nuclear power plants are required to create emergency planning zones within a 10-mile (16-km) radius surrounding their plants. These imaginary 10-mile lines are not walls that hold back the effects of radiation but simply a distance determined by emergency planners. In the event of an accident, the residents of the 10-mile (16-km) zone might not need to be evacuated but could be advised to remain indoors with their windows closed. Nuclear plants also establish smaller zones of two and five miles (three to eight km) surrounding the plants, within which the risk of radiation exposure is much greater.

What happened atThree Mile Island?

Three Mile Island, Pennsylvania, was the site of the United States' worst nuclear accident. Luckily, no radiation was released into the environment and no one was killed. In March 1979, the nuclear reactor at the Three Mile Island plant overheated, breaking the radioactive rods that are inserted inside the reactor. Pennsylvania's governor recommended a voluntary evacuation of pregnant women and preschool children who lived within 5 miles (8 km) of the plant. It was the unexpected self-evacuation of residents in the area that created major problems. The evacuations yielded surprising information about the lack of preparedness of communities for such an event, leading to increased planning and preparedness for nuclear accidents and evacuations.

What is nuclear winter?

A nuclear winter is what would follow a large-scale nuclear war. Radioactive particles, dust, and smoke released into the atmosphere would create a large cloud over the planet, blocking out sunlight and reducing temperatures worldwide. Plants and animals would die due to the extremely low temperatures. An extended nuclear winter could cause the death of millions of people from starvation, cold, and other problems.

What caused the Bhopal disaster?

In December 1984, the U.S.-owned Union Carbide pesticide plant in Bhopal, India, leaked toxic chemicals (methyl isocyanate gas) that killed over 3,800 people. It was the worst industrial accident in history. Union Carbide paid a fine of $470 million to avoid facing criminal charges.

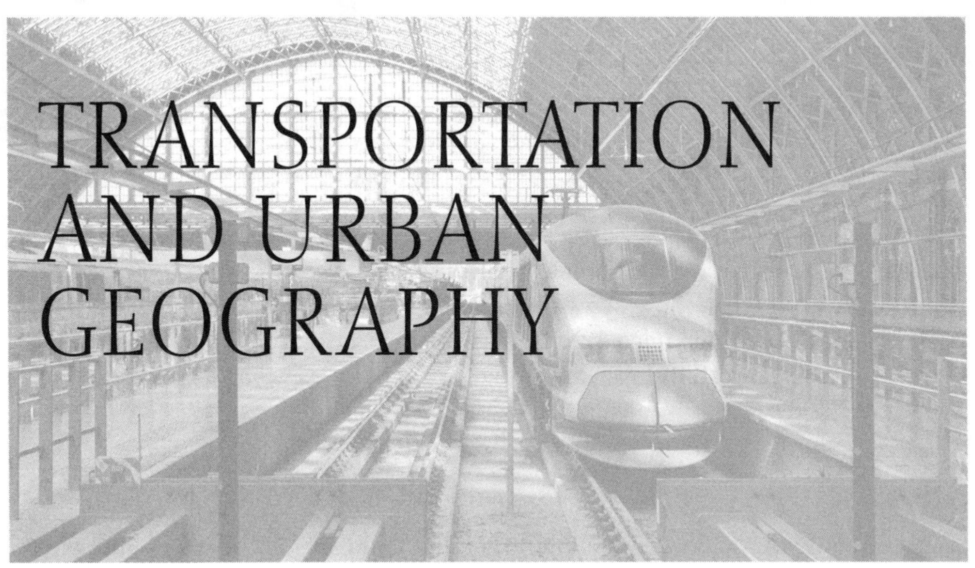

TRANSPORTATION AND URBAN GEOGRAPHY

CITIES AND SUBURBS

What is a city?

In the United States, a city is a legal entity, with a delegated power by a state and county to govern and provide services to its citizens. Cities also have charters, which are somewhat akin to local constitutions, and have specific boundaries.

What was the first city to have more than one million people?

During ancient times, Rome was the world's first city to have a population larger than one million. Rome's population declined during the fall of the Roman Empire in the fifth century, and a city with a population of one million wasn't again seen until the early nineteenth century in London.

What is an urban area?

An urban area consists of a central city and its surrounding suburbs. Urban areas are also known as metropolitan areas. In some cases, urban areas can spread dozens of miles beyond the central city.

How many people in the world are classified as urban dwellers?

According to experts at the United Nations, as of 2014, 54% of the world's population is classified as urban. This number is expected to reach 66% by 2050. Much of the growth in our urban population will come from population growth in India, China, and Nigeria. These countries will account for 37% of the projected growth.

Are there any new discoveries of cities that predate Mesopotamian urban areas?

Every year, scientists uncover evidence of cities, long ago flooded by the rising sea and reclamation of coastline, where archaeological evidence suggests that cities, even older than those in the Sumer Valley, once existed. A recent marine archaeological expedition uncovered traces of a city beneath the sea off the coast of Gujarat, in northwestern India, and found pieces of carved wood that was carbon dated to 7500 B.C.E.

What are some of the oldest, continuously inhabited cities in the world and when were they established?

Some of the oldest cities still inhabited today are Jericho, Palestine (founded 9000 B.C.E.); Byblos, Lebanon (founded 5000 B.C.E.); Damascus, Syria (founded 4300 B.C.E.); Aleppo, Syria (founded 4300 B.C.E.); and Susa, Iran (founded 4200 B.C.E.).

Where did some of the first cities of the world begin?

Scientists generally believe that the Sumer Valley area of ancient Mesopotamia, located in modern-day Iraq in the Middle East, is the origin of the earliest cities.

What is the most populated urban area in the world?

The world has many cities (called "megacities") with populations greater than 10 million inhabitants. The capital city of Tokyo, Japan, still reigns as the most populated urban area in the world, with over 38 million inhabitants. The next largest urban areas include Delhi, India (25 million) and Shanghai, China (23 million). By comparison, the New York/Newark metropolitan area and Cairo, Egypt, have approximately 18.5 million inhabitants each.

What is a megalopolis?

Geographer Jean Gottmann developed the term megalopolis to describe the huge, interconnected metropolitan area from Boston to Washington, D.C. "Boswash," as this original megalopolis has been called, has been joined by such nascent megalopolises as "Chi-Pitts" (from Chicago to Pittsburgh), the Ruhr area in Germany, Italy's Po Valley, and "San-San" (from San Francisco to San Diego).

Are mega-urban area populations growing?

According to experts at the United Nations, global urban populations are expected to grow 1.84% per year between 2015 and 2020, 1.63% between 2020 and 2025, and only

1.44% between 2025 and 2030. Of the top twenty large urban areas in the world, nearly half experience population growth of only 1.5%.

How many cities in the United States have more than one million people?

Ten cities in the U.S.A. have populations in excess of one million people, including New York (8,491,000), Los Angeles (3,928,864), Chicago (2,722,389), Houston (2,239,558), Philadelphia (1,560,297), Phoenix (1,537,058), San Antonio (1,436,697), San Diego (1,381,069), Dallas (1,281,047), and San Jose (1,015,785). In 1890, only three cities in the United States had populations greater than one million people: New York, Chicago, and Philadelphia. Los Angeles recorded only 50,895 people during the 1890 Census.

What are the largest metropolitan areas in the United States?

The New York metropolitan area is America's largest, with 18.897 million people. Los Angeles is second with 12.829 million; Chicago is third with 9.461 million; Dallas is fourth with 6.372 million; and Philadelphia is fifth with 5.965 million people. In 2014, fifty-three U.S. metropolitan areas had populations of one million or more.

How does the population of Tokyo change each day?

The population of Tokyo, Japan, changes each day as 40 million people commute by rail (both above and underground) into the city, riding a combination of trains, subways, and buses for an average commute of two hours per day.

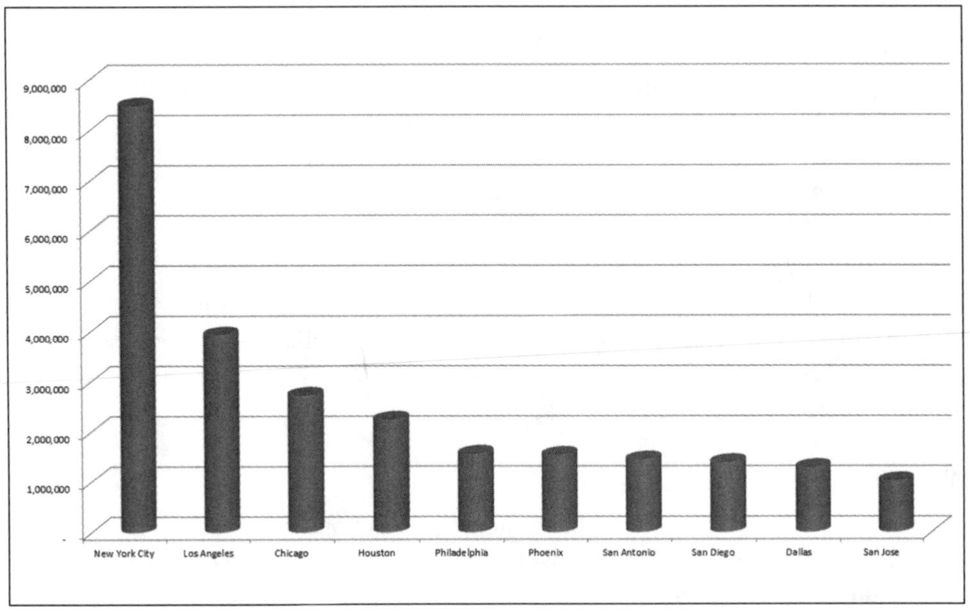

The ten largest U.S. cities all have populations over one million.

What percentage of people live in the megacities of the world?

Only 6.4% of the inhabitants of Earth live in gigantic urban areas. The rest live in areas with populations of less than 10 million people. A total of 453 million people live in the world's megacities, or 12% of the world's urban dwellers.

What major American city loses more people than any other?

Although 298 of the 381 metro areas in the United States gained in population between 2013 and 2014, Wayne County, Michigan (Detroit), leads the nation with a decline of 11,000 people. The next closest American county to lose people is Cuyahoga County, Ohio (Cleveland), with a net loss of more than 4,000 people.

What is a central business district?

A central business district (CBD) of a city is located downtown, often where the city began, and is the primary concentration of commercial buildings. Experts at the American Planning Association believe that a great central business plan includes corroboration on the part of both private people and corporations that are located in, or service, the central business district, as well as public entities.

Who decides where houses can be built?

Almost all American city governments have a department that is responsible for planning the layout of the city. The planning department for each city enforces and delineates zoning, which regulates the location of homes, businesses, factories, and even nuclear power plants.

How many housing units are there in the United States?

According to the U.S. Census Bureau, as of the year 2014, there were approximately 133,957,180 housing units (including apartments and condominiums) in the United States, 8.8% more than in 2013. The homeownership rate is approximately 64.9%. This means that approximately 35.1% of American households are renters.

What is a long lot?

Long lots are long and narrow pieces of property. This type of division of land is common in Europe and places in North America that were initially settled by the French (such as Québec and Louisiana). Each lot has a narrow access to a stream or road but is several hundred feet deep.

When did suburbs become fashionable?

Following World War II, the subsequent housing boom and construction of interstate highways led to the development of low-density housing surrounding cities. These areas of low-density housing are known as suburbs. Suburbs have been extremely popular since the 1950s.

Why do I need a ZIP Code?

ZIP (Zone Improvement Program) Codes were created by the U.S. Postal Service in 1963 to help sort and distribute mail more effectively. Each ZIP Code represents a specific geographic area, with each digit further pinpointing the recipient's address. For instance, the first digit (0—9) of each ZIP Code represents an area of the United States, from "0" in the Northeast to "9" in the West.

Levittown, Pennsylvania, is one example of the large residential developments built in the 1940s through the 1960s that were based on just a couple floor plans for the sake of efficiency.

What was Levittown?

The three Levittowns were large housing developments built from the mid-1940s through the early 1960s by William J. Levitt and his construction company. Levitt invented a process to mass-produce homes by making each home exactly the same. The first Levittown was located in New York and consisted of 17,000 homes. The subsequent Levittowns were built in New Jersey and Pennsylvania. Levittowns were the forerunner of the suburb.

URBAN STRUCTURES

What are some of the largest buildings in the world, in terms of useable space?

The Boeing Everett Factory in Everett, Washington, has over 4.3 million square feet (398,000 square meters) of space. It is followed by the Target Import Warehouse, in Savannah, Georgia, which has 2.02 million square feet (187,664 square meters) of floor area, and the Jean-Luc Lagardère Plant operated by Airbus Industries (the A380 Assembly Plant) near Toulouse, France, with 1,320,000 square feet (122,500 square meters).

What is the largest building in the world in terms of its footprint?

The Aalsmeer Flower Auction, in Aalsmeer, the Netherlands, is the largest building in the world in terms of floor space. It covers 128 acres (518,000 square meters) in area.

What building in the United States is one of the largest office buildings in the world?

The Pentagon, located in Arlington County, Virginia, is one of the largest office buildings in the world. It has approximately 6.5 million square feet (600,000 square meters) of area, with 3.7 million square feet (343,730 square meters) of office space under its roof. The Pentagon provides offices to the 23,000 people who work at the Department of Defense.

What is the largest multipurpose building in the world?

The largest multipurpose building in the world is the New Century Global Center, in the Tianfu New Area of Chengdu, China, with 18 million square feet (1.67 million square meters) of floor space. It is termed "multipurpose" because it houses shopping areas, conference facilities, a commercial center, a university campus, hotels, etc.

What is the largest church building in the world?

The Mormon Church LDS Conference Center in Salt Lake City, Utah, is the largest church building, having over 1.4 million square feet (130,000 square meters) in area.

What are two of the longest enclosed buildings in the world?

Catching a flight from Kansai International Airport in Osaka, Japan, could involve a walk of more than a mile; the airport is 5,580 feet (1.7 km) long. In the country of the United Arab Emirates, at Dubai International Airport, one could walk 1.1 miles (1.774 km) to get to the gate.

What is the largest shopping mall in terms of gross leasable area?

The New South China Mall in Dongguan, China, has more than 7.1 million square feet (659,612 square meters) of leasable space. It is the largest shopping mall in the world in terms of total area, consisting of 9.6 million square feet (892,000 square meters).

What is the tallest self-supporting structure in the world?

In the Sumida District of Tokyo, Japan, stands the Tokyo Skytree Tower, 2,080 feet (634 meters) above the ground. It has been the tallest structure in the world since 2011.

What was the world's first skyscraper?

Completed in 1885 in Chicago, Illinois, the Home Insurance Company Building was the world's first skyscraper.

The Burj Khalifa in the United Arab Emirates is 2,722 feet (829.8 meters) tall.

Why is there such controversy about defining the tallest building in the world?

Most of the controversy about which building is the tallest centers around defining what is a building versus what is a structure. Some buildings, in order to be listed among the tallest buildings, add other nonhabitable structures to the top in order to rise in the rankings. These structures may be in the form of communication towers, which can easily add significant height to the building.

What is the tallest skyscraper in the world?

The Burj Khalifa skyscraper in Dubai, the United Arab Emirates, is currently the tallest skyscraper in the world, standing 2,722 feet (829.8 meters) above the ground.

Why do mosques have domes?

The onion-shaped domes of Islamic mosques and other religious buildings of eastern religions were an architectural style borrowed from the Byzantine Empire. One of the world's most famous onion-domed buildings, St. Basil's Cathedral in Moscow's Red Square, was built in the mid-sixteenth century.

What is a UNESCO World Heritage Site?

The United Nations Educational, Scientific, and Cultural Organization's (UNESCO) stated purpose is to build peace in the minds of mankind. It is one of the most important agencies within the U.N. A World Heritage Site represents our natural or cultural wealth, the most important landmarks in the world that should be shared with everyone. Because of the special significance of these places, they must be protected for us to see so that we may pass on knowledge of our shared culture to future generations. They can be both physical places built by man or natural wonders.

AIR TRANSPORTATION

Who were the Wright Brothers?

Orville and Wilbur Wright were the first two people to fly in a heavier-than-air vehicle, called the *Flyer.* They made their historic flight on December 17, 1903, at Kitty Hawk, North Carolina.

When was the first flight achieved?

In 1783, the Montgolfier Brothers flew the first hot air balloon across Paris, France.

What is the largest airport in the world?

There are three terminals at Dubai International Airport, which occupies 7,200 acres of land (29.14 square km). Of the three, the largest is Terminal Three, which has 18.9 million square feet of floor area (1.76 million square meters) and is the second-biggest building in the world.

What is the busiest airport in the world?

The busiest airport in the world, in terms of numbers of passengers annually, is Hartsfield-Jackson Atlanta International Airport in Atlanta, Georgia, with over 96 million passengers annually.

How many airports are there in the United States?

According to the U.S. Department of Transportation, there are approximately 19,453 airports in the United States, including 5,155 that are designated for public use.

What is the busiest airport outside of the United States?

Beijing Capital International Airport in Beijing, China, is the busiest airport outside the United States, with more than 83.7 million people passing through its gates. Heathrow Airport outside of London, England, has more than 72.4 million passengers passing through the various terminals annually.

ROADS AND RAILWAYS

When were the first roads built?

Archaeologists have found remnants of early roads in Ur, Iraq, and Glastonbury, Scotland, dating back to 4000 B.C.E.

Do all roads really lead to Rome?

Not any longer. During the time of the Roman Empire, the Romans built a massive road network to ensure easy travel in all weather conditions between Rome and the furthest reaches of the empire. The Romans made their roads as straight as possible and paved large sections of them by precisely piecing together cut rock to make a flat surface. Along the 50,000 miles (80,000 km) of Roman roads, markers were placed every Roman mile (just short of a modern mile) so as to indicate either the distance to Rome or to the city where the road originated. After the fall of Rome, the maintenance of the Roman road system was severely neglected, and during the Middle Ages the roads became overused and dilapidated. Though the Romans built these roads over 2,000 years ago, some segments are still in use today.

What is a turnpike?

A turnpike is a toll road. In the late eighteenth century, private companies in the United States and in the United Kingdom built roads and charged users to pass. Beginning in the 1840s, turnpikes had to compete for traffic, and thus profits, with the railroads. The name turnpike is still common on toll highways in the eastern United States, such as the New Jersey Turnpike, the Massachusetts Turnpike, and the Pennsylvania Turnpike.

What road in the United States was known as the National Road?

The Cumberland Road, also known as the National Road, was the first federally funded road in the United States. Though construction began in 1811, the Cumberland Road was not completed until 1852. Stretching 800 miles (1,287 km) from Cumberland, Maryland, to Vandalia, Illinois, the road was built to allow settlers to traverse the Appalachian Mountains and settle in the West. With the advent of the automobile, the road was paved, and in 1926 became part of U.S. Route 40, which stretches across the continent.

What do the Cumberland Road and Cumberland Gap have to do with each other?

Absolutely nothing. The Cumberland Road is more than 100 miles (161 km) from the Cumberland Gap. The Cumberland Gap, which lies near the border of Kentucky, Tennessee, and Virginia, is a pass through the Appalachian Mountains at the Cumberland Plateau. The name "Cumberland" was extremely popular in Colonial America, originating in the name of the British Duke of Cumberland.

What is the difference between a highway and a freeway?

The term highway can be used for any road but most often describes a paved road connecting distant towns. Freeways are multilane highways that use on- and off-ramps, rather than intersections, in order to limit the number of entrance and exit points along the route, hence keeping traffic along the freeway fairly steady.

When was the first freeway built in the United States?

The first freeway (lacking tolls and having limited access) in the United States was the Arroyo Seco Freeway, connecting Pasadena and downtown Los Angeles. It opened in 1940 and is now the Pasadena Freeway, Highway 110. The first urban freeway built in the United States is the Davison Freeway, in Detroit, Michigan, and was opened in 1942. It is 5½ miles long.

What are interstate highways?

President Dwight D. Eisenhower signed the Federal-Aid Highway Act of 1956, which established the system of interstate highways in the United States. Interstate highways are federally funded freeways that allow the rapid transportation of people, goods, and the military across the country.

Did Hitler create the Autobahn?

Though the first modern freeway system in Germany was begun in 1913, Adolf Hitler did create the Autobahn during the Third Reich, from 1933 to 1945. The Autobahn is a freeway system that includes 6,800 miles (10,941 km) of roads across Germany. Though it is widely believed that there are no speed limits on the Autobahn, there are a few segments with marked speed limits.

How are interstate highways numbered?

One- and two-digit interstate highways are numbered according to their direction. Highways that run in an east-west direction are even numbered, while highways that run in a north-south direction are odd numbered. The lowest numbers are in the south and west, while higher numbers are in the north and east. For example, Interstate 10 is an east-west highway that runs from Santa Monica, California, to Jacksonville, Florida; thus, it has an even, low number. Interstate 95 is a north-south highway that runs from Houlton, Maine, to Miami, Florida; thus, it has an odd, high number. Three-digit interstate highways are short spur routes connected to a two-digit interstate.

Why was U.S. President Eisenhower a fan of interstate highways?

In 1919, the young Dwight D. Eisenhower took part in a cross-country military trip from Washington, D.C., to San Francisco. But, due to the state of the highways at that time, the trip took sixty-two days—far too long to defend the country should the need arise. This experience made Eisenhower realize the need for a faster, more efficient mode of transportation across the country. Because of President Eisenhower's support for the Interstate Highway System, it is now officially known as the Dwight D. Eisenhower System of Interstate and Defense Highways. Today, the Interstate Highway System has approximately 47,856 miles (77,017 km) of road.

Why does Hawaii have interstate highways?

Since any freeway funded under the Federal-Aid Highway Act of 1956 is known as an interstate highway, whether it crosses state boundaries or not, Hawaii can have interstate highways. Though they cross no state borders, Hawaii has three interstate highways: H1, H2, and H3.

When was the last interstate highway built?

The construction of new interstate highways came to an end in 1993 with the opening of Interstate 105, the Century Freeway, in Los Angeles, thirty-seven years after construction began on the system. The Century Freeway is an intercity route connecting the coastal community of El Segundo to Interstates 405, 110, 710, and finally 605 in Norwalk.

How many miles of road are there in the United States?

According to the U.S. Department of Transportation, the United States has more roads (paved and unpaved) than any country in the world, with a total of 4,071,000 miles (6,551,639 km).

What is the longest bridge in the world?

The longest bridge of any type is the Danyang-Kunshan Grand Bridge, part of the Beijing-Shanghai High-Speed Railway, and is 102.4 miles (164.8 km) long. The Lake Pontchartrain Causeway, completed in 1956, that connects New Orleans with Mandeville, Louisiana, is the longest bridge connecting a highway system over water, at 24 miles (38.6 km) long. A parallel bridge adding two more lanes was completed in 1969.

What is the Chunnel?

The Channel Tunnel (or Chunnel) is a railroad tunnel under the Strait of Dover in the English Channel. Also known as the Eurotunnel, the Chunnel runs for 31 miles (50 km) between Folkestone (near Dover) in England and Sangatte (near Calais) in France. Opened in 1994, the Chunnel connects England with the rest of continental Europe. In 2014, the Chunnel carried 10,397,894 passengers between the two countries.

What is the most common street name in the United States?

It's not Main Street. Second, Third, First, and Fourth Streets are the most commonly used street names, followed by Park, Fifth, and Main.

The London station for the Chunnel. The train and tunnel connects Britain to the rest of Europe.

Where is the longest Main Street in the United States?

Main Street in Island Park, Idaho, is 33 miles (53 km) long, making it the longest Main Street in the United States.

When was the first automobile built?

Though it had only three wheels, the world's first gasoline-powered automobile was built by Karl Benz in 1885 but patented a few years earlier in 1879.

How many authorized drivers of taxis are in New York City?

There are 52,131 authorized taxicab drivers that service New York City and the surrounding areas, as of 2014.

What countries do New York taxi drivers come from?

In New York City alone, according to data from the New York Open Data Project, as revealed by the *Daily Mail Online*, 23.1% of all taxi drivers come from Bangladesh, 13.2% come from Pakistan, and 9.3% come from India.

Which city has the most taxis?

Congested Mexico City, Mexico, is home to more than 60,000 taxis among its 3.5 million automobiles. The city with the second-greatest number of taxis is Mumbai, India, which has more than 55,000.

How many taxicab drivers are there in the United States?

According to the U.S. Bureau of Labor Statistics, as of 2012, there are more than 233,000 taxicab drivers in the United States. More than 38% of drivers are immigrants to the United States. The median pay for all cab drivers is $22,820 per year.

Where was the first self-service gas station?

In Los Angeles, George Urich opened the first self-service station in 1947.

Who invented the traffic signal?

The red, yellow, and green traffic signal that we are familiar with today was originally invented by Garrett Morgan in 1923. Morgan, who was also the inventor of the gas mask, received numerous awards for his invention.

Who invented the first train?

In 1825, British engineer George Stephenson invented the first train, which was powered by steam. Stephenson's train was introduced to North America in the 1830s and was used until the 1940s when diesel-electric locomotives, which didn't run on expensive coal, replaced steam locomotives.

What is the world's longest subway system?

China's third-oldest subway system is also the world's longest subway system, extending more than 365.4 miles (588 km). Construction began in 1986, and although it was completed in 1993, it continues to grow in length each year.

SEA TRANSPORTATION

Where is the busiest seaport in the world?

Shanghai, China, is the busiest seaport in the world, followed by Singapore and Shenzhen, China. The Port of Los Angeles in the United States is ranked eighteen in terms of container traffic moving through the facilities. The ranking of activity of seaports depends on the weight of the cargo passing through or being loaded, as well as the volume of containers that hold products while they are being shipped.

What does a canal lock do?

Many canals connect two bodies of water that lie at different elevations. Locks are used to gradually move the ships from one elevation to another. Once a ship enters a lock, doors close in front of and behind it. Water is then added or drained from the area to raise or lower the ship to a different elevation. Then, the doors in front of the ship open, and the ship sails down the canal to the next lock or to the open sea.

In which direction do ships sail through the Panama Canal?

Though you would expect them to travel east from the Pacific to the Atlantic Ocean when sailing through the Panama Canal, ships actually travel northwest. Since the Isthmus of Panama lies parallel to the equator, the canal does not lie east-west but rather northwest-southeast.

What is the St. Lawrence Seaway?

Completed in 1959, the 183-mile- (294-km-) long St. Lawrence Seaway was built by deepening and widening the St. Lawrence River between Montréal, Québec, Canada, and Lake Ontario so that large ships could traverse it. The Seaway consists of a series of locks that allow ships to travel from the Atlantic Ocean to the Great Lakes and ultimately on to Chicago. A limiting factor of the Seaway is that ships can only use it between May and November, as it is blocked by ice in the winter.

Why was the Erie Canal built?

The 363-mile (584-km) Erie Canal connects the Hudson River to Lake Erie. Opened in 1825, the canal created a new, shorter route from the northern interior of the United

Six new locks were constructed for the Panama Canal in 2015.

States to the Atlantic Ocean. Prior to the opening of the Erie Canal, goods traveled down the Mississippi River and out to the Atlantic. Since New York City lies on the Hudson River, the canal was responsible for the growth of the city as a major port, helping it to become the largest city in the United States. Once the St. Lawrence Seaway was built in 1959, the Erie Canal became rarely used, since most transportation soon traveled along the seaway.

How does a choke point "choke" a body of water?

A choke point is a narrow waterway between two larger bodies of water that can be easily closed or blocked to control water transportation routes. Though historically the Strait of Gibraltar (connecting the Mediterranean Sea and Atlantic Ocean between Africa and Spain) has been one of the world's most important choke points, the Strait of Hormuz gained significant attention during the Persian Gulf War of 1991. The Strait of Hormuz, bounded by the United Arab Emirates and Iran, connects the Persian Gulf to the Arabian Sea and, thus, to the Indian Ocean. It was feared that if Iraq controlled the Strait, then most of the oil from the region could not be shipped out.

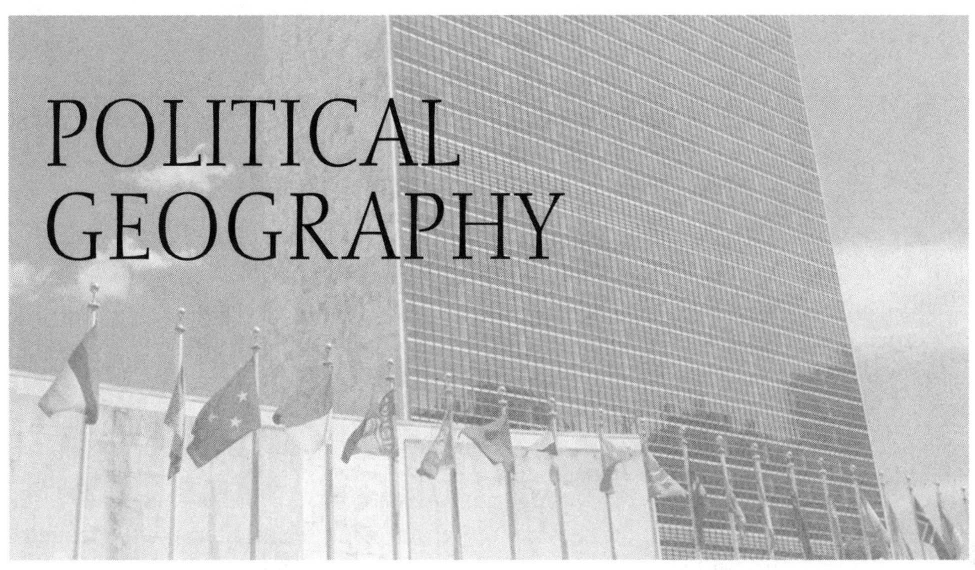

How does geography influence politics?

Geography is a key component in many political decisions and actions. The borders of countries, access to international waters, location of natural resources, access to ports, and the designation of voting districts are a few of the many geographical factors that affect politics.

What is the difference between a country and a nation?

Many people use the terms "country" and "nation" interchangeably. But not all nations are countries, nor are all countries nations. A country is the equivalent of a State and is a political entity. A nation is a group of people with a common heritage and culture. Some nations have a State and are thus called a nation-state. Nation-states include France, Germany, Japan, China, and the United States. Some nations have no State, such as the Kurds and the Palestinians. Some States have multiple nations such as Belgium, which is composed of two multilingual regions, Flanders in the north and Wallonia in the south.

What is the difference between a State and a state?

A State, with a capital "S," is equivalent to a country. A state, with a lowercase "s," is a division of a country, like the states that make up the United States.

Do all countries have states?

While most countries are divided into states, provinces, or departments, there are many that have no political divisions. Large countries without political divisions include Mali, Kazakhstan, Saudi Arabia, and Algeria.

Who controls the world's oil supply?

The Organization of Petroleum Exporting Countries (OPEC) coordinates most of the world's oil production. The members of OPEC meet to coordinate oil policies and prices.

> ### What criteria must an area meet to be considered an independent State?
>
> An independent State must have sovereignty, or the ability to freely exercise all powers of a State under international law, without external dependence. Additionally, a State must be an independent country with its own land, government, permanent population, and economy. Above all, an independent State must have international recognition.

Thirteen countries comprise OPEC: Algeria, Angola, Ecuador, Indonesia, Iran, Iraq, Kuwait, Libya, Nigeria, Qatar, Saudi Arabia, the United Arab Emirates, and Venezuela. Though Russia, the United States, and Mexico are also leading petroleum producers, the three countries are not members of OPEC. Ecuador suspended its membership from 1992 to 2007, and Indonesia left the organization at the beginning of 2009 but rejoined the organization in 2016.

Who else controls the price of gasoline besides oil producers?

There are many factors that influence the price of gasoline. Among them are the rising demand for fuel from developed countries such as the United States; new and growing demand from China and India; market speculation; and public policy. The refining industry, which is under government regulation, is also consolidated into the hands of very few global refiners. This has the potential for removing competitive pricing in the oil industry.

COLONIES AND EXPANSIONISM

Why was it said that the sun never set on the British Empire?

In the early twentieth century, the United Kingdom included colonies located in both North and South America (Canada, British Guiana, and Bermuda), Africa (Egypt, South Africa, and Nigeria), Asia (India, Burma, Hong Kong, and parts of China), and Oceania (Australia, New Zealand, and various islands and territories). Because the British Empire spanned the globe, there was always at least one portion of the empire in daylight.

Why would countries want colonies?

Colonies are a source of raw materials, new land, wider trading opportunities, and militaristic expansion for the mother country. Although the concept of colonization has been practiced for many thousands of years in some form, modern colonies were established around the world from the sixteenth century through the nineteenth century by powerful western nations. After World War II, the concept of colonization was widely attacked as an exploitive policy. Though most colonies were granted independence, several countries still control colonies around the world.

What were some of the earliest colonies?

The Phoenicians, around the year 1000 B.C.E., founded some of the first colonies in Tyre (present-day Lebanon). Colonists from there went on to colonize Carthage (present-day Tunisia) and the coast of Spain. This enabled them to control access to the Atlantic Ocean and trade with the indigenous peoples of what is today Great Britain and France.

How did the Nazis use geopolitics?

During the Nazi era in Germany, 1933–1945, the "science" of geopolitics was utilized to support Germany's concept of Lebensraum, or living space. The Nazi concept of Lebensraum was based on the idea that there was a racial hierarchy that allowed "superior" races to conquer "inferior" races. Adolf Hitler used this sense of geopolitics to invade Czechoslovakia, Poland, and the Soviet Union.

How did irredentism help start World War II?

Irredentism is a term used to describe a situation in which a minority group in one country shares the culture and heritage of people within another country. Members of the minority group may attempt to have their region annexed into the mother country or may be satisfied with the country in which they reside. Adolf Hitler used irredentism as an excuse to invade and conquer Czechoslovakia in 1938. He claimed that the Germans in the Sudetenland, a part of Czechoslovakia, were being treated unfairly, and thus this area should be annexed to Germany. Though Germany's annexation of Czechoslo-

Who owns the world's oceans?

The battle over control of the world's oceans has increased over the past few decades due to the discovery of vast mineral and fuel resources located under the sea. In 1958, the United Nations held the first Conference on the Law of the Sea. Negotiations for the third Law of the Sea Convention occurred between 1973 and 1982. This conference established territorial seas, measuring 12 nautical miles (22.24 km) from the shore of coastal nations that are under the full control of that country. Additionally, countries have mineral, fuel, and fishing rights in an Exclusive Economic Zone (EEZ) that spans 200 nautical miles (370.6 kilometers) from shore. Problems arise when two countries' zones overlap. Median lines between countries have been drawn in most cases, but there are still many areas of disagreement. The Law of the Sea has been ratified by 166 countries and one territory.

vakia did not start World War II, it was the Nazis' first direct aggressive step toward conquering Europe.

THE UNITED NATIONS

When did the United Nations begin?

Established in 1945 at the end of World War II, the United Nations was created for the purpose of maintaining world peace. Its members pledge to work together to solve disputes. The United Nations also oversees many agencies that promote health, welfare, cooperation, and development around the world.

How many countries are members of the United Nations?

There are 193 member states of the United Nations, which is headquartered in New York City. New members include Serbia, Montenegro (from the former Yugoslavia), Tuvalu (an island nation in the Pacific), Timor-Leste (in what was once part of Indonesia), South Sudan, and Switzerland.

Which countries are not members of the United Nations?

While almost every country in the world is a member of the United Nations, there is a short list of countries that are not members: Taiwan, Tonga, and the Holy See (Vatican City). Additionally, Palestine, Abkhazia (part of Georgia), and Kosovo (part of Serbia) are recognized by some members of the U.N. but are not full members as they are not sovereign states.

Why do some countries choose not to join the United Nations?

A country must be willing to give up some of its self-rule for the greater good, which often means the good of the larger, more influential western countries.

How did the League of Nations fail?

The League of Nations, which was created in 1920 and replaced by the United Nations in 1945, failed in its mission to prevent World War II. Even though the league was essentially the creation of President Woodrow Wilson, the isolationist United States never became a member.

The United Nations headquarters is in New York City.

NATO AND THE COLD WAR

What countries are members of the North Atlantic Treaty Organization?

The twenty-eight members of NATO include Albania, Belgium, Bulgaria, Canada, Croatia, Czech Republic, Denmark, Estonia, France, Germany, Greece, Hungary, Iceland, Italy, Latvia, Lithuania, Luxembourg, the Netherlands, Norway, Poland, Portugal, Romania, Slovakia, Slovenia, Spain, Turkey, the United Kingdom, and the United States.

What countries are members of the Commonwealth of Independent States?

The Commonwealth of Independent states comprises of twelve countries that were part of the former Soviet Union, including Azerbaijan, Armenia, Belarus, Georgia, Kazakhstan, Kyrgyzstan, Moldova, Russia, Tajikistan, Turkmenistan, Uzbekistan, and Ukraine. It was created in December 1991.

How did the Soviet Bloc countries respond to the creation of NATO?

In 1955, seven communist countries created the Warsaw Pact to protect against NATO aggression. The Warsaw Pact originally consisted of Albania, Bulgaria, Czechoslovakia, Hungary, Poland, Romania, and the Soviet Union. It disbanded in 1991 with the breakup of the U.S.S.R. and the changes in Eastern Europe.

What did the domino effect have to do with the United States' involvement in the Vietnam War?

American military strategists believed that if one country became communist, it would begin a never-ending succession of countries converting to communism (thus the domino metaphor). North Vietnam, at the time of the American invasion, was comprised primarily of communists and communist sympathizers, whereas South Vietnam was more democratic leaning. Policymakers believed that the United States had to do everything possible to keep every country from falling to communism. This included sending American troops to Vietnam. Though Vietnam fell to the communists, the the-

What is the purpose of NATO now that the Soviet Union is gone?

The North Atlantic Treaty Organization (NATO) was founded in 1949 as an alliance of European and North American noncommunist countries committed to preventing and protecting against communist threats. With the current instability in Russia and Middle East, NATO still sees its role purely as defensive in nature, supporting member states and countries in need of assistance against aggression in the surrounding regions.

ory of the domino effect was proven incorrect because neighboring countries did not fall to communism as predicted. In fact, the U.S. war and policy toward Vietnam caused many adjacent countries to become even more fervently opposed to U.S. policy in the region, and some theorize that this contributed to turmoil and genocide in such countries as Cambodia and Laos. The largest trading partner of the United States in the world today is the communist People's Republic of China.

Who won the Cold War?

The stalemate of the Cold War effectively ended when the people and governments of the former Soviet Union and Eastern Bloc countries decided that they needed a major change in the way in which they organized their government, societies, and economies. The fall of the Berlin Wall in 1989 was only one indicator of the significance of this trend. Popular demonstrations and changes took place in all parts of the former U.S.S.R. and Eastern Europe, leading inexorably to the downfall of the Soviet Union and its control of satellite states.

THE WORLD TODAY

How many communist countries are there remaining in the world today?

There are five remaining communist countries in the world today, four of which are in Asia, including the People's Republic of China, the Democratic People's Republic of Korea (North Korea), the Socialist Republic of Vietnam, the Lao People's Democratic Republic (Laos), and the Republic of Cuba.

How many countries are ruled by monarchs in the world?

Approximately twenty-eight countries in the world have monarchs, from the Kingdom of Swaziland in Africa to the Kingdom of Tonga, a Pacific Ocean island nation. If you include the number of countries or territories that are part of Great Britain's Commonwealth of Nations, the number increases to forty-four.

What are some of the newest countries in the world?

In the 1990s, over two dozen new countries appeared on the map. These included fifteen new countries that were created when the U.S.S.R. broke up in 1991—Armenia, Azerbaijan, Belarus, Estonia, Georgia, Kazakhstan, Kyrgyzstan, Latvia, Lithuania, Moldova, Russia, Tajikistan, Turkmenistan, Ukraine, and Uzbekistan.

In 1990, Namibia split from South Africa to become its own country, while East and West Germany combined to become Germany. The dissolution of Yugoslavia also created several new countries in 1991 and 1992—Bosnia and Herzegovina, Croatia, Macedonia, Serbia and Montenegro, and Slovenia. In 1993, Eritrea became independent of Ethiopia, and Czechoslovakia dissolved into the Czech Republic and Slovakia. That same year, the Pacific island countries of the Marshall Islands, Micronesia, and Palau all became independent. More recently, several new countries have been born, including South Sudan (2011), Kosovo (2008), Serbia (2006), Montenegro (2006), and Timor-Leste (2002).

How many countries does the United States recognize?

The State Department is the official U.S. government agency that recognizes independent countries. It maintains an updated list of the official independent states of the world. As of this writing, there are 203 states listed. Taiwan and Vatican City, though commonly considered countries, are not included on this list. South Sudan, which gained independence in 2011, is the newest country to be recognized.

Why are some borders curvy while others are straight?

There are two primary types of boundaries—geometric and natural. Geometric boundaries are straight and follow lines of latitude, longitude, or a certain compass direction between points. Geometric boundaries were established to divide territories before settlers entered areas. Most of the states in the Western United States have at least a portion of their borders formed by geometric boundaries (especially rectangular-shaped Colorado and Wyoming). Natural boundaries are usually curvy because they follow the crests of mountains or the center of rivers. Natural boundaries are very common in places like Europe, where the region was heavily populated before the countries were created.

Why do third-world countries no longer exist?

The term "third world" was part of the classification of countries during the Cold War. This classification designated those countries aligned with the United States as "first world," those countries aligned with the Soviet Union as "second world," and those countries that were nonaligned as "third world." Over time, the term "third world" came

Why isn't Taiwan recognized by the U.S. government?

The People's Republic of China has had a dispute with Taiwan since the Nationalists, who were fighting the communists on the mainland, fled to Taiwan at the end of World War II. Decades later, as China became more powerful in the eyes of various American political administrations, China gave the United States a choice: recognize China or Taiwan, but not both. The United States chose China. Although officially not recognized, Taiwan is one of the United States' strongest allies in the region and is the United States' tenth-biggest trading partner. Taiwan has very close official and unofficial ties with the United States. The People's Republic of China is the biggest trading partner of the United States, followed by Canada.

to mean a poorer or less developed country. With the dissolution of the Soviet Union and the acceptance of democracy in Russia and Eastern Europe, the classification no longer exists. The preferred terms are now "developed" countries and "less developed" (or "developing") countries.

Why was the border between Yemen and Saudi Arabia always dashed on maps?

In the sandy desert between Saudi Arabia and Yemen, the border between the two countries was in dispute. A treaty signed in Jeddah, Saudi Arabia, in July 2000 resolved the conflict, and maps no longer show a dashed line between the two countries.

What other countries have border disputes?

For years, India and Pakistan have had disputes about the borders of Kashmir, a region located in northern India. Other countries in Central America, like Honduras and El Salvador, have had disputes over very small parcels of land because of ill-defined border markings implemented during the colonial period.

Which countries are surrounded entirely by landlocked countries?

The two countries surrounded entirely by landlocked countries are Uzbekistan and Liechtenstein. Uzbekistan is surrounded by the landlocked countries of Kazakhstan, Kyrgyzstan, Tajikistan, Afghanistan, and Turkmenistan. Liechtenstein is bordered by Switzerland and Austria, neither of which have access to the ocean.

Which countries are the most prosperous, in terms of wealth and well-being of their people?

Experts at the Legatum Institute analyzed data from 142 countries in the world and compared such variables as entrepreneurship/opportunity, education, health, safety/security, personal freedom, and social capital. They found that the top five most prosperous coun-

How is gerrymandering like a salamander?

In 1812, Massachusetts governor Elbridge Gerry signed a law that established an oddly shaped congressional district. It was redrawn by political cartoonists into a salamander-type creature, and thus the term "gerrymander" was born. Gerrymandering is the process of establishing oddly shaped congressional districts in order to include voters from dispersed areas. Gerrymandered districts can be helpful or detrimental to minority groups, depending on who draws the borders. Although U.S. courts have found gerrymandering to be a legal method of establishing congressional boundaries, it has often been used by politicians to ensure that their party has a majority of support in the redrawn district.

tries in 2014 were Norway, Switzerland, New Zealand, Denmark, and Canada. The United States was ranked tenth.

What is the best shape for a country?

Though countries come in various shapes for various reasons, the best shape for a country is compact. A compact country, such as Germany or France, is easier to govern and support than those that are fragmented (such as Indonesia) or elongated (such as Chile). Compact countries are easier to govern because transportation, communication, and internal security are easier to maintain. Also, compact countries have shorter borders to protect. Elongated and fragmented countries are more easily divided and conquered.

THE WORLD ECONOMY

What is GNP?

GNP, or gross national product, is the total value of GDP plus all income from investments around the world.

What is GDP?

GDP, or gross domestic product, is the value of all goods and services produced in a country in a year. Per capita GDP is usually compared between countries and measures the value of GDP divided by the population.

Which country has the highest GDP?

According to experts at the World Bank, the European Community, representing an economic union of the nations of Europe, would have the highest GDP if it were one country. The United States is number one ($16.768 trillion), followed by China ($9.240 trillion) and Japan ($4.919 trillion).

Which country has the highest per capita GDP?

Banking giant Luxembourg in Europe has a per capita GDP of $110,664, followed by Norway ($100,898) and Qatar ($93,714).

Which countries give the highest proportion of their GNP in the form of aid to other countries?

Foreign aid or official development assistance is money given to countries in the form of grants and loans to promote economic development and welfare. According to the Organisation for Economic Co-operation and Development, as a percentage of gross national income (GNI), the following countries give the most money to assist other

Which advanced industrial democracy ranks near the bottom in giving foreign aid as calculated as a percentage of its gross national income?

The United States, although giving the highest amount in dollars of any country (approximately $31.5 billion each year), contributes the least amount of money in foreign assistance as a percentage of its GNP: around 0.19%.

countries in attaining development goals: the United Arab Emirates (1.25%), Norway (1.07%), Sweden (1.02%), and Luxembourg (1.0%)

Which country spends the most money on the military?

According to experts at the Stockholm International Peace Research Institute, the United States leads the world in military spending at $610 billion per year. American taxpayers commit more money than the next seven countries combined (China, Russia, Saudi Arabia, France, the U.K., India, and Germany.) These countries spend a total of $601 billion per year.

Which country is the top exporter of merchandise?

According to analysts at the U.S. Central Intelligence Agency, China leads the world in merchandise exports, at $2.25 trillion. The next biggest exporters include the European Union ($2.173 trillion), the United States ($1.61 trillion), Germany ($1.547 trillion), and Japan ($711 billion).

POPULATION

What are some of the most common surnames used in the world?

Although difficult to confirm, some of the most common surnames come from the world's most populous country, China. Of the approximately 12,000 historically recorded last names, only 3,100 remain in use today. Some of the most widely used surnames include Wang, Li, Zhang (or Chang), Liu, and Chen. These names are shared by hundreds of millions of people.

How many people have ever lived on the planet Earth?

Only a small percentage, anywhere from 5 to 10%, of the humans who have ever lived are alive today. Experts believe that since humans have existed for approximately 100,000 years, the total number that have ever lived is likely in a range between 60 billion and 120 billion people.

How many people live on the Earth?

There are approximately 7.215 billion people on Earth. The planet's population grows by approximately 1% each year, adds 4.3 people each second, and loses 1.8 people each second.

What has the world's population been over time?

Year	Population
0	200 million
1000	275 million
1500	450 million

Year	Population
1750	700 million
1850	1.2 billion
1900	1.6 billion
1950	2.6 billion
1960	3 billion
1975	4 billion
1985	4.85 billion
1990	5.3 billion
1999	6 billion
2010	6.9 billion
2015	7.4 billion

In recent years, how long did it take for the world population to double in size?

In the forty-year period from 1959 to 1999, the population of the world changed from 3 billion people to over 6 billion people.

How fast is the population growing?

The highest growth rate recorded was around 2% in the late 1950s. This rate has been in decline and is now less than 1% per year.

Where is this population growth happening?

Experts at the Population Institute see the majority of the population growth over the next thirty-five years happening in the developing world. By the year 2050, the population of the planet may approach or exceed 9.6 billion people. Many of the people in these

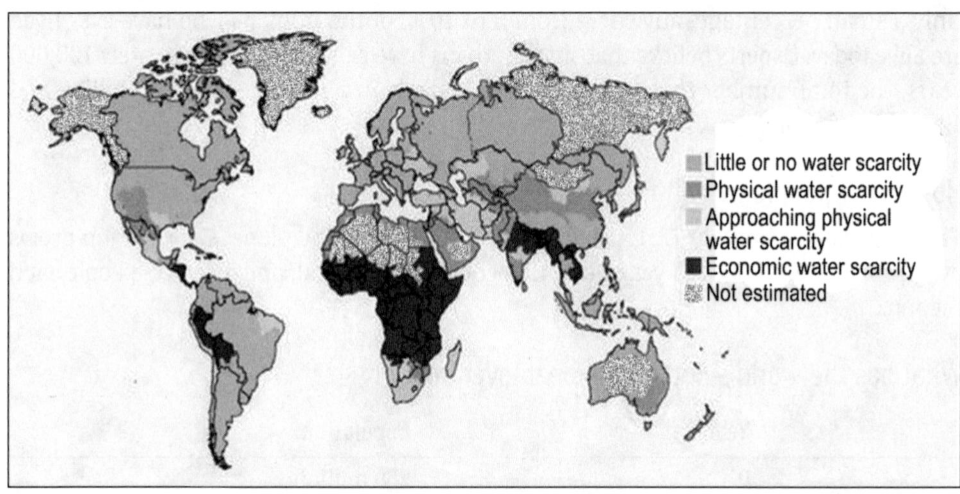

As populations rise and there is a greater need for irrigation and other water needs, water shortages have become an increasing concern across the globe.

What factors affect population growth?

Many factors affect the growth in a population, including an increase in marital age and education level, cultural acceptance of the use of contraceptives, prevalence of disease, access to food and water, civil strife, military action, and genocide.

developing countries struggle today to alleviate poverty and hunger and face political instability, civil unrest, water scarcity, and deforestation of their undeveloped land.

Which country has the world's highest population density?

The density of people across our planet is typically measured as the number of people divided by the amount of land in square miles or kilometers. Some notable countries with high population density include Macau (48,003 people/square mile or 18,534 people/square km), Monaco (43,830 people/square mile or 16,923 people/square km), and Singapore (18,513 people/square mile or 7,148 people/square km). Macau, a former Portuguese colony that is now part of China, consisting of a peninsula and two islands off the southern coast of China near Hong Kong, has the highest population density in the world.

How many people live on the planet?

As of 2015, approximately 7.3 billion people inhabit the Earth. This number is increasing at a rate of approximately twenty-five births every 10 seconds or 1% per year.

Which ten countries have the most people?

Country	2015 Population
China	1.39 billion
India	1.26 billion
United States	325 million
Indonesia	246 million
Brazil	209 million
Pakistan	193 million
Bangladesh	168 million
Nigeria	161 million
Russia	137 million
Japan	128 million

How many people are projected to live on the planet in 2050?

The United Nations Population Division estimates that there will be approximately 8.11 billion people on the planet by the year 2050, even with a declining rate of growth.

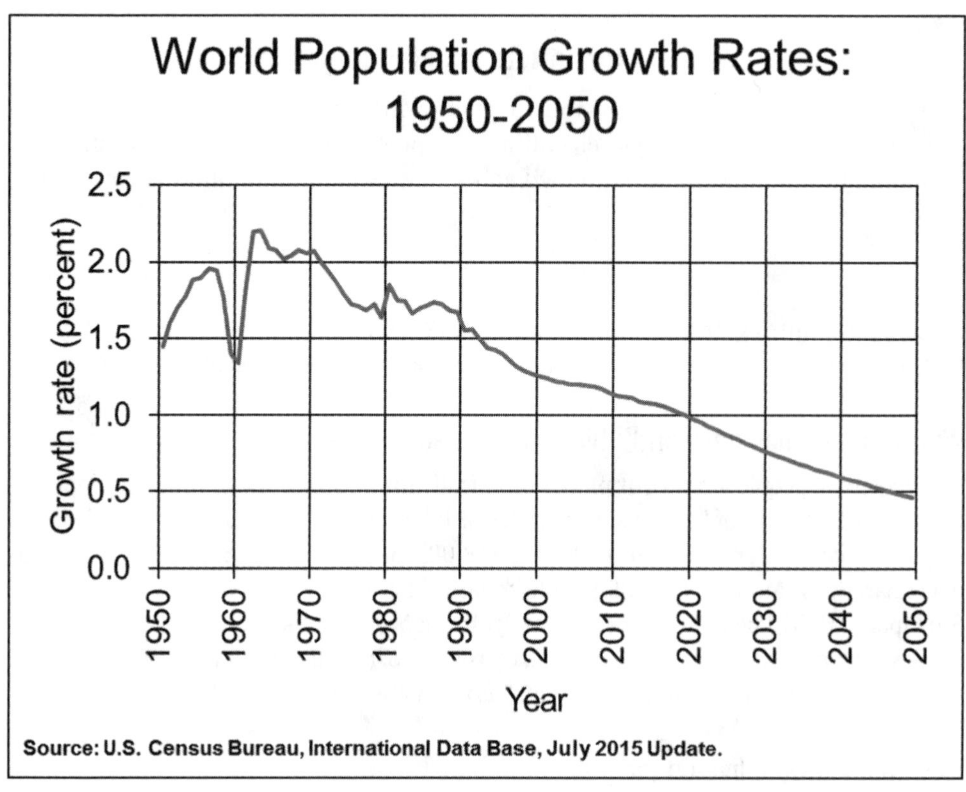

There is some good news about population growth in that rates are expected to decline in the future.

Why is the growth rate declining?

A big reason why the rate of growth is slowly declining is because people are delaying marriage longer. Also, the wide availability of contraception is having a positive impact on decreasing the number of unplanned births.

What top five countries will have the largest population of people in 2050?

Country	2050 Projected Population
India	1.45 billion
China	1.13 billion
United States	357 million
Nigeria	348 million
Indonesia	253 million

How many people on the planet are not literate?

According to experts at UNESCO, while we have improved the literacy rate, we still have a literacy problem in the world today. Approximately 757 million adults, including 115 million kids (two-thirds of whom are females), cannot read or write a simple sentence. An estimated 15% of the world is illiterate, with Sub-Saharan, South Asia, and West Asia among the places in the world with the lowest rates. The literacy rate in places like Afghanistan, Mali, and Senegal have an adult literacy rate of less than 50%.

North Korea is one of several countries that restricts use of the Internet.

What did the average European know about the world in the Middle Ages?

In Europe in the Middle Ages (from the fifth century C.E. to the fifteenth century C.E.), most individuals' knowledge of the world was quite limited. Most people were illiterate, meaning that they could not read or write in their local language, or in Latin, the language of scholarship. Because of this, geographic knowledge developed by the Greeks and Romans (who knew the Earth was a sphere) was all but lost in Europe. Europeans of the time thought of the world as flat and composed of only Europe, Asia, and Africa.

What are the top ten countries that restrict Internet usage or content from their citizens?

Although there are many places around the world with little access to the Internet, there are several countries whose government specifically controls access to the content, blocks access to specific sites, or even filters private email content of users. These countries include

Where are cyberspace and the Internet?

Cyberspace is not space in the old-fashioned sense of the word at all. The Internet is composed of many thousands of computers around the world, which are connected to each other in order to provide information across cyberspace as though there were no global boundaries, mountain passes, or oceans to cross. When you send e-mail to a friend on the other side of the planet, it passes from your computer to that of your Internet service provider and then from computer to computer, making its way to your friend in a matter of seconds. Similarly, when you access a page on the World Wide Web, your computer tells another computer, which tells another computer that you want such and such document delivered to your computer, and it arrives in seconds. Some geographers measure and map cyberspace by looking at where most of the Internet's traffic flows through, to, and from.

China, North Korea, Myanmar, Vietnam, Cuba, Saudi Arabia, Iran, Syria, and Turkmenistan. Other countries that restrict Internet usage in a variety of ways include Bahrain, Belarus, Ethiopia, Pakistan, Russia, the Sudan, the United Arab Emirates, Uzbekistan, and Yemen.

What countries have the most Internet usage?

Country	Internet Users
China	641 million
United States	279 million
India	243 million
Japan	109 million
Brazil	108 million
Russia	84 million
Germany	72 million

What is a census?

A census is an enumeration, or counting, of a population. The information from a census is used to help governments determine where to provide services, based on the demographics of the population. Information about age, gender, number of children, race, languages spoken, education, commuting distance, salary, and other demographic variables are common in a census. This information is compiled and provided to government agencies and is usually accessible to the general public.

How often do countries count their people?

In the United States and most other developed nations, a census takes place once every decade. The Constitution of the United States requires a census to be taken every ten years in order to create districts and determine the number of members of Congress each state is able to send to the House of Representatives.

What was the baby boom?

Due to post-World War II prosperity, there was a boom in American births between 1946 and 1964, now referred to as the "baby boom." During this time, approximately 77 million babies were born in the United States, a very large number compared to that of previous time spans. As the baby-boomers approach retirement age, health and welfare services for the elderly will become a high priority as the country prepares for a higher proportion of older people in its population than ever before.

Are there an equal number of boy and girl babies born?

Though scientists aren't sure why it occurs, there is an average of 105 boys born for every 100 girls. Recent data from the World Bank suggest that globally, this number has increased to 107 boys for every 100 girls born. In the most populated country of China, the ratio is 118 boys for every 100 girls born each year.

What was a leading cause of death of young women in the developed world?

From the beginning of time until the mid-twentieth century, a leading cause of death for young women was complications during childbirth. Now, in most developed countries, there is little risk of death during pregnancy and labor.

Why do some cultures in the world practice infanticide?

Infanticide is the practice of killing an infant. For centuries, various cultures around the globe used infanticide as a form of population control, most commonly because their limited food supply could only feed a certain number of humans. Because of cultural biases, female infants were more often victims of infanticide. The practice of infanticide still occurs today.

Can a woman have multiple husbands?

While some cultures allow men to have multiple wives, there are very few cultures that allow women to have multiple husbands. This practice, known as polyandry, is presently observed by only two cultures: some Tibetans living in remote parts of Nepal and the Nair people of southwestern India. Polygyny, when men have multiple wives, remains legal and practiced in many Islamic and African countries. The collective name for multiple spouses, both polyandry and polygyny, is polygamy.

Why does a firstborn son get everything?

Primogeniture is the system of inheritance in which all inheritable land and property is passed on to the firstborn son. A common worldwide tradition, primogeniture enabled a family's possessions and status to remain intact as they were passed from generation to generation. This practice of the entire inheritances benefiting only the firstborn son resulted in subsequent sons needing to find alternative livelihoods. At times, this ancient custom requires some flexibility. In 1953, because there was no male heir to the Crown of England, Elizabeth II became queen.

Estimates are difficult to establish as to how many homosexual people there are in the world because many people are still reluctant to confess their sexuality, even in an anonymous survey.

How many lesbian and gay people are there in the world?

Experts at Gallup have found that it is a difficult survey question to ask, mainly because many people are uncomfortable with answering the question honestly. Most social scientists believe that between 1 and 10% of the world's population is homosexual. But many organizations believe that approximately 2% of the global population is homosexual. This means that approximately 142.50 million people are gay or lesbian across the world.

Which wealthy country leads the world in child poverty?

Child poverty can be measured in many ways. In one measurement, child poverty is defined as the number of children from households who earn 50% less than the national median. Given this measurement, approximately 37% of children aged zero to seventeen live in poverty in the United States.

How many kids live in poverty in the world?

According to experts at UNICEF, The United Nations Children's Fund, which was created by the U.N. in 1946, estimates the number of kids aged eighteen or under who live in poverty at 569 million, or over half a billion kids. Children constitute nearly half of all people who live in extreme poverty.

How does the United States compare to other wealthy countries, in terms of child poverty?

These same experts at UNICEF also found that, in what is generally believed to be the richest country in the world, the United States, more than one out of every three kids live in poverty, a number that is increasing each year. The United States ranks number thirty-six out of forty-one wealthy countries when comparing kids' poverty levels in similar wealthy countries. By comparison, in wealthy Norway, only 5.3% of kids are living in harsh economic conditions.

LANGUAGE AND RELIGION

What are the most popular languages in the world?

Although it is difficult to estimate, most language experts would agree that the most widely, natively spoken language in the world is Mandarin Chinese, spoken by 1.026 billion people. This is followed by English, with about 841 million native speakers; Spanish, with 489 million native speakers; and Hindi, with 380 million native speakers. Arabic is spoken by approximately 295 million people.

What is the difference between a *lingua franca* and a pidgin?

A *lingua franca* is a language used between people who do not share a common language. English is often used as a *lingua franca* in international business transactions.

Pidgin is a language that has a small vocabulary and is a combination and distortion of two or more languages. For example, pidgin English, a combination between English and indigenous languages, is used in Papua New Guinea between the English-speaking and indigenous people. Most pidgin languages are used as lingua francas, but not all *lingua franca* languages are pidgin languages.

Which religions are practiced by the most people?

According to experts at the Pew Research Center, the number of Christians is estimated at 2.2 billion adherents, Muslims at 1.6 billion, and Hindus at 1 billion. By 2050, because of the relatively high birthrate trend, it is projected that Islam (estimated at 2.8 billion) will have nearly as many followers as Christianity (estimated at 2.9 billion).

How many religions have holy sites in Jerusalem?

Judaism, Islam, and Christianity all regard Jerusalem as a holy city, with many holy sites located within the city. The Western Wall, the remaining wall of the Second Temple, is the holiest site in Judaism. Islam's third-holiest site is the Dome of the Rock and the Al-Aqsa Mosque. The Church of the Holy Sepulchre is a holy Christian site.

DEALING WITH HAZARDS

How widespread was the influenza pandemic of 1918?

In 1918, a deadly flu spread quickly around the world. Within just two years, this influenza pandemic had sickened over a billion people and killed more than 21 million people. Half a million people died in the United States alone.

How did the Black Plague affect the world's population?

Spread by fleas, the bubonic plague, also called the Black Plague, raged through Europe, Asia, and North Africa between the years 1346 and 1350. Though cities attempted to curb the spread of this highly infectious disease by quarantining cities, the fleas easily spread from city to city. Estimates of those killed reach into the tens of millions of people. In Europe and Asia, more than half of the population died of the Black Plague during those four years. Many more died of starvation in the famine that followed because of the staggering depletion of the work force.

Is there enough food to feed the world?

Though there is enough food produced in the world to feed everyone, logistical and political problems make its distribution inefficient. At the current rate of world population growth, we may soon have to change our eating habits and eat more grain and less meat. There is a limited amount of grain that the Earth can produce. Currently, much of this grain is consumed by cattle, rather than humans. If humans were to eat the grain instead of eating the cattle, the calories from the grain would be twenty times more efficient than those calories derived solely from beef.

Thomas Malthus believed that the world's population would grow faster than the food supply, leading to a crisis.

What were Thomas Malthus' ideas on population growth?

In 1798, English clergyman Thomas Malthus wrote "An Essay on the Principle of Population," in which he described the problems of population growth. Malthus argued that the world's population grows faster than the food supply, but there are such checks as war, famine, disease, and disaster that seem to limit the growth in the population.

What revolution attempted to stop world hunger?

Begun in the 1960s, the Green Revolution was an attempt by developed countries and such international organizations as the United Nations to transfer agricultural technology to less developed countries. While the Green Revolution increased agricultural yields, it modified the ecology of traditional agricultural systems (such as through the use of chemical fertilizers and altered seed stock). Crop yields of wheat and rice doubled from the 1960s to the 1990s, but the innovations of the Green Revolution could not keep pace with population growth in the developing world and has yet to cure world hunger.

How many people are abducted and sold into slavery each year?

According to CNN, approximately thirty-six million people are subject to slavery in the world today. Five countries account for 61% of all slaves, or roughly twenty-two million people (India, China, Pakistan, Uzbekistan, and Russia). India is perhaps the worst offender, with an estimated 14.29 million people living in slavery. The African country Mauritania has the highest percentage of their population living in slavery at 4%.

Which countries are responsible for human trafficking?

Many countries participate in human trafficking as accomplices in providing the people to abduct, by not enforcing laws, or by allowing the people to enter their borders. Some of the biggest offenders include Saudi Arabia, Bahrain, Kuwait, Malaysia, Qatar, the United Arab Emirates, Ukraine, Russia, Moldova, Mexico, India, Egypt, and China.

What is a refugee?

A refugee is a person who is forced to flee their homes due to violence, conflict, and fear of persecution. According to the United Nations High Commission for Refugees, there are approximately twenty million refugees, with another sixty million who have forcibly left their home countries today. Most refugees come from developing countries where society is unstable. Refugees usually flee to the closest stable country, so different countries see great variation in the number of refugees based on the political climate of their neighbors. A total of 86% of all refugees live in developing, not developed countries, an increase of 16% since 2005; 51% of all refugees are kids under the age of eighteen.

Is there human trafficking in the United States?

According to the FBI and officials at UNICEF, each year between 14,500 and 17,500 people are trafficked into the United States.

How many people were sold into slavery in the New World and the United States?

From the colonial period until the nineteenth century and the end of the American Civil War, the New World and later, the United States, had a long history of slavery. During the African Diaspora, beginning in 1619, approximately twelve to thirteen million Africans were taken from their homes and sold into slavery in the Americas and the Caribbean. The majority of the slaves who survived the voyage were sent to Brazil. By the beginning of the American Civil War, nearly 240 years after the first slaves arrived in the American colonies, four million people were held as slaves, principally in the South, where slave ownership was legal and was a main source of labor for agriculture.

CULTURES AROUND THE WORLD

What are nomads?

Nomads are tribes that move from place to place in a seasonal circuit over a large region. Though nomadic people often build temporary homes, they consider migratory life within their tribe to be their home. Nomadic tribes are located in marginal areas around the world, from the Sahara Desert to northern Siberia. The nomadic way of life is threatened because of general cultural prejudice against unsettled peoples and urbanization.

Who are Gypsies?

Gypsies are nomadic tribes that travel throughout Europe. Though they once were thought to have originated in Egypt (hence the European name for them), linguistic

Nomads can still be found in some parts of the world, such as this Tibetan man tending to yaks in Nepal.

studies have placed their origin in India. These traveling tribes have been subject to centuries of persecution, including "Gypsy hunts" and extermination at the Auschwitz Death Camp during World War II.

What is brain drain?

When highly educated or highly skilled individuals leave their home countries to go to countries where opportunities are better, the home country experiences "brain drain." This occurs especially in Asian countries, as highly educated Asians (for example, from India, China, and the Philippines) move to the United States, Canada, and Australia for higher-paying jobs. Many people come to pursue a higher education and never return to their home country.

Why don't Americans eat horse meat?

Most religions and cultural groups have some kinds of food taboos. Foods may be avoided entirely or may be avoided on certain days or during certain festivals. Religious food taboos include the avoidance of pork by Muslims and Jews and the avoidance of beef by Hindus. Cultural food taboos also play an important role. For example, Americans don't eat horse meat because it is a cultural food taboo, despite the fact that horse meat is a nutritional and edible type of food. Researchers have also found that when raising horses, they require more feed than cows. Experts in the U.K. found that horse eating was eliminated in the early Christian Church during medieval times because of its connection to paganism.

How many McDonald's restaurants are there in the world?

To some people, McDonald's represents the extreme example of cultural imperialism, an import of one's culture supplanted into another. Today, there are 36,000 McDonald's

Why do some people eat dirt?

The practice of eating dirt or clay, called geophagy or geophagia, is most commonly practiced by pregnant and lactating women and children in the developing world. Since women's bodies require additional nutrients during pregnancy and lactation, the body craves clays and dirts that carry these additional minerals. This practice is most common in the developing world in Africa, but the practice spread to the southern United States, especially in the state of Georgia, where a type of clay named "Kaolin" can be found in nature.

restaurants located in over 100 countries, serving approximately 69 million customers each day.

When did people start eating with forks and spoons?

Though introduced in the fifteenth century, two-tined forks did not come into common use in Europe until the seventeenth century. Prior to that time, people ate with their hands and a knife. Researchers have found that spoons are a much older invention, dating back to Paleolithic times (from 12,000 B.C.E. to 2.9 million B.C.E.). Ancient spoons made of shells or pieces of wood have been found at excavation sites. Further proof of the ancient use of spoons is in the history of words that we use today for spoons, which are derived from the Greek and Latin words "cochlea," meaning "shell," and the Anglo-Saxon word "spon," which means chip of wood, both materials used to make spoons.

TIME, CALENDARS, AND SEASONS

Why did early humans have no need for hours, days, weeks, or months?

Because early humans were hunters and gatherers, they had little need to know the exact time. What was essential to them was an understanding of the seasonal migration of animals, varieties of plant life, and their location.

What time is it at 12:00 A.M.?

A.M. is midnight. In the middle of the night, 11:59 P.M. is followed by 12:00 A.M., not 12:00 P.M., which is noon.

What do A.M. and P.M. mean?

The A.M. is an abbreviation for the Latin words for "before midday" or "ante meridiem," and P.M. is an abbreviation for the Latin words for "after midday" or "post meridiem."

How does military time work?

Twenty-four-hour time, also known as military time, is used in many countries. It begins at midnight with 0000 and the day ends at 2359. The first two digits represent the hour, and the last two digits represent the minutes. Since there are 24 hours in a day, each hour is numbered 00 through 23. For instance, 0100 is 1:00 A.M., 1200 is noon, 1300 is 1:00 P.M., and 2043 is 8:43 P.M.

Does the Earth always rotate and revolve around the sun at the same speed?

No, the Earth's revolution around the sun and rotation on its axis are not perfect. Its daily rotation may vary by approximately four to five milliseconds and is slowing down at a rate of one millisecond each century, due to tidal friction. Additionally, the axis wobbles slightly, and the length of revolution around the sun varies by a few milliseconds.

How long is a day?

A day is the time it takes the Earth to make one rotation, which is 23 hours, 56 minutes, and 4.2 seconds. We round this to an even 24 hours for convenience.

How can I find the exact time?

Today, you can go to many websites on the Internet in order to synchronize your device's time. Just search for the word "time," and you will see links for the exact time. Your devices are also set to automatically update time with various atomic clock time servers, like those maintained by the National Institute of Standards and Technology (NIST). Your device's time is usually accurate to within a few milliseconds.

TIME ZONES

When were time zones established in the United States?

In 1878, Sir Sandford Fleming, a Scottish-born Canadian engineer and inventor, proposed dividing the world into 24 time zones, each spaced 15 degrees of longitude apart. He originally called his concept of a single, world time delineation "Cosmic Time." The contiguous United States was covered by four time zones. By 1895, most states had begun to institute the standard time zones of Eastern, Central, Mountain, and Pacific on their own. But it wasn't until 1918 that Congress passed the Standard Time Act (also known as the Calder Act), establishing official time zones in the United States.

How did trains help establish time zones?

Before trains, many cities and regions had their own local time, which was set based on the sun at their location. The great variation of local times made train schedules confusing. In November 1883, railroad companies across the United States and Canada began to use standard time zones, decades before they came into general use across the United States.

How many time zones does the United States have?

The United States observes nine time zones that are defined by federal law. These time zones are: Atlantic, Eastern, Central, Mountain, Pacific, Alaska, Hawaii-Aleutian, Samoa, and Chamorro.

Which states have multiple time zones?

Florida, Indiana, Kentucky, Michigan (part of the Upper Peninsula), and Tennessee are split into Eastern and Central time zones. Kansas, Nebraska, North Dakota, South Dakota, and parts of Texas are split between Central and Mountain time zones. Part of Nevada is split between Mountain and Pacific time zones.

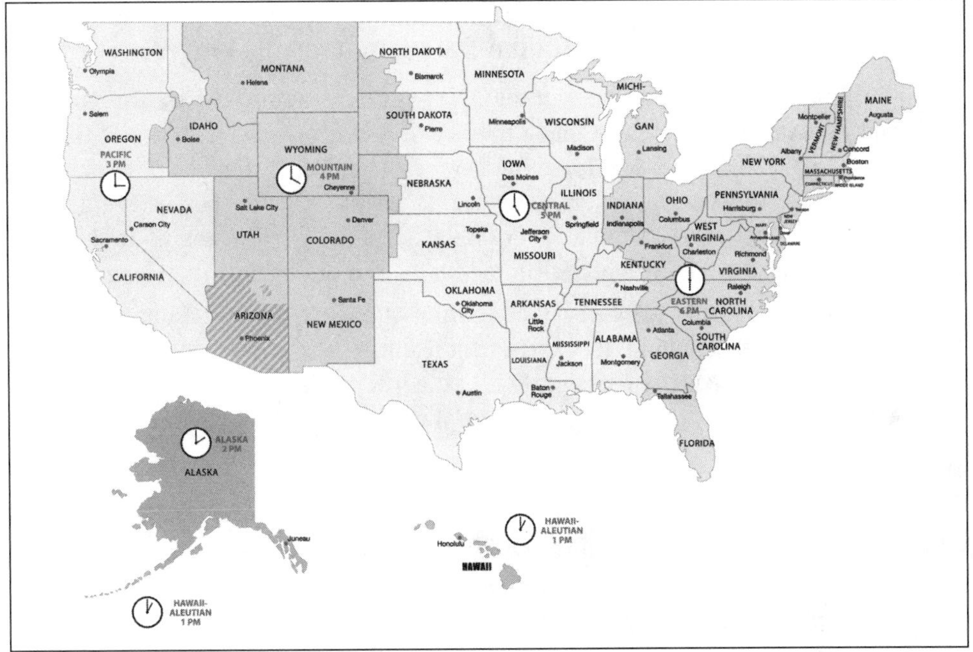

The United States has nine time zones, but most of us are only familiar with Eastern, Central, Mountain, and Pacific.

Which countries use half-hour time zones?

Several countries use half-hour time zones, including Afghanistan, Australia (certain parts), India, Iran, the Marquesas Islands, Myanmar, Sri Lanka, and Venezuela; also, the Canadian province of Newfoundland uses half-hour zones.

If it is 12:30 P.M. in New Delhi, India, what time is it in Kathmandu, Nepal?

If it is 12:30 P.M. in New Delhi, it must be 12:45 P.M. in the neighboring capital of Nepal, Kathmandu. Quarter-hour time zone divisions are used in Nepal, as well as in the Chatham Islands.

How many time zones does China have?

Since China is such a large country, it should span five time zones. Since 1949, the entire country uses only one time zone, which is eight hours ahead of Coordinated Universal Time (UTC) or Greenwich Mean Time (GMT).

What time is it at the South Pole?

Because time zones get narrower the farther you get from the equator, time zones would be very thin near the North and South Poles. To simplify things, researchers living in polar regions, like Antarctica, use one of twelve different time zones, depending on which country manages the research station or base.

Why did the International Date Line recently move?

In 2011, in order to improve economic ties with key trading partners Australia and New Zealand, the Pacific Island nations of Samoa and Tokelau moved their calendars ahead by one day, skipping December 30 entirely. Now they are closer in time to their key trading partners. Before the change, the Independent State of Samoa was twenty-three hours ahead of New Zealand. Now it is one hour ahead. American Samoa, a territory of the United States, did not change its time zone.

In 1994, the country of Kiribati, consisting of twenty-one inhabited islands, straddled the International Date Line, thus dividing the country into two days. In 1995, Kiribati decided to shift its portion of the International Date Line far to the east, so that the entire country could be on the same side of the Date Line.

What happens when I cross the International Date Line?

If you fly, sail, or swim across the International Date Line from east to west, such as from the United States to Japan, you add a day (Sunday becomes Monday). When you travel from west to east, such as from Japan to the United States, you subtract a day (Sunday becomes Saturday).

How fast do you have to travel west to arrive earlier than when you left?

Normally, when flying between London and New York, the trip takes seven hours. Thus, with the five-hour difference in time between the two cities, you arrive two hours "later" than when you left London. If you could have flown on a supersonic passenger jet, traveling at Mach 2 (1,300 miles/hour or 2,092 km/hour), two times the speed of sound, the trip between London and New York would have taken only three hours. Thus, with the five-hour difference in time zones, you would have arrived two hours "earlier" than when you left!

Why is Russia always one hour ahead of the standard time zones for the geographic area?

In an effort to take advantage of the limited amount of light available in winter months, each of Russia's time zones are one hour ahead of the standard time for those zones. Russia also follows Daylight Saving Time and adds an additional hour during the spring and summer months.

What's the difference between Greenwich Mean Time, Coordinated Universal Time, and Zulu Time?

Greenwich Mean Time (GMT), Coordinated Universal Time (UTC), and Zulu Time are three different names for the same time zone. This time zone is situated at the Prime

Meridian, zero degrees longitude, and runs through the Royal Observatory at Greenwich, a section of London, England. It is the Prime Meridian from which other longitudes are determined, east and west.

DAYLIGHT SAVING TIME

Why do we practice Daylight Saving Time?

By moving our clocks forward one hour between spring and fall, we more effectively utilize the light of the sun to keep homes and businesses illuminated, which saves all of us electrical energy.

When did Daylight Saving Time move from the end to the beginning of April in the United States?

The shift of the beginning of Daylight Saving Time from the last Sunday in April to the first Sunday in April took place in 1987, when the Uniform Time Act was amended. In 2007, the beginning of Daylight Saving Time was moved again to the second Sunday in March, and the ending of Daylight Saving Time was moved to the first Sunday in November.

When do countries in the Southern Hemisphere observe Daylight Saving Time?

Because Daylight Saving Time is an effort to save daylight during the summer months, Daylight Saving Time in the Southern Hemisphere occurs from October through March.

When was Daylight Saving Time instituted?

Though Benjamin Franklin suggested a satirical concept of Daylight Saving Time in 1784, it was actually New Zealand entomologist G. V. Hudson who first proposed it in 1895. In 1905, British builder William Willett conceived of the idea of advancing clocks in the summer months, which caught the attention of the British Parliament, but it never passed into law. In the United States, it was not implemented until World War I. Between World Wars I and II, states and communities were allowed to choose whether or not to observe the change. During World War II, Franklin Roosevelt again implemented Daylight Saving Time. Finally, in 1966, Congress passed the Uniform Time Act, which standardized the length of the Daylight Saving Time period. But states and territories can choose not to observe Daylight Saving Time. Arizona, Hawaii, parts of Indiana, Puerto Rico, and some island territories have chosen not to observe Daylight Saving Time.

KEEPING TIME

What is a sundial?

A sundial is an instrument that uses the sun to measure time. A sundial consists of an angled marker, called a gnomon, that casts a shadow on a plate, called a dial plane. On the dial plane, there are marks indicating the hours of the day. During the day, as the sun moves across the sky, the shadow from the gnomon moves across the dial plane, indicating the hour. Sundials were used to measure time before clocks and watches were invented. The Greek philosopher and teacher of Pythagoras, Anaximander, is believed to have introduced the sundial to the Greeks, around the fifth century B.C.E. The idea for the sundial comes from the Babylonians.

When were the first clocks made?

The concept of keeping time in two twelve-hour periods originated in Sumer, present-day Iraq, around 2000 B.C.E. Egyptians, and later Greeks, developed sundials to track the movement of time. Even before 2000 B.C.E., evidence suggests that the Chinese developed water clocks, too. No one knows who first invented the clock, but the archaeological record seems to support the fact that time-keeping devices began to appear in several places, like Sumer and China, around the same time, 2000 B.C.E.

What is a water clock?

Water clocks were the first clocks that didn't rely upon sunlight (as with sundials) to tell time. They operated by dripping water from containers at measured intervals and then measuring it. There were two key types of water clocks: those that measured time by the amount of water remaining in the clock and those that measured time by how much water dripped from the clock, filling a measuring device. Because the invention is so old, there is some debate as to when they first appeared. Some believe that the Chinese, as early as 4000 B.C.E., first created the water clock. Others believe that ancient Sumerians and Egyptians, followed by Greeks, invented and innovated the first water clocks around 2000 B.C.E.

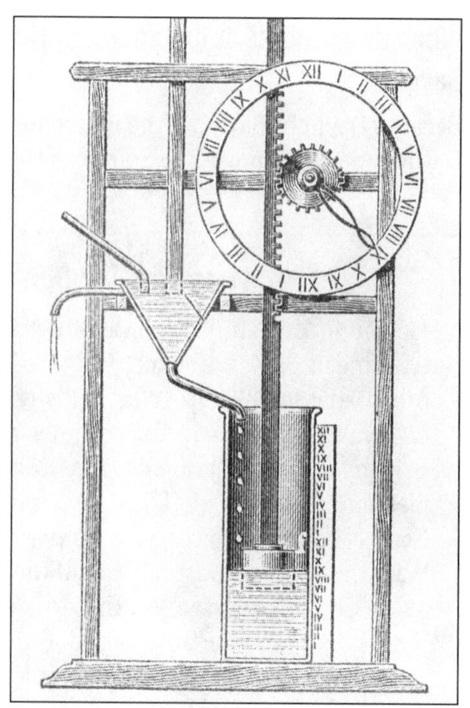

A nineteenth-century illustration of an Egyptian water clock.

When was the first watch made?

In the early sixteenth century, German locksmith Peter Henlein began to produce portable clocks called Nüremberg eggs. At about the same time, another clockmaker from Nuremberg, Germany, named Peter Helle, in an account, was described as having been making small, portable watches for several years. Advances in later centuries led to timepieces that could be worn on the wrist.

What is an atomic clock?

An atomic clock uses measurements of energy released from atoms to precisely measure time. The current model of the atomic clock, created in 1957 by Norman Ramsey, uses measurements from cesium atoms. Atomic clocks are used by NASA, physicists, astronomers, and other scientists who need extremely precise time.

CALENDARS

What do the abbreviations B.C. and A.D. mean?

In our modern calendar, the year 0 represents the year of the birth of Jesus Christ. Years before his birth are known as B.C. or "before Christ." Years since his birth are known as A.D. or "Anno Domini," the "Year of Our Lord."

What do the abbreviations B.C.E. and C.E. stand for?

To secularize the calendar, the terms B.C.E. and C.E. have come into use to replace the abbreviations B.C. and A.D. respectively. B.C.E. means "before the common/current/Christian era" and C.E. stands for "common/current/Christian era."

What is the problem with a calendar based on the cycles of the moon?

The time between two new moons is 29.5 days. After 12 lunar months, a calendar based on the cycles of the moon falls short of a solar year—and thus the cycle of seasons—by 11.25 days. To compensate, the Hebrew calendar, which is based on the moon cycle, has a regulated 19-year cycle in which an extra month is added every two or three years.

How did Julius Caesar fix the calendar?

For years, the Romans had been using a calendar based on lunar cycles. Since each lunar month is 29.5 days, 12 months only adds up to 354 days. But seasons do not follow a lunar cycle, they follow a solar one. A solar year lasts 365 days, 5 hours, 49 minutes. In 45 B.C.E., Julius Caesar implemented a solar calendar so that the seasons would occur at the same times every year. Additionally, Caesar made each year 365 days long, with every fourth year being 366 days long (a "leap" year). Unfortunately, each calendar year was still 11 minutes longer than a solar year, a problem that Caesar did not feel was a big concern at the time.

Why do we have leap years?

We have leap years to keep the calendar accurate with respect to the solar year. It takes the Earth 365 days, 5 hours, 48 minutes, and 46 seconds—just under 365.25 days—to revolve around the sun. If there were no leap years, every 56 years the calendar would be two weeks behind. By adding one extra day every four years, the calendar stays accurate.

When was January 1 chosen as the beginning of the year?

In Caesar's calendar modifications of 46 B.C.E., he decreed that the year would begin on January 1 instead of March 25, as it had in the past. At that time, Caesar also designated the number of days in each month, unchanged to the present day.

What was the longest year in history?

The year 46 B.C.E. was decreed by the Roman Emperor Julius Caesar to be 445 days long in order to correct the calendar, which was 80 days off, based on the seasons. The year was also known in Latin as *annus confusionis*, or "the year of confusion."

Why were ten days lost from the year in 1582?

In 46 B.C.E., Julius Caesar implemented the Julian calendar, which was eleven minutes longer than a solar year. By 1582, those eleven minutes each year had added up to ten days. Pope Gregory XIII aligned the calendar with the solar year by declaring October 5, 1582, to be October 15, 1582, in the Catholic regions of the world, thus correcting for the ten lost days.

Pope Gregory XIII reworked our calendar to fix the Julian calendar.

What is the Gregorian calendar?

In addition to moving the calendar forward by ten days in 1582, Pope Gregory XIII also corrected the error of the Julian calendar. He declared that years ending in "00" would not be leap years, except those divisible by 400 (such as the year 2000). The Julian calendar, with Pope Gregory's correction, is known as the Gregorian calendar, which most of us use today.

When was the Gregorian calendar adopted in the United States?

Though Catholic countries switched to the Gregorian calendar in the sixteenth century, Protestant countries, such as England and its colonies, refused to switch from the Julian to the Gregorian calendar at that time. It wasn't until 1752 that Britain and its colonies, including the colonies that later became the United States, switched to the Gregorian calendar. By that time, there was an eleven-day difference in time, so September 3, 1752, became September 14, 1752.

Is the Gregorian calendar accurate?

Almost! It is still twenty-five seconds longer than the solar year. Therefore, after about 3,320 years we will be a full day ahead of the solar year. The keepers of time will have to deal with this problem when the time comes.

What type of calendar did the French use between 1793 and 1806?

In 1793, during the French Revolution, an entirely new calendar was established by the National Convention. The calendar was designed to help rid French society of its Christian influences. Within this new calendar there were 12 months, each consisting of three decades. Each decade was composed of 10 days. Five days (six during a leap year) were added at the end of the year to add up to 365 (or 366) days. Napoleon reinstated the Gregorian calendar in 1806.

What type of calendar was used in the Soviet Union between 1929 and 1940?

The Soviets created the Revolutionary calendar, which had five days in a week (four for work, the fifth as a day off) and six weeks in a month. Five days (six during a leap year) were added at the end of the year to add up to 365 (or 366) days.

Which celestial bodies are the days of the week named after?

The names of the days of the week come from Roman or Norse names for the planets:

Day	Celestial Body (Roman/Norse)
Sunday	Sun/Sol
Monday	Moon
Tuesday	Mars/Tui
Wednesday	Mercury/Woden
Thursday	Jupiter/Thor
Friday	Venus/Frygga
Saturday	Saturn

When did the twenty-first century begin?

The twenty-first century began at 12:00 A.M. on January 1, 2001. Since the first century, which spanned the years 1 to 100, centuries have been counted beginning with the year ending in "01" rather than "00." For instance, the twentieth century consists of the years 1901 through 2000.

THE SEASONS

Why do we have summer and winter?

Since the Earth is tilted 23.5 degrees, the sun's rays hit the Northern and Southern Hemispheres unequally. When the sun's rays hit one hemisphere directly, the other hemisphere receives diffused rays. The hemisphere that receives the direct rays of the sun experiences summer; the hemisphere that receives the diffused rays experiences winter. Thus, when it is summer in North America, it is winter in most of South America, and vice versa.

Where on the planet is it light 24 hours a day in the summer?

In the extreme north and south parts of the Earth (north of 66.5 degrees north latitude and south of 66.5 degrees south latitude), it is light 24 hours a day during the summer and dark 24 hours a day during the winter. Northern cities like Reykjavik, Iceland, and Murmansk, Russia, have nearly 24 hours of daylight for a short period of time during the summer months.

How do people cope with continual light or darkness in high latitudes?

Murmansk, Russia, is the largest city north of the Arctic Circle. The city receives no sunlight for several months out of the year, making it one of the most psychologically extreme environments on the planet. Residents of the city (about 307,257) walk along artificially lit streets that give the appearance of sunlight, undergo artificial sun treatments (much like tanning booths), and often suffer from the condition known as Polar Night Stress. Polar Night Stress symptoms include fatigue, depression, vision problems, and susceptibility to colds and viruses. Other large Arctic cities include Norilsk, Russia (population 177,506), Tromso, Norway (population 71,590), and Vorkuta, Russia (population 70,548).

Where are the Arctic and Antarctic Circles?

The Arctic Circle is located at 66.5 degrees north of the equator and the Antarctic Circle is located at 66.5 degrees south of the equator. Areas north of the Arctic Circle and south of the Antarctic Circle have 24 hours of light during the summer and 24 hours of darkness during the winter.

What are the Tropics of Cancer and Capricorn?

The two tropics are the lines of latitude where the sun is directly overhead on the summer solstices. The Tropic of Cancer is at 23.5 degrees north and passes through central

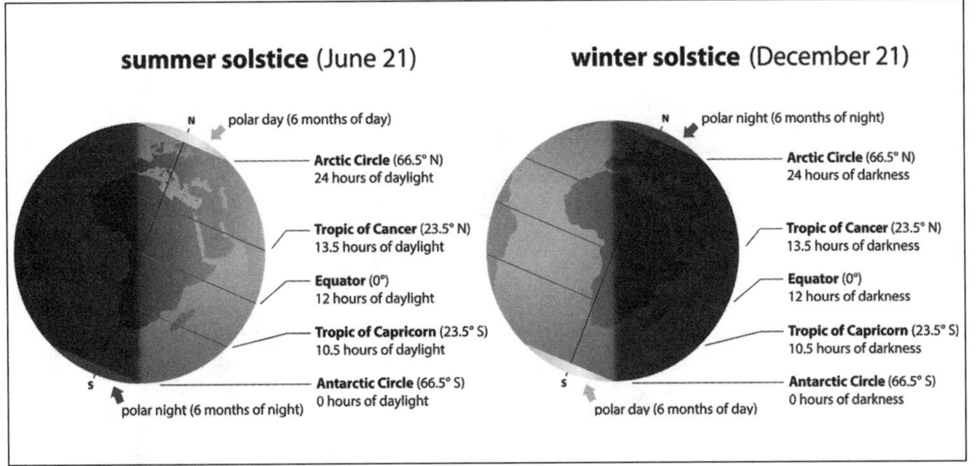

During the summer solstice, the North Pole is tilted toward the sun, and it is tilted away during the winter solstice.

Mexico, northern Africa, central India, and southern China. The Tropic of Capricorn is at 23.5 degrees south and passes through central Australia, southern Brazil, and southern Africa.

What are the solstices?

There are two solstices—one on June 21 and the other on December 21. On June 21, the sun is directly above the Tropic of Cancer at noon and heralds the beginning of summer in the Northern Hemisphere and the beginning of winter in the Southern Hemisphere. On December 21, the sun is directly above the Tropic of Capricorn at noon and heralds the beginning of winter in the Northern Hemisphere and the beginning of summer in the Southern Hemisphere.

What are the equinoxes?

There are two equinoxes—one on March 21 and the other on September 21. On both equinoxes, the sun is directly over the equator. March 21 heralds the beginning of spring in the Northern Hemisphere and the beginning of fall in the Southern Hemisphere. September 21 heralds the beginning of fall in the Northern Hemisphere and the beginning of spring in the Southern Hemisphere.

Can you stand an egg on end only on the spring equinox?

It is a common legend that an egg can be balanced on its end only on the spring equinox (March 21). Actually, there's nothing magical about gravity on the spring equinox that would allow an egg to stand on end. It can happen at any time of the year with patience and perseverance.

EXPLORATION

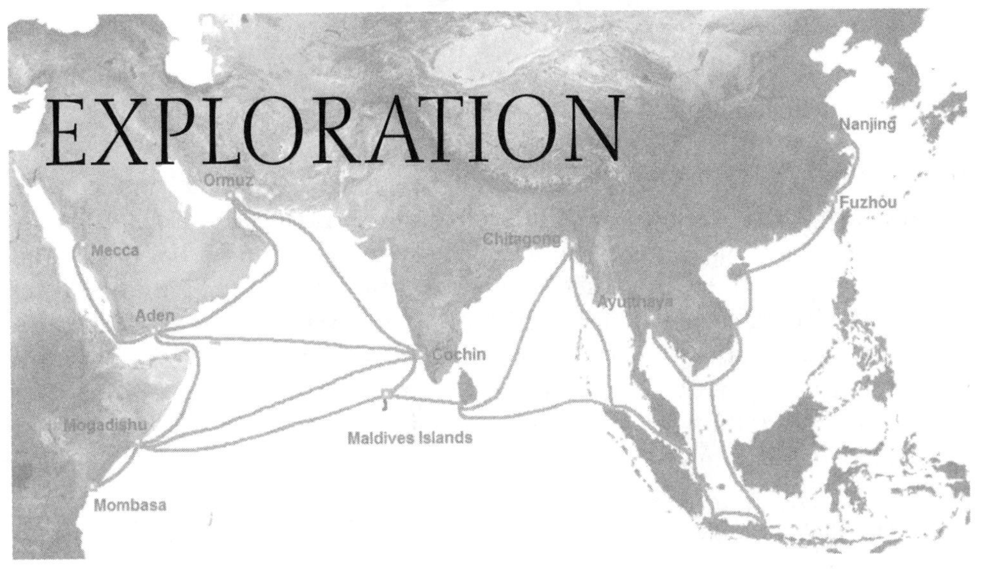

EUROPE AND ASIA

How did people from the Arctic region of Siberia populate South America?

Many scientists believe that around 14,000 B.C.E., small groups of people, the original explorers, emanating from what is now Siberia in Russia, crossed a land bridge to what is now Alaska. From there, they moved as far south as South America. Considerable genetic evidence, archaeological finds, and skeletal remains support this theory.

When did seafaring Phoenician explorers begin discovering colonizing the Mediterranean Region?

As early as 1000 B.C.E., at the high point of their strength and influence, Phoenicians began colonizing much of the coastal areas around the Mediterranean Sea, including Algeria, Cyprus, Sardinia (Italy), Sicily (Italy), Libya, Malta, Mauritania, Morocco, Portugal, Spain, Tunisia, and Turkey. Ancient Greek historians Strabo and Herodotus believed that the Phoenicians originated from what is now Bahrain. We owe much of our language to the Phoenicians, an assemblage of coastal societies of people that existed for more than 3,000 years.

When did the Chinese Empire begin naval exploration?

During the Song Dynasty (960–1270 C.E.), the Chinese began to build seafaring trade ships. It was during the Yuan Dynasty (1271–1368 C.E.) that Chinese traders began to appear in the ports of Sumatra, Ceylon (Sri Lanka), India, and as far west as Africa. They were seeking goods for the royal kingdom, such as spices, ivory, medicine, and tropical woods.

What were Marco Polo's contributions to exploration?

Though Marco Polo did not actually discover anything, his writings in "The Travels of Marco Polo" served as Europe's introduction to the East and spurred interest in exploration. Marco Polo, born in the mid-thirteenth century in Venice, traveled with his father and uncle to China. During his stay, Polo served Emperor Kublai Khan as an ambassador, governor, and in a host of other diplomatic positions. In his thirties, he returned to Venice and fought against the city-state of Genoa that was at war with Venice and was eventually captured. While imprisoned in Genoa, he dictated the story of his travels to a fellow prisoner, creating the somewhat exaggerated memoir, "The Travels of Marco Polo." He was the first person to chronicle his travels to the Far East and inspired many travelers after him.

What did Marco Polo note in his journals about the Chinese fleet when he arrived in the thirteenth century?

Marco Polo noted in his journals that Chinese ships had crews of more than 300, cabins for 60 people, and four sailing masts, which were far bigger than anything he had seen at the time.

Which explorer was named the Grand Imperial Eunuch by the emperor of China?

The Chinese explorer Zheng He (a.k.a. Cheng He) helped Emperor Yung-lo come to power in 1402 C.E., and in 1404 C.E., the emperor named Zheng the Grand Imperial Eu-

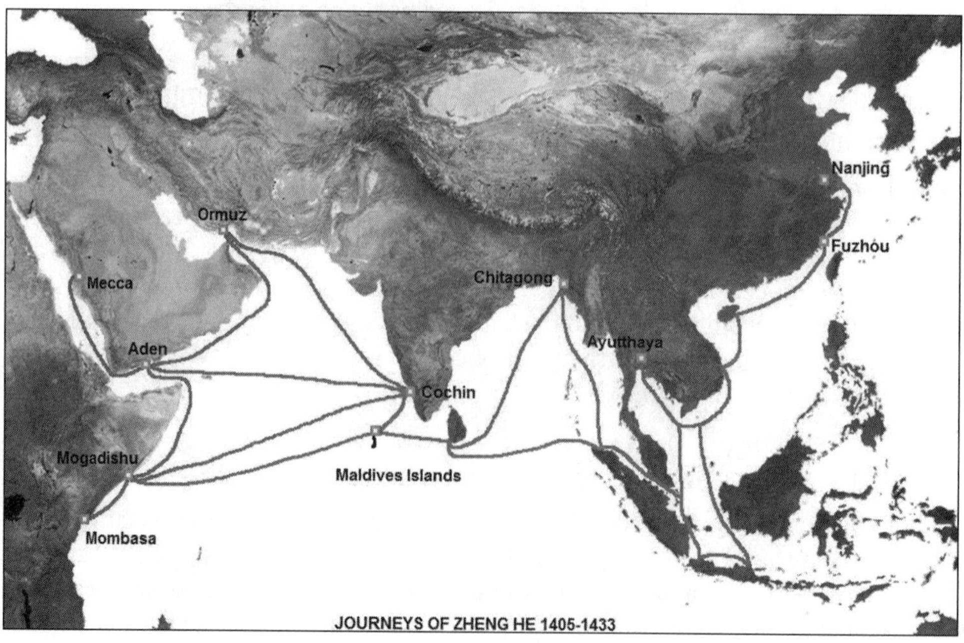

Chinese explorer Zheng He explored much of the area around the Indian Ocean in the early fifteenth century.

nuch. In 1405 C.E., he set sail on the first of his seven voyages, which spread Chinese influence and knowledge throughout South Asia and Africa. China moved toward a period of isolationism after Emperor Yung-lo died.

What did Genghis Khan conquer?

Genghis Khan, the ruler of the Mongol Empire, conquered an area stretching from China to western Russia to the Middle East. Khan created the world's largest empire, which began to dissolve following his death in 1227 C.E. Geneticists believe that millions of people throughout the former empire, living today, are direct descendants of Genghis Khan.

Who was the greatest explorer of the Arab world?

Known as the "Muslim Marco Polo," Ibn Battuta (1304–1369) was a Moroccan explorer who explored much of Africa and Asia over a period of 30 years. In his lifetime, he traveled over 75,000 miles (120,000 km), gaining the reputation as the most-traveled man on Earth and one of the greatest travelers of all time.

What were the Crusades?

From the late eleventh through the thirteenth centuries, groups of armed Christian Europeans invaded the Middle East to take the Holy Land from the Muslims and reclaim it for Christianity. These acts were supported by the Catholic Church. The Crusaders ruthlessly murdered and pillaged throughout their long journeys to the Middle East and continued their brutality once there. Though the Crusades were a horrific era, the knowledge of the world gained by the Crusaders spurred a better geographic understanding.

Who was Prester John?

During the period of the Crusades (late eleventh to thirteenth centuries C.E.), a legend arose that a king allegedly named Prester John needed assistance in fending off the Muslims. In the twelfth century, a letter arrived for the pope, and several other European leaders, claiming to be from "Prester John," the leader of a Christian kingdom in the east that was in danger of being overrun by infidels. Prester John reportedly asked for help from European brethren. Though Prester John and his kingdom were never discovered, his mysterious letter sparked travels and explorations for centuries in an attempt to rescue the kingdom.

Who disguised himself as a Muslim to travel to Mecca?

Since non-Muslims are not allowed into the sacred city of Mecca, British explorer Sir Richard Francis Burton disguised himself as an Afghan pilgrim in order to enter the city in 1853. Burton, having learned various languages in the military, explored India, the Middle East, Africa, and South America. Also a prolific writer, Burton published many accounts of his journeys and is perhaps most famous for translating the book

1001 Arabian Nights into English and his ability to speak twenty-nine languages. He was educated at Trinity College, Oxford University.

AFRICA

Who was John Hanning Speke?

John Hanning Speke (1827–1864) was an officer in the British Indian army. In 1856, Speke was sent by the British Royal Geographical Society to discover lakes believed to exist in eastern Africa. In 1858, Speke and fellow explorer Sir Richard Francis Burton discovered Lake Tanganyika. Speke and Burton split, and while traveling on his own, Speke discovered Lake Victoria and claimed it to be the source of the Nile River. Though many did not believe the lake was the source, Speke returned to the lake in 1860 and proved Lake Victoria was indeed the source of the Nile River.

Whose body was preserved and then hand carried for nine months to the coast of Africa?

Dr. David Livingstone, the world-famous explorer of Africa, died while exploring the area now known as Zambia. His body was embalmed with sand, and his heart was buried under a nearby tree. His body was then wrapped in cloth and covered with tar to waterproof it. Loyal servants carried his body for nine months, all the way to the eastern coast of Africa, where the body was then transported to Britain on the HMS *Vulture*. On April 18, 1874, his body was buried in Westminster Abbey.

Who was Captain Kidd?

Though Captain William Kidd (c. 1645–1701) was hired by the British to fight pirates, in so doing, he became a pirate himself. After he set sail for Africa, an area swarming with

Which explorer fought for both the Union and Confederate armies in the U.S. Civil War and went on to discover a famous missing African explorer?

Though born in Britain, Sir Henry Morton Stanley sailed to the United States and worked there for several years before the start of the Civil War. He joined the Confederate Army but was captured in 1861 at the Battle of Shiloh. He then joined the Union Army. Stanley is best known for his 1871 search for the missing African explorer David Livingstone and his greeting upon finding him: "Dr. Livingstone, I presume?"

pirates, reports returned to Britain that Kidd himself had captured several ships. Upon learning of a warrant for his arrest, Kidd sailed to Boston to meet a benefactor, who then had Kidd arrested and sent back to Britain. On May 23, 1701, Kidd was hanged for piracy.

British privateer Captain William Kidd was hired to fight pirates and became one himself, a crime for which he was hanged.

THE NEW WORLD

Who was the first European to see the Pacific Ocean from its eastern shore?

In 1513, Vasco Núñez de Balboa crossed the Isthmus of Panama and became the first European to see the Pacific Ocean from its eastern side. Wearing full armor, he walked straight into the ocean and claimed it and all the land it touched for Spain. He named the discovery the "Mar del Sur" (the South Sea). He settled on the island of Hispaniola (modern-day Dominican Republic and Haiti) and founded the first permanent European settlement in the Americas, in present-day Panama (1510 C.E.).

What is the Strait of Magellan?

The Strait of Magellan is a narrow, east-west water passage that separates mainland Chile and Argentina from the region of Tierra del Fuego to the south. Ferdinand Magellan discovered the strait that bears his name in 1520 during his voyage that circumnavigated the Earth. Magellan used the strait as a shortcut around the southern tip of South America. The waters of the Strait of Magellan are violent and surrounded by dangerous rocks.

What was the intent of Magellan's expedition?

Ferdinand Magellan left Europe in 1519 and was intending to reach what was known as the Spice Islands (modern-day Indonesian archipelago of the Moluccas), by way of South America, in order to gain exclusive European rights to the trade in spices (including cloves, nutmeg, and mace). He succeeded in reaching the Philippines in 1521, where he was later killed in a skirmish with native people. Though five ships and 241 men left Europe on September 20, 1519, only one ship, the *Victoria*, returned to Spain with eighteen men on September 6, 1522. Despite Magellan's death on April 27, 1521, the Magellan Expedition successfully circumnavigated the globe. When *Victoria* returned home, it was stuffed with 381 bags of cloves, which was worth enough money to pay for the expedition in its entirety.

Who was Alexander von Humboldt?

Alexander von Humboldt (1769–1859) was a German geographer who explored much of Latin America, including Venezuela, Peru, Ecuador, Colombia, and Mexico. He also explored Europe and Russia. Von Humboldt traveled deep within the Amazon rain forests, developed the first weather map, and wrote a five-volume encyclopedia in which he sought to describe all of human knowledge about the Earth. Von Humboldt was the world's last great polymath (one with encyclopedic learning). His books, which were published over twenty-one years, inspired Charles Darwin.

Who was Leif Ericsson?

In the early eleventh century C.E., an explorer likely born in Iceland named Leif Ericsson was the first European to set foot in North America. According to legend, Leif Ericsson, a Norse explorer, visited Helluland, Markland, and Vinland, which are believed to be Baffin Island, Labrador, and Newfoundland, respectively. He was the son of the explorer Eric the Red, who founded the first Norse settlement in Greenland (986 C.E.).

Why is Christopher Columbus credited with discovering the New World?

Despite the fact that Christopher Columbus was neither the first person nor the first European to reach the Americas, Columbus's discovery was important in that it prompted mass exploration and colonization of the New World. Though the idea of Columbus "discovering" the New World is extremely Eurocentric, Europeans have credited Columbus with the discovery of, and the enthusiasm surrounding the exploration of, the New World.

Did everyone during Christopher Columbus's time think that the world was flat?

No, they did not. Though the common perception is that Christopher Columbus had to convince King Ferdinand and Queen Isabella that the world was a sphere, Columbus actually had to convince the king and queen of the distance around the world. Though the ancient Greeks had discovered that the Earth was a sphere, centuries passed before this was generally accepted. By the fifteenth century, however—the time of Columbus's sailing—most educated people believed the world to be round.

Is it true that Columbus deliberately fudged the circumference measurement so as to make a better case for his trip?

Though most scholars believed that the circumference of the Earth was approximately 25,000 miles (40,000 km), Columbus used an estimate of 18,000 miles (29,000 km) to push his case in order to make his trip seem more achievable and the costs more reasonable. Columbus used Posidonius's smaller estimate, rather than Eratosthenes's larger and more accurate estimate, to make the trip appear shorter.

Who was Ponce de Leon?

The Spanish conquistador Juan Ponce de Leon searched for the fountain of youth. In 1493, he sailed with Christopher Columbus on his second voyage to the New World. In 1506, he discovered gold on an island and returned as its governor for two years. He named the island Puerto Rico. Two years later, he was replaced by Columbus's son. In 1513 during his travels in search of the mythical fountain, reportedly located on the legendary island of Bimini, Ponce de Leon discovered Florida near the present-day city of St. Augustine.

The eponymous source of the name America comes from the Italian Amerigo Vespucci, the first person to recognize that the New World was not part of Asia.

How did America get its name?

Though Christopher Columbus discovered the New World, he always believed that he had reached Asia, not realizing that he had encountered new continents. The Italian explorer Amerigo Vespucci, who had explored the New World and published accounts of his travels, was the first person known to have distinguished the New World from Asia. The German cartographer Martin Waldseemüller, who had read of Vespucci's travels, published a map of the New World in 1507 with what is now known as South America, using the name "America," in honor of Amerigo. The name stuck.

How fast did the *Mayflower* sail?

In 1620, the Pilgrims sailed on the *Mayflower* from Plymouth, England, to the New World in sixty-six days. Though the *Mayflower* relied upon intermittent wind for propulsion, it averaged two miles (3.2 km) per hour across the Atlantic Ocean.

What did James Cook not discover?

During the eighteenth century, James Cook was sent on several expeditions of discovery, one of which was to the southern Pacific Ocean in search of the legendary landmass "Terra Australis Incognita." Though the continent now known as Australia had already been discovered, a centuries-old belief foretold of another huge continent in that area. Cook traveled to the southern Pacific Ocean and disproved the legend of "Terra Australis Incognita." On another expedition, Cook was sent to find a water route north of North America from Asia to Europe. As Cook sailed, he discovered the Sandwich Islands (Hawaiian Islands) and determined that a northwest passage was not feasible because of

> ### Was George Washington a geographer?
>
> George Washington manifested the abilities of a cartographer and a surveyor at an early age. When he was 13, Washington made his first map, which was of his father's property, Mt. Vernon. At the age of 17, Washington was appointed surveyor of Culpepper County, Virginia. At age 21, Washington entered military service, and the rest is history. His career as a surveyor and cartographer spanned more than 50 years. During this time, he surveyed more than 200 tracts of land and held title to more than 65,000 acres of land. His cartographic work is housed at the Library of Congress, in Washington, D.C.

ice. On his way back from this nondiscovery, Cook was killed in the Sandwich Islands during a struggle over the theft of one of his boats.

Who was John Wesley Powell?

Though he lost an arm in the Civil War, John Wesley Powell became one of the leading surveyors of the nineteenth century. Through the winter of 1868, Powell lived with a Native American tribe, the White River Ute. In 1869, Powell explored the Grand Canyon. Traveling on a boat along the Colorado River, he faced dangerous rapids, hostile Native Americans, and weather extremes. In 1880, Powell was appointed the second director of the U.S. Geological Survey.

Who was Vancouver, Canada, named after?

In the 1790s, George Vancouver, who had previously accompanied James Cook on his explorations for "Terra Australis Incognita" and the Northwest Passage, explored and mapped the Pacific Coast of North America. Vancouver, a British Royal Navy officer and explorer, circumnavigated Canada's Vancouver Island, and it and the city of Vancouver (founded 1881) were named for him. He was also the first European to set foot upon the coast of what is now British Columbia.

How early were the islands of the Pacific explored?

Polynesians, the people who settled many of the islands in the South Pacific, brought their people, culture, and knowledge to the islands before 1500 B.C.E. Experts at celestial navigation, reading the currents, flight patterns of birds, and how to travel in light, canoelike rafts, the ancient Polynesians left New Guinea and traveled first to the areas we know of as the Solomon Islands, then on to present-day Vanuatu. As the distance between islands became even greater, they refined their boatmaking ability to create ships with double hulls that were able to carry animals, people, and trading supplies as far east as Hawaii and even Easter Island, which is 2,237 miles (3,600 km) west of Chile. By the year 1000 C.E., Polynesian culture could be found in a gigantic triangle—the Polynesian Triangle—spreading across thousands of miles of ocean.

THE POLES

Who was first to reach the North Pole?

Though American explorer Robert Edwin Peary is credited as the first to reach the North Pole, it is likely that he only came within 30 to 60 miles (48 to 80 km) of 90° North during his expedition in 1909. Who actually reached the North Pole first is still being debated.

Who was one of the great African American explorers?

Matthew Henson, an African American born in 1866, joined Peary's first Arctic expedition and spent seven years there, covering more than 9,000 miles. Many experts believe that Henson arrived at the North Pole forty-five minutes ahead of expedition leader Peary, and he actually is the first person to find and stand on the North Pole.

Matthew Henson was part of Peary's Arctic expedition.

Who was the first person to reach the South Pole?

In 1911, the Norwegian explorer Roald Amundsen and the British explorer Robert Scott were racing against each other to be the first to reach the South Pole. On December 4, 1911, Amundsen and his crew of four reached the South Pole at 90° South. Approximately one month later, Scott and his team arrived at the pole. Depressed from their defeat and with inadequate supplies of food, Scott and his team died while trying to return to their base camp. His frozen remains were discovered eight months after he died, in November of 1912, with only two of his four-person team.

THE UNITED STATES OF AMERICA

PHYSICAL FEATURES AND RESOURCES

What is the highest point in the United States?

Alaska's Denali (known as Mt. McKinley from 1917 to 2015) is the highest point in the United States at 20,320 feet (6,194 meters). In the contiguous 48 states, the highest point is California's Mt. Whitney at 14,495 feet (4,418 meters), on the east part of the Great Western Divide. It is less than 100 miles (161 km) from North America's lowest point, Death Valley, which is 282 feet (86 meters) below sea level.

What is the highest point east of the Mississippi?

North Carolina's Mt. Mitchell is the tallest point east of the Mississippi River at 6,684 feet (2,037 meters). It can be found only 35 miles northeast of Asheville.

What is the largest lake above 7,000 feet in North America?

Yellowstone Lake in Yellowstone National Park is the largest lake above 7,000 feet (2,100 meters) in North America. It is situated at 7,735 feet (2,358 meters) above sea level.

What is the deepest lake in the United States?

Crater Lake in Oregon, lying within the collapsed crater of an ancient volcano, is the nation's deepest lake at 1,932 feet (589 meters). Crater Lake has no feeder streams and is filled solely by precipitation. The area surrounding the lake receives an average of 44.42 feet (13.54 meters) of snow per year.

What is the Continental Divide?

The Continental Divide is simply the line that divides the flow of water in North America. Precipitation east of the divide flows toward the Atlantic Ocean, while precipitation west of the divide flows toward the Pacific Ocean. The divide follows the line of the highest ranges of the Rocky Mountains and is not a distinct mountain range.

How many Great Lakes are there?

There are five Great Lakes: Huron, Ontario, Michigan, Erie, and Superior. The acronym "HOMES" can help you to remember the names of the five Great Lakes. All of the lakes, except for Lake Michigan, lie on the U.S./Canada border.

What is the largest freshwater lake by area in the world?

Lake Superior is the largest freshwater lake by area in the world. It is 31,820 square miles (82,414 square km) in area and is approximately 383 miles (616 km) long. By volume of water, it is larger than all of the other Great Lakes combined.

Where are some of the world's shortest rivers?

Some of the world's shortest rivers are in the United States. The world's shortest river is the Roe River, in Montana, which has been measured to be between 58 and 200 feet long (17.68 meters–60.96 meters). It is also notable because it is fed by Giant Springs, which is the

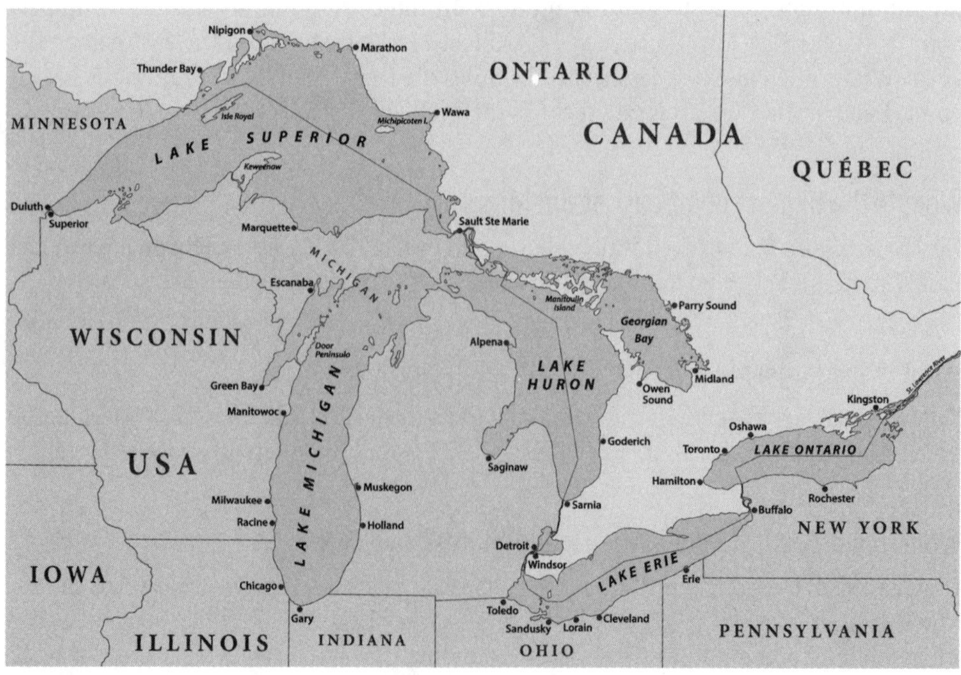

The Great Lakes, when taken together, form the largest source of fresh water on our planet.

Where do the Great Lakes rank in terms of size?

Lake Superior is 31,820 square miles (82,414 square kilometers) and is the world's largest freshwater lake by area; Lake Huron is 23,000 square miles (59,6000 square kilometers) and the third-largest freshwater lake by area in the world; Lake Michigan is 22,000 square miles (58,000 square kilometers) and the fourth-largest freshwater lake by area in the world; Lake Erie is 9,930 square miles (25,719 square kilometers), the world's tenth-largest freshwater lake by area; and Lake Ontario, 7,520 square miles (19,477 square kilometers), is the twelfth-largest freshwater lake by area in the world.

largest freshwater spring in the United States. Oregon's D River, which is a mere 120 feet long (36.6 meters), connects Devil's Lake to the Pacific Ocean near Lincoln City, Oregon.

Which state has the most lakes?

Though Minnesota is known for its "10,000 lakes" (the slogan on its license plates), it actually has 11,842 lakes that are greater than ten acres in size. But it is not the state with the most lakes. Neighboring Wisconsin has even more, with about 15,000 lakes. Alaska is the definite winner, with over three million unnamed natural lakes. Only 3,197 of these lakes are named.

Why is Coney Island called an island even though it's not?

Though now a peninsula of Long Island, Coney Island was actually an island at one time. The popular amusement park of the early twentieth century is now attached to Long Island due to the silting up of Coney Island Creek, which once separated the two islands. The creek was filled in the 1920s and early 1960s.

THE STATES

What are the five largest states in the United States of America?

The five largest states are Alaska, covering an area of 665,384 square miles (1,723,337 square km), Texas with 268,596 square miles (695,662 square km), California with 163,694 square miles (423,967 square km), Montana with 147,039 square miles (380,831 square km), and New Mexico with 121,590 square miles (314,917 square km).

What are the five smallest states?

The five smallest states are Rhode Island, covering an area of 1,555 square miles (4,001 square km), Delaware with 2,489 square miles (6,446 square km), Connecticut with

Why is Rhode Island called an island even though it's not?

Rhode Island's official name is "Rhode Island and Providence Plantations" and includes not only the area of the mainland (the location of the city of Providence), but also four major islands. The largest of these islands is named Rhode Island, from which the state gets the first part of its name.

5,543 square miles (14,357 square km), New Jersey with 8,723 square miles (22,591 square km), and New Hampshire with 9,349 square miles (24,214 square km).

What are the five most populated U.S. states?

The five most populated states are California (37.3 million), Texas (25.2 million), New York (19.4 million), Florida (18.8 million), and Illinois (12.8 million).

What are the five least populated U.S. states?

The five least populated U.S. states are Wyoming (563,626), Vermont (625,741), North Dakota (672,591), Alaska (710,231), and South Dakota (814,180).

How many states have four-letter names?

There are three states that tie for having the shortest name: Utah, Ohio, and Iowa.

How many state names end in the letter "a"?

Twenty-one of the fifty states end in the letter "a."

How was the Delmarva Peninsula named?

The Delmarva Peninsula, located on the East Coast, contains all of Delaware and portions of Maryland and Virginia. The 180-mile (290-km) peninsula was named by combining the abbreviations of each of those three states: Del., Mar., and Va.

What do the Sandwich Islands and Hawaiian Islands have in common?

The Sandwich Islands and Hawaiian Islands are actually the same set of islands. In 1778, when Captain James Cook discovered the islands, he named them the Sandwich Islands. Gradually, the islands began to be known by their indigenous name of Hawaii. Cook named the islands after his supporter, John Montagu, the fourth Earl of Sandwich.

What is the largest island in the United States?

The Island of Hawaii (also known as the Big Island) is the largest island in the United States at 4,021 square miles (10,414 square km). Puerto Rico, a U.S. territory, is the second largest at 3,435 square miles (8,897 square km).

Where is the world's largest mountain?

Hawaii's Mauna Kea is the world's largest mountain. It begins on the sea floor and rises 33,480 feet (10,205 meters). By comparison, Mt. Everest only rises 29,000 feet (8,839 meters) from its base. Mauna Kea's peak reaches 13,680 feet (4,170 meters) above sea level.

How many islands does Hawaii include?

There are a total of 122 islands in the Hawaiian chain. The eight main islands include Ni'hau, Kaua'i, O'ahu, Moloka'i, Lana'i, Kaho'olawe, Maui, and Hawai'i. The southernmost island, Hawaii, also known as the "Big Island," is the largest with 4,000 square miles (10,360 square km) in area. The two westernmost islands are known as the Midway Islands and, while they are part of the United States, they are not part of the state of Hawaii.

Which Hawaiian island was once a leper colony?

On the island of Moloka'i, authorities established a secluded leper colony on an inaccessible cove. Its founder, Father Damien (1840–1889 C.E.), served and cared for the inhabitants. He eventually died of the disease, and the Catholic Church made him a saint for his work.

Which state has the highest annual divorce rate?

The state with the highest number of divorces per 1,000 people is Nevada, with 7.1 divorces per 1,000 people. The District of Columbia has the lowest divorce rate at 2.4 divorces per 1,000 people, and it is followed by both Massachusetts and Georgia, with 2.5 divorces per 1,000 people.

What is the driest state?

Nevada averages only 7.5 inches (19 centimeters) of rainfall annually, making it the driest state in the Union. In 2013, California was the driest state, beating a record of sparse precipitation that was set in 1898, with only 7.38 inches (18.75 cm) of precipitation.

What is the official neckwear of Arizona?

Like an official state flower or animal, each state can add additional official symbols. Since 1971, the official neckwear of Arizona is the bolo tie.

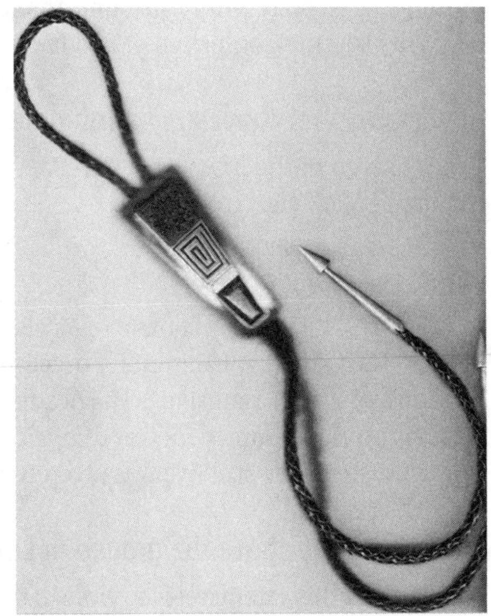

Bolo ties are rather like small, ornamental ropes that resemble a necklace more than a tie.

What is the only state with a diamond mine?

Arkansas is the only state with a diamond mine, located in the town of Murfreesboro, in the southwestern part of the state. Although the mine is no longer in commercial production, it is now Crater of Diamonds State Park. It is the only public diamond mine in the world and allows visitors to take any diamonds that they find.

Which state borders only one other state?

Bordering just New Hampshire, the state of Maine is the only state that borders only one other state.

Which states border the most states?

Both Tennessee and Missouri are adjacent to eight states. Tennessee borders these eight states: Arkansas, Missouri, Kentucky, Virginia, North Carolina, Georgia, Alabama, and Mississippi. Missouri, a state with many neighbors, borders the following eight states: Iowa, Nebraska, Kansas, Oklahoma, Arkansas, Tennessee, Kentucky, and Illinois.

Which state has the longest coastline?

Alaska has more than 6,640 miles (10,686 km) of coastline. The next largest is Michigan, which has 3,288 miles (5,291 km) of coastline, far longer than both Florida with 1,197 miles (1,926 km) and California with 840 miles (1,352 km).

What state has only one legislative body?

Since 1937, Nebraska has had a unicameral legislature. All of the other forty-nine states have a legislature comprised of two houses, or a bicameral legislature.

How many states have land north of Canada's southernmost point?

Twenty-seven of the fifty states have land north of Canada's southernmost point, Pelee Island in Lake Erie.

Where is Acadiana?

In 1755, the British took control of Acadie, New Brunswick, Canada, and exiled the city's French descendant inhabitants. Forced to leave their homes, these people, known as Acadians, moved to Louisiana. The Acadians are still present in Louisiana, and their culture has produced famed food and music, called Cajun. The area in southern Louisiana where the Acadians still live was given the moniker "Acadiana" in the mid-1950s.

Why are graves above the ground in Louisiana?

In Louisiana, the water level is so close to the surface of the ground that coffins, rather than being buried, are placed in tombs above ground to avoid the possibility of coffins floating out of place. A Louisiana cemetery looks like a miniature city, with tombs and alleyways.

Do Louisiana and Alaska have counties?

No. Louisiana is divided into sixty-four parishes rather than counties. These parishes are no different than counties, other than in name. The word parish comes from the parish system of the Catholic Church and thus shows the French and Catholic influence on Louisiana.

Alaska has twenty boroughs and census areas to delineate their land area, instead of using counties.

Is Puerto Rico a state?

No, Puerto Rico is a commonwealth of the United States. In 1898, the United States acquired the island of Puerto Rico from Spain, but Puerto Rico did not become a commonwealth until 1952. Though Puerto Ricans pay no federal income tax and cannot vote for president, they are citizens of the United States and can move freely around the country. Puerto Rico is allowed one representative to the House of Representatives, but that member cannot vote. With approximately 3.548 million residents, Puerto Rico is currently the largest colony in the world.

How did the United States obtain the Virgin Islands?

The United States purchased the Virgin Islands, which includes three main islands (St. Croix, St. John, and St. Thomas) along with fifty smaller islands, from Denmark in 1916 to help defend the Caribbean Sea. Approximately 104,170 people live on the islands, which are now a territory of the United States.

CITIES AND COUNTIES

How many cities and towns are there in the United States?

There are approximately 19,508 incorporated (with local governing control) cities and towns in the United States, the vast majority of which have fewer than 25,000 people. There are approximately 752 cities with populations of 50,000 people or more.

What are the fastest-growing, large U.S. cities in terms of population growth from 2010 to 2013?

From 2010 to 2013, the fastest-growing, large U.S. cities have been Austin, Texas, with 12% growth; Charlotte, North Carolina, with 8.4% growth; and Denver, Colorado, with 8.2% growth.

Measured by area, what is the largest city in the United States?

Sitka, Alaska, is the largest city in the United States, comprising over 4,811.4 square miles (12,460.8 square km). But Sitka is a small town, with only approximately 10,300 residents.

How much did the population of New Orleans decline after Hurricane Katrina struck?

Although many residents returned in the aftermath of Hurricane Katrina, New Orleans's population had declined by more than 53% since the 2005 disaster. By mid-2014, the population recovered to approximately 79% of what it was before the hurricane struck.

What cities have the most crime in the United States?

According to the FBI, the most dangerous cities in America, by number of violent crimes per 1,000 residents, are Detroit, Michigan; Oakland, California; and Memphis, Tennessee.

What are the northernmost, southernmost, easternmost, and westernmost cities in the United States?

Barrow, Alaska, is the northernmost; Hilo, Hawaii, is the southernmost; Eastport, Maine, is the easternmost; and Adak, Alaska, is the westernmost city in the country.

Where is the center of the U.S.A.?

The geographic center of the lower forty-eight states is located at 39°50' North, 98°35' West, approximately four miles (6.5 km) northwest of Lebanon, Kansas.

Where is the center of North America?

The geographic center of North America (including Canada, the United States, and Mexico) is located six miles (10 km) west of Balta, North Dakota, at 48°10' North, 100°10' West.

What is the highest settlement in the United States?

The city of Alma, Colorado, is located at an elevation of 10,355 to 11,680 feet (3,156 to 3,560 meters) and is the highest settlement in the United States.

What is the oldest continually occupied city in the United States?

St. Augustine, Florida, was established in 1565 by Spanish explorer Pedro Menéndez de Avilés. St. Augustine is located on Florida's eastern (Atlantic) coast and now has a population of approximately 13,000. It is the oldest continually occupied city in both the United States and all of North America.

The downtown square of old St. Augustine, Florida, a city that was first settled by the Spanish in 1565.

Which is farther west—Los Angeles, California, or Reno, Nevada?

Though Nevada is California's eastern neighbor and Los Angeles sits on the Pacific Coast, Reno is farther west than Los Angeles. Reno is located at 119.82° West, while Los Angeles is located at 118.25° West.

What city is named for a game show?

The popular radio and television game show *Truth or Consequences* (1940–1988) offered to host its tenth-anniversary show in a city that would change its name to the show's title. In 1950, Hot Springs, New Mexico, changed its name, and the tenth-anniversary show was broadcast from Truth or Consequences, New Mexico. The city still holds the name today. The city is located in southern New Mexico, about 140 miles (225 km) south of Albuquerque and 120 miles (193 km) north of El Paso, Texas, and the Mexican border.

What city is known as the Earthquake City?

Charleston, South Carolina, claims the nickname "Earthquake City." On August 31, 1886, Charleston suffered from the largest earthquake in history to strike the east coast of the United States. Sixty people were killed in the quake, which had an estimated Richter magnitude of 6.6. The city has had approximately twenty earthquakes of magnitude V or greater since then.

Where does Los Angeles get its water?

Not much of the water in Los Angeles comes from local sources. Most of it is brought from hundreds of miles away. Large aqueducts, man-made channels used to transport water, were built to carry water from Owens Valley (in east-central California), from the Colorado River, and from the rivers of Northern California to Los Angeles. Though this method has brought fresh water to a region that desperately needs it, it has also drained and damaged the ecologies that once depended on the water now being sapped from its supply.

How many counties are in the United States?

There are 3,143 counties in the United States. The state with the fewest number of counties is Delaware, with three. Texas has the most counties, with 254.

What are the largest and smallest counties in the United States?

San Bernardino County in California is the largest, at 20,105 square miles (52,072 square km). It stretches from metropolitan Los Angeles to the Nevada/Arizona border. The smallest county in the United States is Kalawao, Hawaii, at 12 square miles (33.7 square km) of land. It has a very small population of only ninety people.

What county in the lower forty-eight contiguous states of the United States has both the lowest point and the highest point?

California's Inyo County has both the continental United States' highest and lowest points, which cannot be seen from one another. Inyo County is situated east of the city of Fresno, California, and southwest of Yosemite National Park. Inyo County is also home to some of the oldest life forms on Earth, the bristlecone pines, which are estimated to have sprouted in 3500 B.C.E.

PEOPLE AND CULTURE

How big is the population of the United States?

The U.S. population is approximately 321 million people, with one person being born every seven seconds, one person dying every thirteen seconds, and one international immigrant arriving every thirty-three seconds, yielding a net gain of one new person every eleven seconds. Within the United States, more than thirty-eight million people live in California, and nearly thirty million people live in Texas.

What percentage of the U.S. population are legal immigrants from other countries?

A total of 13% of the U.S. population are immigrants. Although there are many places in the world that people may choose to live, approximately 20% of all people who leave their countries choose to live in America. Of the total U.S. population, 80 million people are either first- or second-generation Americans.

Where do most legal immigrants to the United States come from?

Mexican-born immigrants account for approximately 28% of the estimated 41.3 million foreign-born U.S. residents. The rest are composed of immigrants from India, China, the Philippines, Vietnam, El Salvador, Cuba, Korea, the Dominican Republic, and Guatemala. Immigrants from these countries comprise nearly 60% of the U.S. immigrant population.

How many illegal immigrants are there in the United States?

Some estimates take into account responses to the U.S. Census Survey, pegging the number at approximately 11.3 million people, which has stabilized over the past five years. They comprise 5.1% of the entire U.S. labor force.

Where do the illegal immigrants come from?

Approximately 52% of illegal immigrants come from Mexico.

Where do these immigrants live?

Sixty percent of all illegal immigrants live in six states: California, Texas, Florida, New York, New Jersey, and Illinois.

How many Internet users are there in the United States?

There are nearly 280 million Internet users in America, making up approximately 87% of the population. Approximately 10% of all Internet users in the world are located in the United States.

Smartphones have become omnipresent in American society, sometimes much to the annoyance of people who want to watch a concert or parade.

What percentage of American adults use a smartphone?

According to experts at the Pew Research Center, 64% of American adults use a smartphone, which is up from 35% in 2011. More than 85% of younger adults aged 18–29 use smartphones today.

What percentage of American households regularly give to charitable organizations?

Approximately 95.4% of American households give to charitable organizations. On average, they give $2,974 each year, donating $358.38 billion annually in total.

What percentage of Americans volunteer each year?

According to the U.S. Bureau of Labor Statistics, 25.3% of Americans volunteer every year. This means approximately 62.8 million people volunteer through or for an organization at least once each year.

What is the leading cause of death in the United States?

Approximately 23.5% of all deaths in America are attributed to cardiovascular disease, making it the country's leading cause of death. Among all deaths in America, 22.5% are attributed to cancers, which are the second-leading cause of death in America.

What is the average life expectancy in the United States?

The average life expectancy in the United States is 78.8 years.

What is the oldest college in the United States?

Harvard University is the oldest college or university in the United States. It was founded in 1636 in Cambridge, Massachusetts, just outside of Boston.

How valuable is a college education in the United States?

Workers in the United States with at least a bachelor's degree earn an average of $48,707 per year. High school graduates can expect to earn $30,000 per year, while those who did not complete high school earn $23,900 on average. Workers with an associate's degree earn an average salary of $37,500 per year, while workers who attained an advanced degree can make on average $59,600 per year.

How many educational institutions are there in the United States?

There are more than 98,000 primary and secondary schools, nearly 31,000 private schools, and more than 7,200 universities, colleges, and postsecondary schools in the United States.

One famous landmark in Boston Common Park is the statue of George Washington.

How many foreign students attend universities and colleges in the United States?

There are more than 524,000 foreign students currently attending schools in the United States. More than half of the foreign students come from these cities: Seoul, Beijing, Shanghai, Hyderabad, and Riyadh. And 45% of foreign students stay after graduation to work in the United States.

What is the most popular national park in the United States?

Tennessee and North Carolina's Great Smoky Mountains National Park is the most popular national park in the United States, drawing over 10 million visitors each year. Located near the eastern border of North Carolina and Tennessee, the park is home to over ten thousand species of flora and fauna, is an International Biosphere Reserve, and is a World Heritage Site.

What is the oldest public park in the United States?

In 1634, William Blackstone sold fifty acres of his land to the town of Boston. This pastureland was set aside for common use and called the Boston Common, making it the oldest public park in the United States. The Common is situated in front of Massachusetts's State House and has always been open public land.

What is the most popular theme park in the United States?

The Magic Kingdom at Walt Disney World in Orlando, Florida, has more than 19 million visitors annually. It is also the most popular theme park in the world.

How many visitors come to the United States every year?

Approximately 75 million people visit the United States each year from abroad. Nearly 30.1% of the tourists are coming from Canada. Tourists and other visitors contribute $220.8 billion per year to the U.S. economy.

Where was the world's first monument to an insect established?

On December 11, 1919, Enterprise, Alabama, dedicated a monument to the boll weevil. This tall statue of a woman with raised arms holding a boll weevil declares, "In profound appreciation of the boll weevil and what it has done as the herald of prosperity." The boll weevil, a beetle that attacks bolls of cotton, spread across the South at the beginning of the twentieth century, wiping out cotton crops. Residents of Enterprise switched from cotton crops to peanut crops, thus discovering a new era of prosperity.

Who carved Mt. Rushmore?

Gutzon Borglum, an American sculptor, designed this national memorial located in the Black Hills of South Dakota. Construction began in 1927 and was nearly complete when Borglum died in March 1941. Borglum oversaw construction of the 60-foot-tall (18-meter-tall) heads of Presidents George Washington, Thomas Jefferson, Abraham Lincoln, and Theodore Roosevelt. After Borglum's death, his son Lincoln completed the work on the unfinished Roosevelt by the end of 1941.

How many automobiles are there in the United States?

There are approximately 253 million automobiles in use in the United States, with an average age of 11.4 years old.

Where was the first commercial air flight?

On January 1, 1914, the first scheduled commercial air flight took passengers from Tampa Bay, Florida, to St. Petersburg, Florida. The service, which lasted only a few weeks, took one person at a time over the 22-mile (35-km) route. It was piloted by Tony Jannus.

Where is the Rust Belt?

The Rust Belt is a term used to describe the United States' declining manufacturing region of the Northeast and Midwest. The region begins roughly in south-central New York and moves eastward through Pennsylvania, West Virginia, Ohio, Michigan, Indiana, northern Illinois, and southeastern Wisconsin. Factory closures, especially those of automotive and related factories, steel mills, and textile mills, have resulted in high unemployment and declining population in key Rust Belt cities. One way to observe this condition is by looking at the loss of manufacturing jobs in this region over a period of time.

Where is the Sun Belt?

The Sun Belt, known for its warm temperatures, is a geographic area, roughly south of the 36th parallel north latitude, that spreads across the southern and southwestern states of the United States. It has been an area of high population growth over the past few decades as more families and individuals have moved to Sun Belt states like California, Arizona, Texas, and Florida.

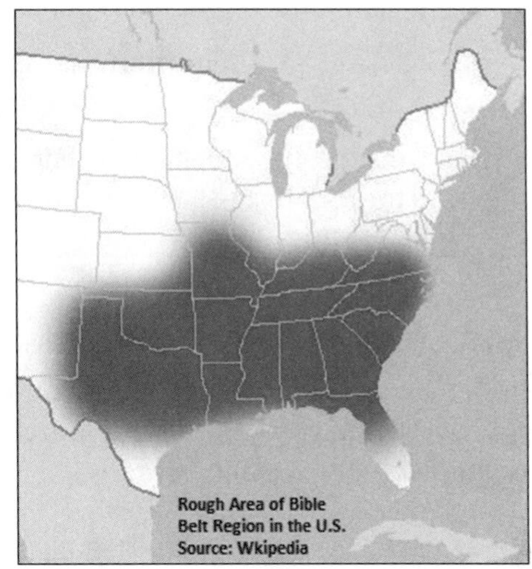

The region known as the "Bible Belt," because of its concentration of conservative Christians, encompasses much of the southeastern United States.

Where is the Bible Belt?

The Bible Belt, a region noted for its high proportion of fundamentalist Christian-believing population, is located roughly in the southern and south-central United States, running from about Texas eastward to Virginia and southward to northern Florida.

What did the Mason-Dixon line originally divide?

While the Mason-Dixon line commonly refers to the division between the "North" and "South" in the eastern United States, it was originally used to help resolve a boundary dispute involving the colonial states of Maryland, Pennsylvania, and Delaware. It was surveyed by Charles Mason and Jeremiah Dixon between 1763 and 1767. During the Civil War, the boundary between Pennsylvania and Maryland was extended westward to represent the line between the slave and nonslave states. Today, parts of the Mason-Dixon line demarcate the borders of West Virginia, Pennsylvania, Maryland, and Delaware.

How many countries have seen U.S. military action since the beginning of the year 2000?

The United States has sent military staff to, bombed, invaded, or occupied twelve foreign countries since 2000, including Macedonia, Afghanistan, Yemen, Philippines, Colombia, Iraq, Liberia, Haiti, Pakistan, Somalia, Syria, and Libya.

Why was there a Russian outpost in California?

Established in 1812, Fort Ross was a Russian outpost located in Sonoma County, California. The outpost was started so that Russian fur traders could explore and exploit the area. A Russian presence was maintained at the fort until 1841.

Where was the first shopping mall in the United States?

In 1922, Country Club District opened in the suburbs of Kansas City, Kansas, and was the first shopping mall designed for shoppers commuting by automobile in the country.

What is the oldest continuously published newspaper in the United States?

In October of 1764, while Connecticut was still a colony, Thomas Green founded the *Hartford Courant* in Hartford, Connecticut. It is the oldest continuously published newspaper in the United States.

How many businesses are started each year in the United States?

Each year, more than 400,000 new companies are started in America, joining the more than 26 million companies that make the American economy the largest in the world. Most companies (99.9%) have fewer than 500 employees, and approximately 6 million of these companies are actively conducting business.

HISTORY

When was the first permanent British settlement established in the United States?

In 1607, the colony of Jamestown was established in Virginia by British colonists transported there by the London Company. Though subjected to many attacks by Native Americans, the colony was ultimately destroyed in 1676 by its own rebelling colonists in Bacon's Rebellion.

How was the area that comprises the United States formed?

The United States began as thirteen British colonies on the Atlantic Coast. In 1783, the United States gained the Northwest Territory, the area encompassing what is now Ohio, Indiana, Illinois, Michigan, and Wisconsin. Spain and the United States agreed on the northern boundary of Florida in 1798, and the United States then took control of the Mississippi Territory. In 1803, the Louisiana Purchase (which included most of the area west of the Mississippi) doubled the size of the country. In 1845 the independent Republic of Texas was annexed, and Spain ceded Florida to the United States. In 1846, the Oregon Territory (which included Oregon, Washington, and Idaho) was officially designated with a treaty between the United States and the United Kingdom. The Mexican War of 1846–1848 led to the secession of California, Utah Territory, and New Mexico Territory to the United States. The Gadsden Purchase of 1853 added southern Arizona. Alaska was purchased from Russia in 1867, and Hawaii was annexed by the United States in 1898.

How did the Monroe Doctrine protect the Americas?

In 1823, President James Monroe gave a speech that declared the area, known as the Americas, off limits to all European powers. These policies became known in the 1840s as the Monroe Doctrine. Since 1823, the United States has used the Monroe Doctrine not only to prevent intervention by Europeans but also to further its own expansionist goals.

What was Manifest Destiny?

First used by newspaper columnist John Louis O'Sullivan in an 1845 editorial, Manifest Destiny was the phrase used to describe the assumption that American expansion to the Pacific Ocean was inevitable and ordained by God. The phrase was used to defend the annexations of Texas, California, Alaska, and even the Pacific and Caribbean islands.

What were Lewis and Clark looking for?

President Thomas Jefferson sent Meriwether Lewis and army officer William Clark to search for a northwest passage, a waterway that would connect the Atlantic and Pacific oceans. Jefferson asked Lewis to explore the Missouri River and the Columbia, Oregon, and Colorado rivers to see if any of these waterways connected the two oceans for commerce. Beginning in May 1804 and lasting through September 1806, the two men and their expedition party, named the Corps of Discovery, traveled through the uncharted Louisiana Territory and the Oregon Territory. Though they did not locate a northwest passage, Lewis and Clark documented the geography, biology, and some inhabitants of the West.

When was the most territory added to the United States at one time?

In 1803, the United States purchased over 800,000 square miles (2,000,000 square km) of land from France for $15 million. This territory, known as the Louisiana Purchase, extended the United States from the Mississippi River to the Rocky Mountains and doubled the size of the United States.

What was Seward's Folly?

Seward's Folly, also known as Seward's Icebox, was the derogatory nickname given to the area known as Alaska, purchased by the United States from Russia in 1867. The $7.2 million purchase, heavily encouraged by Secretary of State William Seward, was criticized by many, thus dubbed "Seward's Folly." The Alaskan Gold Rush of 1900 proved Seward to be a very wise man. In 1959, Alaska became the forty-ninth state. Alaska is bigger than Texas, California, and Montana combined and is 20% of the size of the lower forty-eight states of the continental United States.

Aside from the Louisiana Purchase, how was the American West obtained?

There were several other purchases and wars fought to gain the land west of the Louisiana Purchase. These included the Gadsden Purchase of land from Mexico and sections of the west ceded to the United States by Mexico, Texas, and the Oregon Country.

What was the Oregon Trail?

The Oregon Trail was a 2,200-mile-long (3,500 km) pioneer trail that extended from Independence, Missouri, to Portland, Oregon. Migrants traveled along this trail in an effort to reach and settle the sparsely populated American West. It took migrants approximately six months to traverse the Oregon Trail and reach Oregon. The trail was heavily used from 1846 to 1869. Portions of the trail are still visible today in places such as the Whitman Mission National Historic Site in Washington State.

How many Native American tribes are there in the United States?

There are 567 recognized tribes in the United States, with approximately 1.9 million members.

How many Native Americans were killed during the period of European colonization in the Americas and the subsequent American expansion?

Some historians suggest that approximately 90% of the population was decimated, but they cannot agree on the number of indigenous people inhabiting North America at the time of Columbus. The range in population estimates could be from 2.1 million to 18 million people. The great majority of Native Americans who had made what is now America their home for tens of thousands of years were killed outright by European colonists and, later, U.S. military forces. Diseases brought to the Americas from Europe, including smallpox and the plague, also decimated the indigenous tribes. Furthermore, intertribal warfare, spurred on by the political interests of Europeans, enslavement, and the mass killing and forced relocation in the nineteenth century of the hundreds of thousands of Native Americans who still managed to survive all conspired to nearly de-

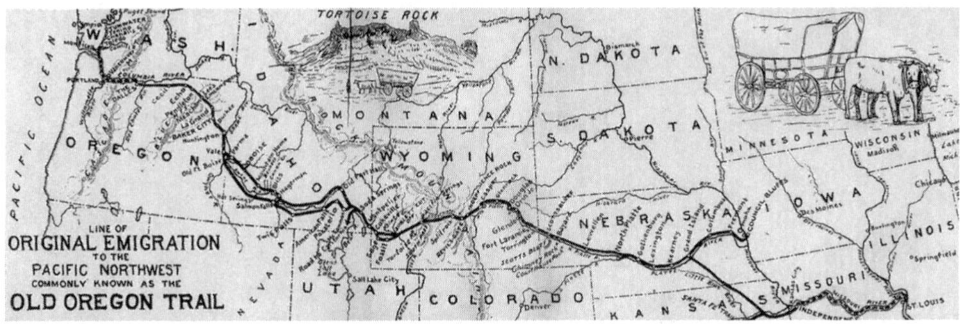

Originally blazed by early fur trappers and mountain men, the Oregon Trail became the preferred route for many pioneers heading toward the American West.

What happened to the Arawak indigenous people after Christopher Columbus arrived in the New World?

Many experts assert that when Columbus arrived on the shores of the islands of the Caribbean, there were between 250,000 and 1 million Island Arawaks (also known as Lokono or Taino indigenous people). By the middle of the sixteenth century, disease, slavery, and outright killing brought the number down to fewer than 500. Today, about 10,000 Lokono people are living in such places as Venezuela, Guyana, Suriname, and French Guiana, and many more descendants living in the neighboring region.

stroy Native tribes and their cultures. Over 300 years of American history, from 1622 to 1924, there were approximately ninety-seven military conflicts with Native Americans.

How many Native American reservations are there in the United States?

There are approximately 326 reservations on 56.2 million acres of land held in trust by the United States. Although the land is owned and administered by the federal government under the Department of the Interior's Bureau of Indian Affairs, each reservation has limited sovereignty, including its own legal system. The largest area of land is the 16-million-acre Navajo Nation, which is spread across Arizona, New Mexico, and Utah.

What was the Trail of Tears?

In 1838, the United States rounded up approximately 15,000 members of the Cherokee Nation and forced them from their homes in Tennessee to live on a reservation in Oklahoma. The removal of the Cherokee Nation from their land was done so that citizens of the United States could use the fertile lands in Tennessee. Along the "trail," approximately one-fourth of the Cherokees died from malnutrition, disease, and government inefficiency.

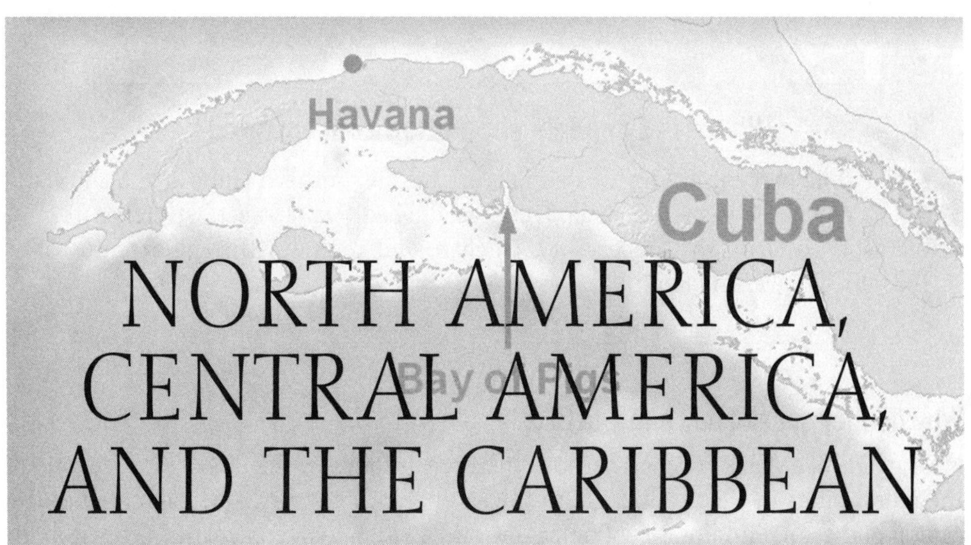

NORTH AMERICA, CENTRAL AMERICA, AND THE CARIBBEAN

What's NAFTA?

In 1994, Canada, the United States, and Mexico entered into the North American Free Trade Agreement (NAFTA), which reduces tariffs and economic controls on imports and exports of products between the three countries.

Has NAFTA helped trade between Canada, Mexico, and the United States?

From 1993 to 2012, trade between the NAFTA countries has quadrupled, from $290 billion to more than $1.1 trillion.

CANADA

What are the largest Canadian urban areas?

Toronto (5.132 million people), Montréal (3.407 million), and Vancouver (2.135 million) are Canada's largest urban areas.

How many provinces and territories comprise Canada?

Canada is divided into ten provinces and three territories. The ten provinces are Alberta, British Columbia, Manitoba, New Brunswick, Newfoundland and Labrador, Nova Scotia, Ontario, Prince Edward Island, Québec, and Saskatchewan. Canada's territories are Yukon Territory, the Northwest Territories, and Nunavut.

Where are the Prairie Provinces?

The grassy, central Canadian states of Manitoba, Saskatchewan, and Alberta are called the Prairie Provinces. The area produces wheat, petroleum, and natural gas.

Where is Canada's population concentrated?

Sixty-one percent of Canada's population lies in southern Ontario and southern Québec in eastern Canada. This area, dubbed "Main Street," stretches from Windsor, Ontario, to Québec City, Québec, and includes the major cities of Toronto, Ottawa, and Montréal.

Why do they speak French in Québec?

Most of Québec speaks French because the French founded the city of Québec, one of the oldest cities in North America, in the seventeenth century. Québec served as the capital of the surrounding New France until the British took control of the territory in 1763. Even though it was ruled by the British from the eighteenth century onward, the region that was once held by France has remained a center of French culture and language in North America.

How different from the rest of Canada is the province of Québec?

The differences in culture between the province of Québec and the rest of Canada have been so extreme that there are strong secessionist forces within Québec that want Québec to become its own country. Though two referendums for Québec's independence have been voted upon in Canada-wide elections, both the 1980 and 1995 referendums failed to pass. But these failures have not quieted Québec's secessionist forces, which began in 1968 with formation of the political party "Parti Québécois," and it is likely that future referendums will be held.

What is Nunavut?

Nunavut, Canada's third territory, whose capital is Iqaluit, is home to Canada's indigenous people, the Inuit. This new territory, which entered the dominion in 1999, covers approximately one-fifth of Canada's land area but contains less than 1% of Canada's population, with just over 36,687 people.

When did Canada have a transcontinental railway?

After nearly a decade of setbacks, Canada completed its first transcontinental railroad in 1885. The Canadian Pacific Railway opened western Canada to settlement and greatly helped the city of Vancouver, as the terminus, to grow.

What was the Klondike Gold Rush?

In 1896, gold was discovered in an area of western Canada known as the Klondike, located in the Yukon Territory where the Klondike and Yukon rivers meet. Once the news of the discovery spread, tens of thousands of people headed west, creating the Klondike Gold Rush.

A photo of the Dyea Waterfront (March 1898), a place for gold prospectors to stock up and prepare for life in the Klondike.

What did the phrase "fifty-four forty or fight" mean?

In the mid-nineteenth century, many Americans wanted to see the United States' territory expand northward into the area now known as Canada. The phrase "fifty-four forty or fight" referred to the desire to move the boundary northward to 54°40' North, which would have encompassed much of southern Canada. Ultimately, the boundary was fixed at 49° North, where it sits today.

What is the deepest lake in North America?

In Canada's Northwest Territories, Great Slave Lake is the continent's deepest, at 2,015 feet (614 meters). The lake is named after the indigenous people who live near the lake, the Slave.

How long are the Rockies?

The Rocky Mountains (Rockies) extend over 2,000 miles (3,200 km), from the Yukon Territory in Canada to Arizona and New Mexico in the southern United States.

How is northern Canada gaining elevation?

During the ice ages, thick ice sheets covered northern Canada, including Hudson Bay, pushing the continent down by the sheer force of the extreme weight of the ice. Ever since the melting of the ice sheets at the end of the ice age, the ground has been rising a few inches each year.

How was Niagara Falls stopped?

The water from the Niagara River falls over two waterfalls, divided by Goat Island. Only 6% of the water from the Niagara River falls over American Falls, while Horseshoe Falls (or Canadian Falls) carries the majority of the water. In 1969, a temporary dam was built to divert the water from American Falls to Horseshoe Falls for several months in order to study the erosion endemic to both waterfalls.

GREENLAND AND THE NORTH POLE REGION

How did Greenland get its name?

In 982 C.E., Greenland was named by its first colonizer, Eric the Red. Having been banished from Iceland, Eric the Red established a colony on Greenland and gave it a pleasant-sounding name in order to attract other colonists. In reality, Greenland is not very green. Though small coastal areas are habitable, most of the island is mountainous and covered by ice sheets. Greenland is an autonomous territory of Denmark.

How thick is Greenland's ice sheet?

Greenland's ice sheet is generally more than 1.2 miles (2 km) thick and 1.9 miles (3 km) thick at its deepest point.

What is the northernmost landmass in the world?

Kaffeklubben Island, at the very northernmost point of Greenland's Peary Land peninsula, is the northernmost landmass in the world. It is located at 83°40' North and is 2,460 feet (750 meters) farther north than Cape Morris Jesup, also located in Peary Land. The cape is at 83°38' North.

Is there land at the North Pole?

Since the North Pole lies in the Arctic Ocean, which is mostly covered by a large icecap year-round, there is no land near the North Pole. Though there is no land, animals such as the polar bear live upon the icecap.

How big is the area of the North Pole ice region?

The Arctic ice mass covers an area of 2.24 million square miles (5.79 million square km). The current size of the Arctic ice shelf is 521,200 square miles (1.35 million square km) lower than the average area from 1981 to 2010.

Is global warming melting the North Pole?

Yes, many scientists believe so. From 1979 to 2005, the North Pole region lost 2.05 million square miles of total area (5.31 million square km), which is equivalent to an area

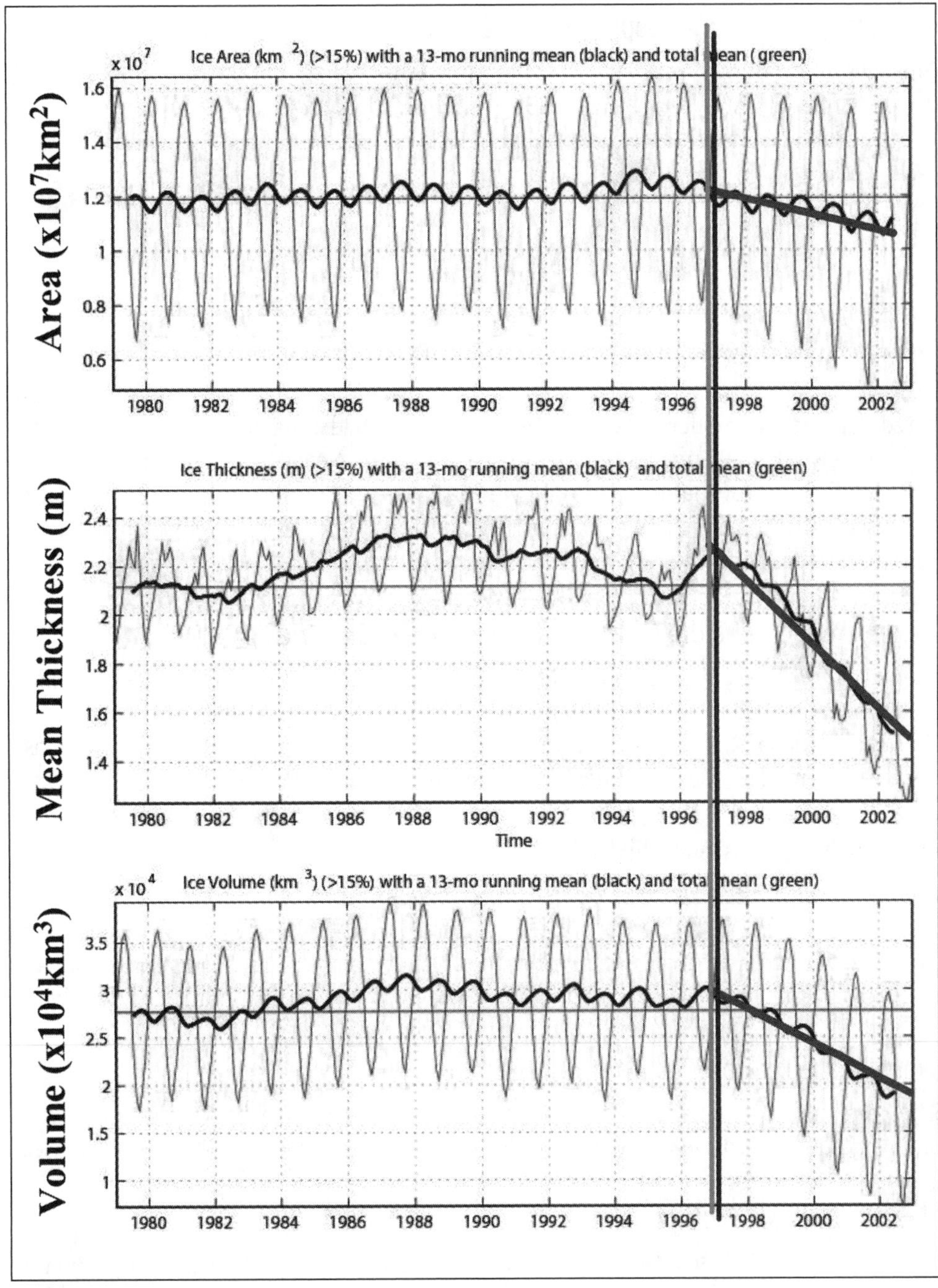

These graphs illustrate how much Arctic ice has been lost by the year 2002. In more recent years, that rate has accelerated considerably.

the size of California and Texas combined. These changes have not been seen since between 6,000 and 125,000 years ago.

How many indigenous people live in the Arctic?

More than four million people live in the Arctic. They are spread across national boundaries in such places as Alaska, Canada, Greenland, Sweden, Norway, and Siberia. There are over forty separate ethnic groups living in this region.

MEXICO

What was the biggest and most populated city in the New World?

From approximately 100 B.C.E. through 800 C.E., the city of Teotihuacan flourished. Located northeast of modern Mexico City, Teotihuacan had a maximum population of approximately 200,000 and was the first major city in the New World. The city was graced with the Pyramid of the Sun and the Pyramid of the Moon. Teotihuacan should not be confused with the Aztec city of Tenochtitlan, which was built nearly six centuries later in the area that is now known as Mexico City.

Where are Sierra Madre Occidental and Sierra Madre Oriental?

Sierra Madre Occidental and Sierra Madre Oriental are two mountain ranges in Mexico. The ranges' names stem from the meanings of the words "occidental" (western) and "oriental" (eastern). Thus, the Sierra Madre Occidental lies along Mexico's west coast, and the Sierra Madre Oriental lies along the east coast.

How many states are in Mexico?

Mexico, or the United Mexican States, as it is formally known, is divided into thirty-one states plus the federal district of Mexico City. The largest state is Chihuahua, located in northern Mexico, with 95,540 square miles (247,460 square km). The most populated state is the state of Mexico, located west of Mexico City, with 15.175 million people.

Where is the Yucatan?

The Yucatan is the name of a large peninsula in southern Mexico. The Yucatan Peninsula "points" toward Florida and separates the Gulf of Mexico from the Caribbean Sea.

How much of Mexico's population lives in Mexico City?

Approximately 17.1% of Mexico's population lives in the capital Mexico City's metropolitan area. It is the largest metropolitan area in the Western Hemisphere and in the Spanish-speaking world, with a population of 21.2 million.

What is the world's largest gulf?

The Gulf of Mexico, which is approximately 615,000 square miles (1.6 million square km), is the world's largest gulf. The Gulf of Mexico is surrounded by Mexico, the southern coast of the United States, and Cuba and is connected to the Caribbean Sea to the east.

How do maquiladoras help clothe the United States?

Maquiladoras are Mexican factories owned by foreign (usually U.S.) corporations. Most often located along the Mexican–U.S. border, maquiladoras receive raw materials from the United States and produce finished goods for export. Maquiladoras commonly produce clothing and automobiles.

How much does an employee of a maquiladora earn?

The average wage per hour of people working in border factories is $5.18 per hour. Maquiladoras prefer to employ women, who tend not to challenge poor working conditions and can be paid as little as one-sixth of what workers can make north of the border in the United States.

How much business is transacted between Mexico and the United States each day?

The trade in goods and services between Mexico and the United States is approximately $1.5 billion every day.

CENTRAL AMERICA

What is the difference between the terms Central America and Latin America?

When we use the term Central America, it includes the countries that connect North and South America and are located between Mexico and Colombia. The seven countries of Central America are Guatemala, Belize, El Salvador, Honduras, Nicaragua, Costa Rica, and Panama. Latin America is a much broader term and includes Central America as well as Mexico and all of the countries of South America.

What is the population of Latin America and the Caribbean?

The population of Latin America and the Caribbean is approximately 613 million. Of the world's thirty megacities, five are in Latin America, including Mexico City (22.2 million people), São Paulo (21.25 million people), Buenos Aires (16.5 million people), Rio de Janeiro (14.45 million people), and Bogotá (12 million people).

What is the primary religion throughout Latin America?

Due to Spanish and Portuguese colonization, 84% of Latin Americans are raised Catholic, but only 69% currently identify themselves as Catholic. Protestants make up about 9% of the religious affiliation of the region, and the rest are atheists, nonreligious, animists, or other religions.

What was the last Central American country to obtain its independence?

In 1981, Belize, formerly known as British Honduras, became the last Central American country to obtain independence.

Are there lots of mosquitoes on the Mosquito Coast?

The Mosquito Coast is an area 64 miles (103 km) wide and approximately 250 miles (400 km) long along Nicaragua's eastern shore. Though the Mosquito Coast receives an average of 100–250 inches (254–635 centimeters) of rain annually, making it a perfect breeding place for mosquitoes, the Mosquito Coast was named after the indigenous people of the area, the Mosquito Indians.

Who owns the Panama Canal?

In 1903, the U.S.-backed revolutionaries in western Colombia revolted and created the independent country of Panama, which was immediately recognized by the United States. The newly independent Panama gave the United States use of a 10-mile- (16-km-) wide strip of land across the Isthmus of Panama, where the United States built the Panama Canal. The United States maintained control of the Panama Canal and its surrounding land, called the Canal Zone, until 1999, when a 1977 agreement between the United States and Panama officially took effect and turned the canal over to the Central American nation.

What monetary unit is used in Panama?

The currency in Panama is called the Balboa, after the explorer Vasco Núñez de Balboa, because Balboa established the first European settlement in Panama. It is tied 1:1 to the U.S. dollar, which is also used as an official currency in the country.

Balboa coins like these are used in Panama, as well as the U.S. dollar.

Who is a peon?

A peon is a farm laborer in Central America who works on large farms known as haciendas.

What is the world's second-longest barrier reef?

The second-longest barrier reef in the world lies just off the Caribbean Sea coast of Belize, on the northeastern corner of Central America, and consists of Lighthouse Reef and Glover's Reef. Belize's reefs are only a few dozen miles long while Australia's Great Barrier Reef, the longest reef in the world, is hundreds of miles long.

CARIBBEAN

How far away are the East and West Indies from each other?

The East and West Indies are separated by half the planet. The West Indies are islands in the Caribbean, including the Greater Antilles, the Lesser Antilles, and the Bahamas; the East Indies include islands that encompass Indonesia, Malaysia, and Brunei. When Christopher Columbus reached the New World in 1492, he believed that he had actually found a shorter route to the East Indies. Thus, Columbus thought the islands he had reached made up a portion of the Indies and considered the islands' inhabitants to be "Indians."

Where are the Windward Islands?

The Windward Islands are located in the Caribbean Sea and are exposed to the northeast trade winds (northeasterlies) of the Atlantic Ocean. Because of their vulnerability to these winds, the islands were named the Windward Islands. The Windward Islands include Martinique, St. Lucia, St. Vincent, the Grenadines, and Grenada.

Where are the Leeward Islands?

The Leeward Islands are also located in the Caribbean Sea and are less exposed to the northeasterlies. Because these islands are "lee," or away, from the wind, they were named the Leeward Islands. The Leeward Islands include Dominica, Guadeloupe, Montserrat, Antigua, Barbuda, St. Kitts, Nevis, Anguilla, and the Virgin Islands.

Was Cuba ever a part of the United States?

The United States went to war against Spain in 1898 to assist Cubans who were rebelling against Spanish rule. The United States took control of Cuba during the Spanish-Amer-

What is the difference between the Greater and Lesser Antilles?

The Greater Antilles refers to the four largest Caribbean islands: Cuba, Hispaniola, Puerto Rico, and Jamaica. All smaller Caribbean islands make up the Lesser Antilles.

ican War in 1898 and held it until 1902, when Cuba was granted independence. The three-year military occupation by the United States ended with an agreement that the United States would be allowed to lease Guantanamo Bay, which the United States still uses as a naval base.

How did Cuba become a communist country?

Having been an independent country for fifty-seven years, the Cuban government, run by the dictator Fulgencio Batista y Zaldívar, fell to the communist leader Fidel Castro in 1959. Because of Cuba's communist government, the United States severed its relationship with Cuba, forcing the island to ally itself with the Soviet Union. In October 1962, the presence of this nearby communist country caused extreme concern in the United States when the U.S.S.R. attempted to place nuclear missiles within Cuba. The "Cuban Missile Crisis" is thought to be the closest the Cold War ever got to a real nuclear war.

Where is the Bay of Pigs?

The Bay of Pigs is a bay in southwestern Cuba. In 1961, the bay became the location of an attempted coup against the Cuban government by revolutionaries trained and financed by the U.S. Central Intelligence Agency. After the attempted coup failed, the United States abandoned the revolutionaries, most of whom were killed or captured in the days following the coup attempt.

Where is the oldest church in the Americas?

The oldest church, the Cathedral Basilica Menor de Santa Maria, was built by Columbus's son Diego, in Santo Domingo, the Dominican Republic. The first stone was set in 1514.

What was the first independent country in the Caribbean?

Haiti was the first independent country in the Caribbean. In 1791, the slaves in Haiti revolted, which led to Haiti's independence from France in 1804. Though Haiti once oc-

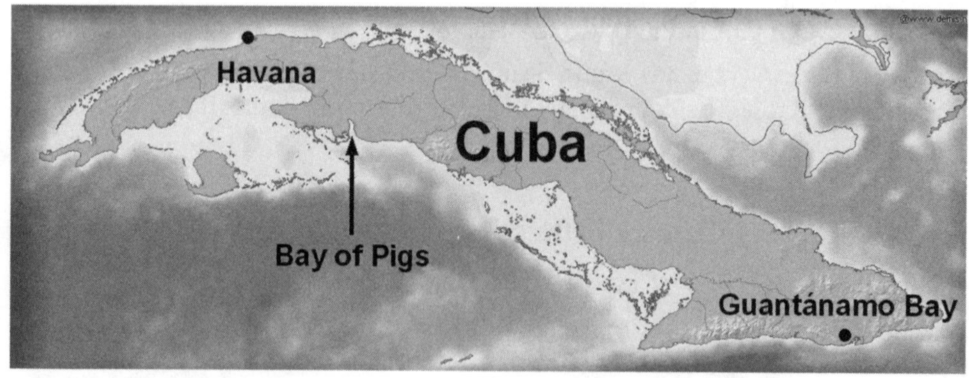

The Bay of Pigs, site of the disastrous U.S. invasion of Cuba in 1961, is located on the south side of the island nation.

Where is the Western Hemisphere's oldest university?

Founded in 1538 by a papal decree, the Autonomous University of Santo Domingo in the Dominican Republic is the oldest university in the Western Hemisphere.

cupied the entire island of Hispaniola, Haiti now shares the island with the Dominican Republic.

What two Caribbean islands are shared by four different countries?

The island of Hispaniola is shared by the Dominican Republic and Haiti. The island of St. Martin is shared by both France (Saint-Martin) and the Netherlands (Sint-Maarten).

How many people visit the Caribbean each year?

Approximately twenty-five million people visit the Caribbean's beautiful countries and islands.

Which Caribbean country leads the region in tourism?

The Dominican Republic is the Caribbean country most visited by tourists, attracting approximately 4.7 million people annually. The beautiful beaches, clear seas, and tropical climate lure tourists from around the world, especially from Europe. Cuba is the second most visited Caribbean country, with 2.8 million tourists annually, followed by Jamaica with 2 million annual visitors.

Do things really disappear in the Bermuda Triangle?

The "Bermuda Triangle," or "Devil's Triangle," is a popular legend that suggests a supernatural or paranormal reason for a supposedly large number of missing aircraft and seagoing vessels within its area. The legend generally places the area of the Bermuda Triangle in the Atlantic Ocean, with its three corners located at Bermuda, Puerto Rico, and Miami, Florida. But you won't be able to find the Bermuda Triangle on a map since it is not a geographically or politically defined area, and its location is solely designated by the legend.

What really caused Flight 19 to disappear in the Bermuda Triangle?

The legend of the Bermuda Triangle seems to assert little evidence that this area is subjected to anything but natural hazards and human error. Most of the evidence for the phenomena in the Bermuda Triangle stems from the disappearance of five aircraft of Flight 19 in December 1945, as well as a search plane that was sent to find them. Though the popular version of the disappearance of Flight 19 assumes a mysterious end, a mixture of missing navigational apparatus, human error, low fuel, and choppy seas most likely led to the squadron's disappearance and ultimate demise.

PHYSICAL FEATURES AND RESOURCES

What is the Pan-American Highway?

Begun in the 1930s, the Pan-American Highway is the result of an international effort to create a highway stretching from Fairbanks, Alaska, to Buenos Aires, Argentina. In 1962, a bridge known as the Bridge of the Americas was built over the Panama Canal to continue the highway over the canal. A 100-mile (161-km) stretch of the highway called the Darién Gap, in eastern Panama, still remains unfinished.

Which city is more east: Santiago or Miami?

Even though it lies on the west coast of South America, Santiago, Chile, is actually farther east than Miami, Florida. Though it is common to envision South America as directly south of North America, South America actually lies southeast of North America.

Which river carries more water than any other in the world?

Though the Amazon River is the second longest in the world at 4,000 miles (6,400 km) long, it carries more water to the ocean than does any other river in the world, approximately 7,380.8 cubic feet (209,000 cubic meters) per second.

What is the highest navigable lake in the world?

Lake Titicaca, located on the border between Peru and Bolivia, is the highest navigable lake in the world, with a surface elevation of 12,507 feet (3,812 meters). Though there are higher lakes in the world, Lake Titicaca is the highest one in which commercial boats are used. Lake Titicaca was the center of Incan civilization.

Where is the world's tallest waterfall?

Angel Falls, along the Gauja River in Venezuela, is the world's tallest waterfall at 3,212 feet (979 meters). American pilot Jimmy Angel discovered the waterfall and named it after himself in 1935. At the time of the discovery, the falls were known to indigenous peoples such as the Pemon for thousands of years. They called the falls Kerepakupai Merú, which means "waterfall of the deepest place."

Is the Strait of Magellan crooked?

Yes, it is! The Strait of Magellan is a winding waterway between South America and the islands of Tierra del Fuego at the southern tip of South America. This strait was discovered by the explorer Ferdinand Magellan in 1520 and has been used as a shortcut to avoid having to sail around Cape Horn, the southern tip of South America. It took Magellan thirty-eight days to pass through the strait.

What are the Andes?

The Andes are a mountain chain that runs along the entire west coast of South America, from Panama (at the southern tip of Central America) to the Strait of Magellan (at the southern tip of South America). This chain is about 4,300 miles (7,000 km) long and contains high plateaus and one of the driest deserts on the planet, the Atacama Desert. The tallest mountain in South America, Aconcagua, at 22,834 feet (6,960 meters) in elevation, is located in the southern Andes, on the border between Chile and Argentina. The ancient Inca city of Machu Picchu is located in the Andes of Peru. It is the world's highest mountain range outside of Asia.

What are the four climatic regions of the Andes?

The Andes are known for their four defined climatic zones, which are based on elevation. The lowest zone, *tierra caliente* (hot lands), is ascribed to the area from the plains to 2,500 feet (762 meters) and is where most of the population resides. The second zone is *tierra templada* (temperate land), which is from 2,500 to 6,000 feet (762 to 1,829 meters). The third zone is *tierra fria* (cold land), which is from 6,000 to 12,000 feet (1,829 to 3,658 meters). Above 12,000 feet (3,658 meters) is the fourth zone, *tierra helada* (frozen land).

What are the Cordilleras of Colombia?

In Colombia, the Andes are split into three separate mountain ranges. They are the Cordilla Occidental (western range), the Cordilla Central (central range), and the Cordilla Oriental (eastern range). The city of Cali is located in the valley between the Occidental and Central ranges, while Bogotá is located between the Central and Oriental ranges.

San Rafael Falls in Ecuador is just one of many beautiful sights to be found in the Amazon rain forest.

What is the Atacama Desert?

The Atacama Desert is the driest nonpolar desert on Earth, located along the Pacific Coast of northern Chile. It is completely barren of plant life. The town of Calama, which is located in the Atacama, has never received rain. The Atacama is a source for nitrates and borax. Some evidence shows that parts of the desert had not received any significant rainfall from 1570 to 1971. It may also be the oldest continuous desert on Earth, originating approximately 3 million years ago.

What is the world's largest tropical rain forest?

The Amazon rain forest is the world's largest tropical rain forest. It occupies one-third of Brazil's land area and averages over 82 inches (210 centimeters) of rain a year. The rain forest loses approximately 15,000 square miles (38,850 square km) of forest each year because of clear-cutting. The Amazon rain forest is home to about 90% of the Earth's animal and plant species and is a major producer of the world's oxygen. Over the past 40 years, 20% of the rain forest has been cleared, which is more than the previous 450 years.

Which country is the world's leading copper producer?

Chile produces 31% of the world's copper concentrates annually and controls approximately 28% of the world's copper reserves. The world's total copper production is 18.7

million tons (17 million metric tons); thus, Chile's production is 5.8 million tons (5.3 million metric tons).

HISTORY

How was the New World divided between Spain and Portugal?

In 1493 Pope Alexander VI divided the New World into Spanish and Portuguese spheres of influence. A line was placed "100 leagues" or about 300 miles (480 km) west of the Azores, located several hundred miles west of Portugal in the Atlantic Ocean. Everything in the New World to the east of this demarcation line, which lay off the east coast of South America, belonged to Portugal, while the lands in the west belonged to Spain. Since this division provided little land for Portugal, the Portuguese were dissatisfied. The Treaty of Tordesillas established a new line about 800 miles (1,300 km) to the west of the old line. Pope Julius II approved the line in 1506.

Who was Simón Bolívar?

Simón Bolívar was a Venezuelan military and political leader, educated at a young age in Spain. In the early nineteenth century, Bolívar led the fight in South America for independence of various territories from the monarchy of Spain. He is revered as a hero among South Americans for his role in the independence of Venezuela, Colombia, Ecuador, and Peru. Bolivia was named in honor of Bolívar. His legacy, both positive and negative, still endures today.

Who was Che Guevara?

Ernesto "Che" Guevara (1928–1967) was an Argentine Marxist revolutionary, physician, author, and guerrilla leader. He formed his political ideas after seeing the extreme poverty in Latin America, believing that the cause was economic inequality due to monopolistic capitalism, neocolonialism, and imperialism. He was instrumental in helping Cuba's Fidel Castro overthrow the U.S.-backed dictator Fulgencio Batista.

How did Brazil become a Portuguese colony?

Most of the present-day country of Brazil was east of the line drawn in the Treaty of Tordesillas, a treaty signed by both Spain and Portugal, in 1494. The treaty provided for a line of demarcation between lands discovered by Columbus to the west, which were given to Spain, and lands to the east, which were given to Portugal. Brazil was claimed by Portugal after explorer Pedro Cabral arrived, but it was not until 1532 before Portugal established a colony. Brazil's official language is Portuguese, making it the only Portuguese-speaking country in South America.

Which South American country was the first to gain independence from colonial rule?

In 1816, Argentina gained independence from Spain. International recognition of the independent country, then called the United Provinces of the Plate River, did not come until 1823, when the United States recognized the new state.

Who is Alberto Fujimori and what is his legacy in Peru?

Alberto Fujimori, who was president of Peru from 1990 to 2000, was credited with ending terrorism in Peru and turning around a devastated economy. Some believe, however, that he trampled on the rights of individuals and indigenous people during his authoritarian rule. He later was convicted on charges of abuse of power in ordering the illegal search of the apartment of his security chief's spouse and a variety of other crimes. He has been imprisoned since 2007 for a maximum of twenty-five years.

What is Machu Picchu?

Machu Picchu is an ancient Incan city constructed at an elevation of 8,000 feet (2,438 meters) above sea level and located about 43 miles (69 km) northwest of Cuzco in Peru. It was built by the Incan ruler Pachacuti Inca Yupanqui between 1460 and 1470. It is comprised of more than 200 buildings, which are visited by thousands of tourists each year. Tourists reach the city either by bus or by a ritualistic twenty-mile hike to the summit. It was rediscovered by a Yale University team, headed by Hiram Bingham, in 1911. Some argue that other explorers may have discovered the site earlier, including a German businessman named Augusto Bern in the 1880s.

The ruins at Machu Picchu bear evidence of the once amazing Incan city in the Andes Mountains.

What was Gran Colombia?

Gran Colombia or "Great Colombia" was the name given to describe the geographic territory of an area that encompasses modern Colombia, Venezuela, Ecuador, Panama, northern Peru, western Guyana, and northwest Brazil. After many wars against Spain for independence, Gran Colombia, led by Simón Bolívar, became an independent country in 1821. In 1830, Gran Colombia was split into Colombia (which included Panama), Ecuador, and Venezuela.

PEOPLE, COUNTRIES, AND CITIES

What percentage of the world's poor live in South America?

Approximately 4.7% of all of the world's poor people (approximately 585.5 million) live in South America. Poor is defined as people having incomes of less than $1.25/day.

What are the most heavily urbanized countries in South America?

Uruguay, Argentina, Chile, and Venezuela all have an urbanization level (the percentage of population who live in urban areas) of more than 85%.

What is MERCOSUR?

The Southern Cone Common Market, also known as MERCOSUR, is a trade group that includes Brazil, Argentina, Paraguay, and Uruguay. A treaty was drafted in 1991, but it was not until 1995 that the trade group was operating. The main mission of MERCOSUR is to reduce trade barriers between member countries and to promote economic unity.

Which South American countries are members of OPEC?

Ecuador (526,000 barrels/day) and Venezuela (2.7 million barrels/day) are both members of OPEC (Organization of the Petroleum Exporting Countries). In terms of oil production among all OPEC members, Ecuador is the smallest.

What is a plaza?

Most Latin American cities have an open public square at the center of the downtown called the plaza, usually comprised of three institutions: the cathedral, administrative center, and court of law. The plaza is a great public area that is used for festivals and ceremonies and is typically surrounded by shopping areas.

Which place in South America is part of the European Union?

Although not a country, French Guiana, lying just north of Brazil and one of the twenty-six departments of France, is part of the European Union. It uses the euro as its national currency.

Which South American territory has yet to gain its independence?

French Guiana, on the northeast coast of South America, has been a colony of France since 1817. It is officially a department (state) of France and is used as a launch site of the European Space Agency.

What is Devil's Island?

Devil's Island, located off the coast of French Guiana, became the overseas prison of France in the middle of the nineteenth century. France used the facility as a penal colony from 1852 until 1953.

Where has one-third of the population of Suriname emigrated to since 1975?

Suriname was a Dutch colony until it gained independence in 1975. Since then, between 350,000 and 455,000 of its residents have emigrated to the Netherlands.

What is the actual name of the city of Bogotá?

Bogotá, Colombia, is officially called Bogotá, Distrito Capital. It was originally known as Santa Fe, and from 1991 to 2000, the actual name became Santa fe de Bogotá. Today, the Colombian capital has approximately nine million people living in the city, with an additional four million residing in its metropolitan area.

What is a cartel?

A cartel is an organization made up of businesses that band together to eliminate competition, collude to maintain high prices, and control supply and production of a product or service. In South America, the word refers to the drug cartels of Colombia, most notably the Medellin and Cali cartels, both of which, by 2011, were suppressed by the Colombian government. Later, government members and lieutenants of former cartel operatives stepped in and created their own cartels. Today, they are still manufacturing and distributing cocaine and its derivatives into the United States, which is still the biggest consumer of cocaine in the world.

Where does cocaine come from?

Cocaine is produced from the coca plant, which was originally domesticated by the Incas. Coca paste from the coca plant is refined to make cocaine. Illegal cartels and individuals in Colombia, Peru, and Bolivia are major producers of cocaine. Colombia produces approximately 42% of the world's cocaine.

What percentage of Americans use cocaine?

Approximately .5% of the U.S. population uses cocaine. To demonstrate the current downward trend of usage, 1% of the population used cocaine in 2006.

Where did Charles Darwin develop his theory of natural selection?

Charles Darwin graduated from Cambridge University in 1831 and spent the next five years of his life as a naturalist on board the HMS *Beagle*. The *Beagle* traveled around the world, including among its destinations the Galapagos Islands, 621.4 miles (1,000 km) west of South America, near what is now the country of Ecuador, where Darwin spent six weeks collecting data from which he developed his theories of natural selection, published in a book entitled *On the Origin of Species by Means of Natural Selection* in 1859.

Charles Darwin (shown here in 1854) significantly advanced the case for the theory of evolution. The concept of evolution had been around for centuries before his landmark book was published.

Why is Mt. Chimborazo interesting to geographers?

Mt. Chimborazo, near the capital city of Quito in Ecuador, is interesting to geographers because it is the highest point from the center of the Earth. This is because the Earth is not a perfect sphere but really is an oblate spheroid shape, meaning that it bulges outward around the equator. The mountain, which lies in the Andes mountain range, is 20,564 feet high (6268 meters) and is a dormant volcano that last erupted around 550 C.E.

Why did Peru and Ecuador fight three wars in the twentieth century?

When Ecuador split off from Gran Colombia in the nineteenth century, it signed a border agreement with Peru, defining its boundaries along the Marañón River. In 1941, Peru invaded Ecuador and occupied half the country for ten days. Afterward, a peace treaty was brokered and guaranteed by the United States, Brazil, Argentina, and Chile. The United States mapped the border, leaving approximately 48 miles (78 km) of a line in the Cordillera del Cóndor area unmarked. The area became the center for disputes between the countries in 1941, 1981, and again in 1995.

How large is Brazil?

Brazil makes up just under 50% of the land area of the entire South American continent. It is the world's fifth-largest country, with 3.29 million square miles (8.514 million square km) of land and 21,411 square miles (55,455 square km) of water.

Who designed and planned the capital of Brazil?

Brasilia was created in 1956 by two people: urban planner Lúcio Costa and architect Oscar Niemeyer. The landscape architect for the project was Roberto Marx.

When did Brazil move its capital city?

In 1960, Brazil moved its capital city from Rio de Janeiro to a brand-new city in the center of the country, called Brasilia. Brasilia was designed and constructed on empty land near the center of the country in the 1950s. The capital was moved to reaffirm Brazilian independence, exchanging a colonial capital on the coast for a new, interior capital. The interior, underdeveloped location of the new capital allowed for a fresh start, as well as an opportunity to develop the region.

What statue overlooks Rio de Janeiro?

The 124-foot- (38-meter-) high statue of "Christ the Redeemer" stands, arms outstretched, over the city of Rio de Janeiro. The statue of Jesus Christ, with its base on top of Corcovado Mountain at 2,340 feet (713 meters), was built in commemoration of the one-hundredth anniversary of Brazilian independence. It was completed in 1931 after nine years of construction.

The statue "Christ the Redeemer" is an iconic symbol overlooking the city of Rio de Janeiro, Brazil.

What is the largest city in the Amazon River Basin?

Manaus, Brazil, is the largest city in the basin, with a population of over 2 million people. Manaus is the capital of Brazil's largest state, Amazonas, and is a major trading center for the region, with its Free Economic Zone. When the Amazon basin was the only known source of rubber, Manaus experienced a boom but subsequently declined in importance due to the planting of rubber in other regions of the world.

What country is crossed by both the equator and a tropic?

Brazil is the only country crossed by the equator at 0° and the Tropic of Capricorn 23.5° South.

What South American city has more people of Japanese descent than any city outside of Japan?

São Paulo, Brazil, has more people of Japanese descent than any other city outside of Japan. It is estimated that between 1.4 and 1.5 million people are of Japanese descent living in the city. Another approximately 275,000 Japanese–Brazilians live in Japan. The original settlers, 791 farmers, traveled to Brazil from Kobe, Japan, in 1908.

What is Mardi Gras?

The Catholic festival Mardi Gras literally means "fat Tuesday" in French. Parades, dancing, and carnivals are all part of this pre-Ash Wednesday celebration. Known as Carnival, it is very popular in Rio de Janeiro, Brazil, and New Orleans, Louisiana. The Brazilian festival is a significant source of tourism-related income for the country. In 2015, 977,000 Mardi Gras tourists generated $782 million in revenue for Rio.

What makes Brazilian automobiles run?

Over half of Brazilian automobiles use alternatives to petroleum known as biofuels, including gasohol and ethanol. Gasohol is made from sugarcane, and ethanol is made from alcohol. The two fuels are much less expensive than petroleum-based gasoline. In 2016, Brazil produces approximately 8.1 billion gallons (30.68 billion liters) of ethanol and exports 357 million gallons (1.35 billion liters) per year.

Which country is the world's leading coffee producer?

Brazil produces more than 5.986 billion pounds (2.721 million metric tons) of coffee beans each year. Another South American country, Colombia, is third in the world and produces 1.531 billion pounds (696,000 metric tons) annually.

What are the capitals of Bolivia?

Bolivia has two capitals. La Paz is the administrative capital, while Sucre is the constitutional and judicial capital. Several countries divide national functions between cities.

What is an altiplano?

Altiplanos are high plains located among the mountains that are suitable for habitation. In the fifteenth and sixteenth centuries, the civilization of the Incas developed in the altiplanos of the Andes Mountains. The Bolivian capital of La Paz is also located on an altiplano.

What is the world's highest capital city?

La Paz, Bolivia, is the world's highest capital city. La Paz is located high in the Andes Mountains at an elevation of 11,975 feet (3,650 meters). It was founded in 1548 by Spanish explorers and is now home to approximately 877,000 people.

What port does landlocked Bolivia use?

Having no access to the sea itself, Bolivia made an agreement in 1992 with Peru to lease and develop its port at the Pacific Coast city of Ilo.

Who owns Easter Island?

Easter Island, or Rapa Nui National Park, located 2,237 miles (3,600 km) west of Chile, is owned by Chile. The island contains over 100 large rocks (called moai) carved into the

How long is Chile?

Chile stretches approximately 2,700 miles (4,344 km) along the western coast of South America. At its widest, it is only 217 miles (350 km) across. Chile is a classic example of an elongated country, which makes governing difficult.

shape of heads, complete with facial features. These large heads vary in size from 10 to 40 feet (3 to 12 meters) in height and are made out of a soft, volcanic rock.

Who's fighting over the Falkland Islands?

The Falkland Islands (also known as Islas Malvinas), located 300 miles (480 km) east of the southern tip of South America, have long been a source of conflict between the United Kingdom and Argentina. Though the islands have been occupied by the British since 1833, Argentina has claimed the islands as its own since the eighteenth century. In 1982, Argentina invaded the islands, but the British regained possession within a matter of weeks. Argentina still claims the Islas Malvinas and is pursuing its claim through diplomatic channels. The capital is Stanley.

What is the world's southernmost city?

Ushuaia, in southern Argentina, is the world's southernmost city. Ushuaia sits on Tierra del Fuego Island, south of the Strait of Magellan. It has a population of approximately 57,000 people.

WESTERN EUROPE

PHYSICAL FEATURES AND RESOURCES

Which river touches more countries than any other?

The Danube River, which begins in Germany, passes through or borders ten countries in Europe, more than any other river in the world. On its journey, the Danube River encounters Germany, Austria, Slovakia, Hungary, Croatia, Serbia and Montenegro, Romania, Bulgaria, Moldova, and Ukraine.

Where are the Highlands?

The island of Great Britain is traditionally divided into highlands and lowlands along the Tees-Exe line, which runs from the mouth of the River Exe, in the southwest, to the River Tees in the northeast. To the southeast of this line lie the flat plains of England, while to the northwest lie the Scottish Highlands.

What are the Alps?

The Alps are Europe's most famous mountain chain, running east-west for approximately 750 miles (1,200 km). The mountain chain crosses Austria, France, Germany, Italy, Liechtenstein, Monaco, Slovenia, and Switzerland. The Alps include Mont Blanc, at 15,781 feet (4,810 meters), the highest point in Western Europe.

Where is the southernmost glacier in Europe?

The southernmost European glacier, named Calderone glacier, is also high atop the Apennine mountain range in Italy, near Corno Grande, the highest point in the Apennines. Due to recent global climate change, it has lost a significant amount of its mass and may disappear entirely by 2020.

What are the Apennines?

The Apennines are a mountain range extending from northern to southern Italy for approximately 750 miles (1,200 km). The highest point is a place called Corno Grande, which reaches a summit at 9,554 feet (2,912 meters).

How much oil is produced by Europe?

Europe produces 3,320,000 barrels of oil per day, which is approximately 3.5% of the world's total production of 95,960,000 barrels per day.

Who is the biggest producer of oil in Europe?

Norway wins as Europe's biggest oil producer, producing approximately 1.9 million barrels per day. The United Kingdom and Denmark are the second- and third-biggest oil producers in Europe.

What is the dividing line between Europe and Asia?

Though Europe and Asia are actually part of one large landmass, tradition has split the region into two continents along the Ural Mountains, a mountain chain that runs north-south for 1,600 miles (2,500 km) from the Arctic Circle in western Russia to the Ural River in northwestern Kazakhstan. Formed between 250 and 300 million years ago, they are far older than many famous mountains, including the Alps, Himalayas, Andes, and Rocky Mountains.

HISTORY

Which country had the world's first legislature?

Though Iceland had been settled about sixty years earlier by the Norwegians, Iceland's parliament, the Althing, was created in 930 C.E.

Who settled Denmark?

Surprisingly, Denmark was not settled by Europeans from the continent directly to its south but was settled in the eighth century by Danes from nearby islands, southern Sweden, and the Scandinavian Peninsula.

When did the European Union begin?

In 1951, six Western European countries joined together in the European Coal and Steel Community. As more members joined, the organization grew in scope and soon became an organization that helped mend and meld the economies of Europe. In 1993, the European Community was renamed the European Union (EU). Today, there are twenty-seven member states: Austria, Belgium, Bulgaria, Cyprus, Czech Republic, Denmark, Estonia, Finland, France, Germany, Greece, Hungary, Ireland, Italy, Latvia, Lithuania, Luxembourg, Malta, the Netherlands, Poland, Portugal, Romania, Slovakia, Slovenia, Spain, Sweden, and the United Kingdom. The European Union has a flag, an anthem, and in 1999 began using a single monetary unit (the "Euro").

What was the Potsdam Conference?

Nine weeks after the end of World War II, the leaders of the United States, the United Kingdom, and the U.S.S.R. met at Potsdam, Germany, from July 17 to August 2, 1945, for a conference to determine how to punish and control Germany and other eastern territories. The Potsdam Conference divided Germany and Austria into Soviet, French, American, and British zones of control.

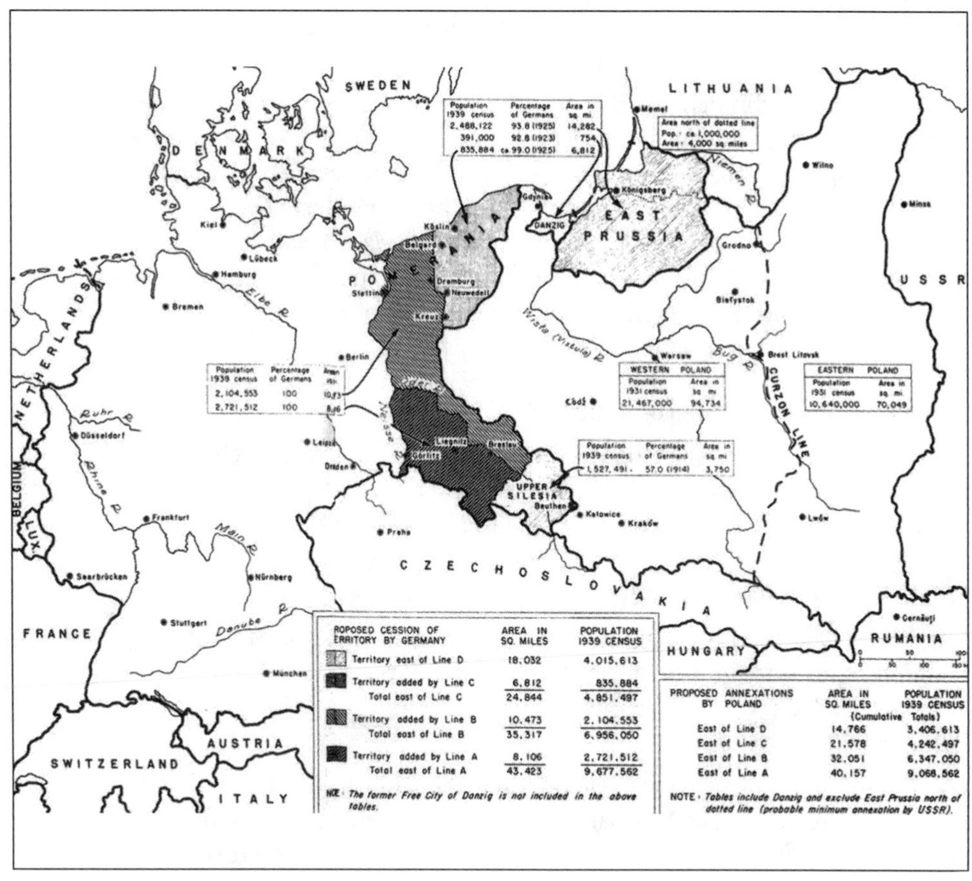

A demographic map used during the 1945 Potsdam Conference to determine new borders. The borders of Germany and Poland, especially, received significant revision.

What is a Reich?

The word "reich" literally means "realm" in German. The First Reich is considered to be the Holy Roman Empire from 962 to 1806 C.E. The Second Reich was Germany, united under Otto von Bismarck, was from 1871 to 1918. In 1933, Adolf Hitler established the Third Reich, the Nazi regime, which lasted until the defeat of Germany at the end of World War II in 1945.

What was the Berlin Wall?

At the end of World War II, Germany was divided into four zones, each occupied separately by the United States, the United Kingdom, France, and the U.S.S.R. The city of Berlin, while located entirely within the Soviet-occupied zone, was itself divided into four zones. Soon thereafter, the Soviets stopped cooperating with the other Allied powers. The three zones occupied by the United States, United Kingdom, and France joined together to create West Germany, while the Soviet zone became East Germany. A similar split occurred in the city of Berlin.

The city of Berlin held the dichotomy of east versus west, communist versus capitalist. Many people who lived in East Berlin could see that those in West Berlin generally had a higher standard of living. It is estimated that over two million East Germans fled to the West within Berlin. In August 1961, the communist government, determined to stop this mass exodus, began to build the Berlin Wall, a wall that physically divided East and West Berlin. On the west side, the wall became the location of spray-painted messages that voiced free opinions; on the east side of the wall lay a deserted area of barbed wire and armed guards called "No Man's Land."

When did the Berlin Wall come down?

For decades, the Berlin Wall stood as the physical version of the psychological "iron curtain" that separated east from west. On November 8, 1989, the Berlin Wall came tumbling down, and soon thereafter the era of the Cold War also ended.

Where was Checkpoint Charlie?

Checkpoint Charlie was a famous crossing point on the Berlin Wall between East and West Berlin, used mainly by tourists and U.S. military personnel.

What was the Maginot Line?

The Maginot Line was a defensive zone that was built in the 1930s to defend France against the possibility of a German invasion. The zone consisted of underground tunnels, artillery, antitank obstacles, and many other defensive structures and stratagems to slow down invading Germans. The Maginot Line stretched for approximately 200 miles (322 km) near the French–German border.

Who was Ötzi the Iceman?

In 1991, two German tourists were hiking in the Ötzal Alps, on the Italian side of the border between Italy and Austria, when they happened upon what appeared to be a corpse buried beneath the ice. Astounded, they alerted authorities. It was found that Ötzi (as the body was named) was a forty-five-year-old traveler himself who was more than 5,300 years old. It was likely he faced what investigators believe was a violent death. On his body were more than fifty tattoos, jewelry, and weapons, all of which gave researchers clues into the lives of our Copper Age ancestors. Various artifacts, as well as a full-size sculpture of what he may have looked like, are on display at the South Tyrol Museum of Archaeology in Bolzano, Italy.

During World War II, when the Germans invaded France, the Germans bypassed the Maginot Line by storming through neutral Belgium. Thus, the Maginot Line had failed its one great test because it was too short. The line was also rendered obsolete by the fact that it did not provide defense against the new, modern warfare that included aircraft.

PEOPLE, COUNTRIES, AND CITIES

What are the most populous cities of Europe?

Western Europe's largest city is London, with 8.6 million inhabitants. Berlin (3.6 million), Madrid (3.2 million), Rome (2.9 million), and Paris (2.3 million) are the next four largest cities.

What are some of the most expensive cities in the world?

According to the *Economist* magazine's recent World Cost of Living Survey, half of the world's most expensive cities to live in are found in Europe, including Paris, France; Oslo, Norway; Zurich, Switzerland; Geneva, Switzerland; and Copenhagen, Denmark.

What are four of the smallest countries in Europe by population?

Some of the smallest countries in Europe include Vatican City, with 839 residents; San Marino, with 32,831 residents; Liechtenstein, with 37,370 residents; and Monaco, with 37,800 residents.

What is Iceland's leading export?

Over 30% of Iceland's exports are fish and fish products. The fish industry employs 7% of the nation's workforce, and the country is economically vulnerable to fluctuations in world fish prices.

Where is the Jutland Peninsula?

The Jutland Peninsula, also called the Cimbrian Peninsula, extends to the north from Germany and is home to the continental portion of the country of Denmark.

Where is the Black Forest?

Located in southwestern Germany, the Black Forest is a densely forested, mountainous region that is a popular location for vacationing, with its many health resorts and wilderness trails. The Black Forest is the source of the Danube River and is renowned for its cuckoo clocks.

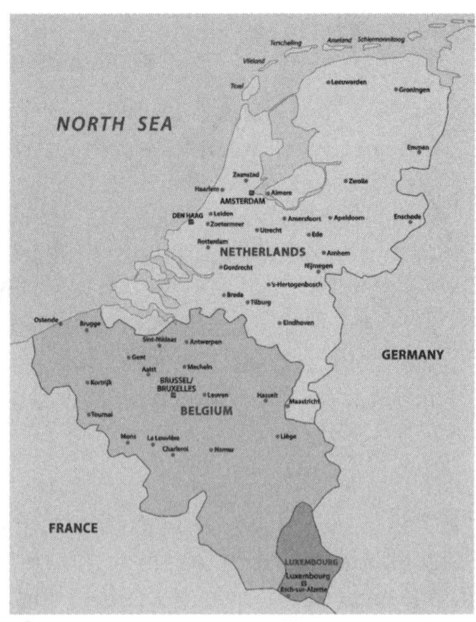

The Benelux countries are the Netherlands, Belgium, and Luxembourg.

What is Benelux?

Benelux stands for Belgium, the Netherlands, and Luxembourg, and it represents an economic alliance between the three that was formed in 1944. At the time, Belgium was primarily industrial, and the Netherlands was primarily agricultural; the two countries' economies complemented each other, and the economic union strengthened their relationship. Luxembourg, which has a varied economy and is extremely small, has long been closely affiliated with its two larger neighbors and thus also benefited from the union. Today, the countries are referred to as Benelux whenever referred to in terms of market analyses, commerce, and trade.

Where are the "low countries"?

Belgium, the Netherlands, and Luxembourg are known as the "low countries" because of their low elevation.

What are the two cultural groups that make up Belgium?

The Walloons in southern Belgium, called Wallonia, are descendants of the Celts, comprise 41% of the population, and speak French. The Flemings in northern Belgium, called Flanders, comprise 59% of the population, are descendants of German Franks, and speak Flemish, a language similar to Dutch.

How do the Netherlands keep getting bigger?

For hundreds of years, the Dutch have been expanding the size of their country by building dikes and reclaiming land. These lands, known as polders, have greatly expanded the size of the Netherlands and are now considered one of the seven wonders of the modern world.

What is the Randstad?

The Randstad is a region of the Netherlands that includes the metropolitan areas of Amsterdam, The Hague, Rotterdam, and Utrecht. The urban area of the Randstad holds 7.1 million people.

What is The Hague?

The Hague is a city on the west coast of the Netherlands with an approximate population of 516,000. The Hague is the seat of the government of the Netherlands but not its capital (Amsterdam). It is the home of many international organizations, such as the International Court of Justice and the International Criminal Court.

What's the difference between England, Great Britain, and the United Kingdom?

Northeast of France lie two large islands: Great Britain to the east and Ireland to the west. On the island of Great Britain there are three regions: England in the southeast, Wales in the southwest, and Scotland in the north. The other island, Ireland, is divided into two political divisions: the region called Northern Ireland in the north and the country of Ireland in the south. The United Kingdom is a country that includes all three regions on the island of Great Britain (England, Wales, and Scotland) and the one northern region on the island of Ireland (Northern Ireland).

What are the British Isles?

The British Isles are composed of the two large islands of Great Britain and Ireland (separated by St. George's Channel) and many small islands nearby. The British Isles include two countries: the United Kingdom and Ireland.

What is Hadrian's Wall?

Hadrian's Wall was built under the direction of the Roman Emperor Hadrian in 122 C.E. Located in northern Great Britain, it was intended to keep out the Caledonians of Scotland. Built of mud and stone, the Wall stretched nearly 75 miles (120 km), from Solway Firth in the west to the Tyne River in the east (near Newcastle).

What is the Commonwealth of Nations?

The Commonwealth of Nations, formerly known as the British Commonwealth, consists of fifty-three member states including the United Kingdom and now-independent

Is Scotland a country?

While Scotland does have limited self-rule, it is still part of the country of the United Kingdom. Scotland occupies the northern portion of the island of Great Britain.

former countries and territories of the British Empire. The Commonwealth is not a policymaking body but is solely a loose voluntary association between countries that were formerly under British control. It is an influential organization because its members occupy approximately 11.5 million square miles (29.9 million square kilometers) across the continents, which is nearly a quarter of all land on the planet.

What is Land's End?

Land's End has quite an appropriate name, as it is a cape at the southwestern tip of Great Britain that is the westernmost point of England; the "end of land" in the west.

What is a moor?

A moor is uncultivated pasture land. You'll find moors in the United Kingdom. In the United States, most people call them fields or prairies.

Where is Camelot?

Although the stories of King Arthur originated in twelfth-century France, the legendary sixth-century castle of King Arthur is located somewhere in Great Britain. Camelot was not only the home of King Arthur and Queen Guinevere but was also the location of the Round Table and its famous knights.

Did people bathe in Bath?

The Roman baths found in Bath, England, were originally built on a Celtic holy site over a period of 300 years, beginning in 60 C.E. The structures include a hot bath, a tepid or medium bath, and a frigid bath complex. Though the ancient city lies buried beneath it, the modern city of Bath is also renowned for its hot springs, which once warmed the Romans and now offer a relaxing bath in this spa town.

How many Irish left Ireland during the Great Famine?

In the mid-nineteenth century, Ireland suffered from the "Great Famine." From 1845 to 1850, a fungus ravaged the potato crops of Ireland, destroying the primary food source of Irish peasants. Though many have called this tragic event the "Great Potato Famine," the mass starvation of the Irish people was caused more by the lack of assistance from the British government than by the famine itself. It is estimated that over one million people died during these catastrophic times, and approximately twice that number left their homeland in an effort to begin a new life elsewhere.

Where is Catalonia?

Catalonia is an autonomous region in northeastern Spain. Because Catalonia is home to more than 7.5 million Spanish Catalans, who have their own language and culture, many in the region would like the territory to become an independent nation. Spain does not want Catalonia to secede, as Catalonia is responsible for a sizable portion of

Spanish economic production. Catalonia's capital is Barcelona, the host city of the 1992 Summer Olympics. In September 2015, a coalition of Catalan nationalist political parties won a majority of seats in the regional assembly elections and will push for independence, against the wishes of the Spanish government in Madrid.

Catalonia is a region in northeastern Spain.

Where is the Rock of Gibraltar?

The Rock of Gibraltar is a limestone mountain located on the Gibraltar peninsula in southern Spain. The city of Gibraltar, located on this same peninsula, is actually a British Overseas Territory with approximately 30,000 residents. Much of the area is used as a British naval air base. This is the perfect location from which to control the Strait of Gibraltar, the small waterway that connects the Mediterranean Sea with the Atlantic Ocean. Spain has continually advocated a claim for this area, which was won during a war in 1704, but has been unable to retrieve this vital piece of land.

How wide is the Strait of Gibraltar?

The strait, which connects the Mediterranean Sea to the Atlantic Ocean between Africa and Spain, is 8.9 miles (14.3 km) wide at its narrowest points.

What are Ceuta and Melilla?

On the opposite side of the Strait of Gibraltar, at the northern tip of Morocco, Spain has its own autonomous community, consisting of the cities of Ceuta and Melilla, which are also strategically located to control the Strait of Gibraltar. Ceuta is 7.5 square miles (18.5 square km) in area, and Melilla is 4.7 square miles (12.3 square km) in area. Although the two cities are autonomous and part of Spain, Morocco considers the cities to be occupied territories.

Who rules Andorra?

Since 1278, the tiny country of Andorra, nestled in the Pyrenees between France and Spain, has been jointly ruled by two people who live outside the country: the president of France and the bishop of La Seu d'Urgell in northeastern Spain. France and Spain jointly take responsibility for the defense of Andorra.

In what country do people have a very high life expectancy?

If you are lucky enough to be born in Andorra, your life expectancy will be on average eighty-one years.

Where is Atlantis?

Atlantis, the legendary underwater utopia supposedly located west of the Pillars of Hercules (the land on either side of the Strait of Gibraltar), was first described by Plato in the fourth century B.C.E. as a magnificent civilization that was swallowed by the sea. Though Plato believed Atlantis to have been destroyed, the legend has grown over the centuries to describe this civilization as an underwater kingdom. Researchers now believe that the legend of Atlantis was based on the ancient Minoan civilization that lived on the Greek islands of Thira and Crete, which disappeared after a volcanic eruption in the sixteenth century B.C.E. Thus, the Minoan civilization in Thira and Crete fits the approximate date of Atlantis's destruction but not its supposed location.

Where is Gaul?

Gaul was a region in ancient Europe that included much of what we know today as France, Luxembourg, Belgium, most of Switzerland, and parts of the Po Valley in Northern Italy, as well as the parts of the Netherlands and Germany on the west bank of the River Rhine. This area was originally settled by Celtic people. The Romans conquered the region during the second and first centuries B.C.E. Afterward, other empires took control of Gaul, and it eventually became the kingdom of the Franks in the fifth century C.E.

What was the French Community?

The French Community, also known as Communauté française, was an association of former colonies and territories that was created in 1958. The union sought to allow for a system of limited self-government of former colonial territories controlled by France and incorporate them into one French union in an attempt to subdue uprisings between settlers and locals in various French territories throughout the world.

During the Refugee Crisis of 2015, how many refugees from the war-torn Middle East region came to France seeking asylum?

Since the Refugee Crisis began in early 2015, more than 430,000 people from the Middle East have come to Europe seeking asylum, but only 24,000 (5.6%) of them will remain in France.

Where is the French Riviera?

The French Riviera, also known as Côte d'Azur, is located in southeastern France, near the border with Italy, along the Mediterranean Sea. The French Riviera is a major vacation spot for Europeans, with its mild Mediterranean climate and beautiful scenery. The tiny country of Monaco is located within the French Riviera and adds to the Riviera's

The last leg of the Tour de France is raced down Paris's famous Champs-Élysées.

image of luxury with the multitude of casinos and hotels at Monte Carlo. Although there is no official known boundary defining the area, it generally begins at the border with Italy and extends westward along the coast to Toulon. The city of Nice (population 344,000) is the largest city on the Riviera.

Where did the Tour de France begin and end in 2015?

The Tour de France changes its course each year, but the last leg is always along Paris's famous boulevard, the Champs-Élysées. The bicycle race is approximately 2,000 miles (3,200 km) long and takes twenty-five to thirty days to complete. In 2015, it began in the city of Utrecht, the Netherlands.

Which European country produces the most nuclear energy?

France produces the second most nuclear energy in the world, generating 418 terawatt-hours per year. This is more than half of the 833.6 terawatts that all European countries produce in a year. Of the 130 operational nuclear power plants in Europe, fifty-eight are located in France.

What was one of the earliest tunnels through the Alps?

The Fréjus Rail Tunnel (also known as the Mont Cenis Tunnel) was one of the first tunnels through the Alps and the first major railroad tunnel in the world. Opened in 1871, the tunnel spanned 8.5 miles (13.7 km) and connected France and Italy.

What is the European country most visited by tourists?

Over 83.7 million people visit France each year. Spain is the next most popular in Europe with 65 million visitors.

What is the Giro d'Italia?

The Giro d'Italia is the second most important multistage bicycle race in Europe. It began in 1909, and instead of a yellow jacket for the fastest stage winner, recipients wear a pink-colored jersey reminiscent of the newspaper publisher and founder of the race's use of pink newsprint for his newspapers.

How long does someone born in Italy live?

Italy has one of the highest life expectancies in the world. People live on average eighty-three years. Italian women may live an average of eighty-five years.

What are the seven hills of Rome?

The city of Rome sits upon seven hills named Capitoline, Quirinal, Viminal, Esquiline, Caelian, Aventine, and Palatine. According to ancient legend, the first settlement in the area, the city of Romulus, was built upon Palatine Hill.

What is Europe's oldest independent state?

San Marino claims to have been founded in the year 301 C.E. Its first constitution was established in 1600. San Marino is located on Mt. Titano, on the northeastern side of the Apennine Mountains in Italy, and, at 24 square miles (62 square km) in area, is one of the world's smallest independent countries and is said to have more cars than people. San Marino has an estimated population of 32,000.

Where is the longest tunnel in the world?

The Gotthard Base Tunnel and Ceneri Base Tunnel, located beneath the Swiss Alps, is 35.4 miles (57 km) long in both directions and is the longest tunnel in the world. It began in 1993 and is expected to be completed in 2016, at a cost of $10 billion. When it opens, trains will travel through it at 160 mph (250 kph).

Which country has the highest unemployment rate in the European Union?

Of all twenty-eight members of the European Union (EU), member country Greece has the highest unemployment rate, with approximately one-quarter of the available work-force either out of work or seeking employment elsewhere.

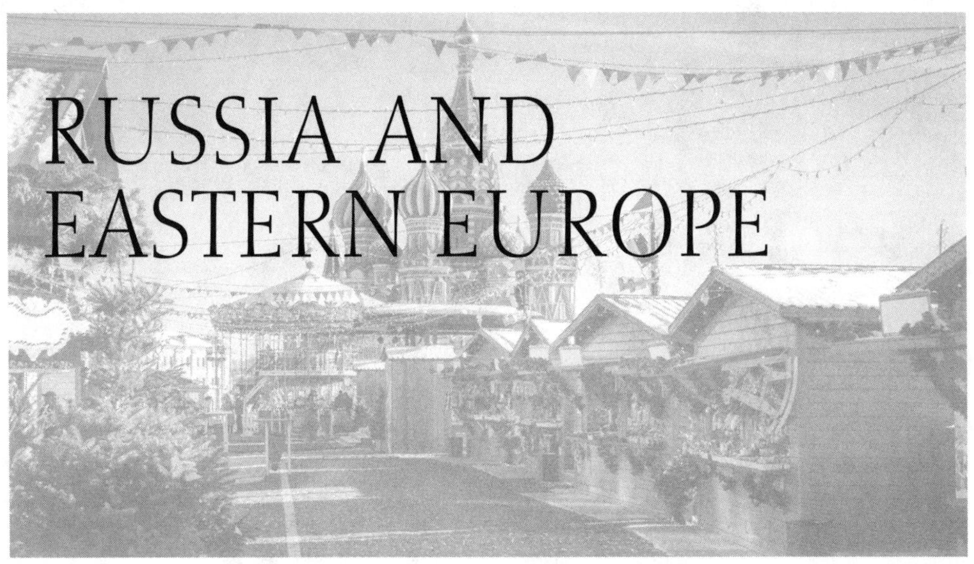

RUSSIA AND EASTERN EUROPE

RUSSIA AND THE FORMER SOVIET STATES

What do the names St. Petersburg, Leningrad, and Petrograd have in common?

St. Petersburg, Leningrad, and Petrograd were three names for the same city. Located in northwestern Russia along the Gulf of Finland, the city was originally founded as St. Petersburg in 1703 by Tsar Peter the Great. Since "St. Petersburg" sounded too German to be the capital city of Russia, the city's name was changed to Petrograd in 1914. After the death of communist leader Vladimir Lenin in 1924, the city's name was again changed, this time to Leningrad. After the dissolution of the Soviet Union in 1991, Leningrad once again became St. Petersburg.

Do Caucasians come from the Caucasus Mountains?

The term "Caucasian race" was first used in 1785 by German philosopher Christoph Meiners. In the latter half of the nineteenth century, scientists attempted to categorize the world's peoples in terms of race. Each race was defined by the color of its skin, a process that was laced with stereotypes. These scientists used the term "Caucasian" for "white" people because they believed that "white" people originated in a region of the Caucasus Mountains of Southwest Asia and that the people from this region represented the archetype. But this thesis was not based upon any real scientific evidence. Since that time, we have learned that there is but one human race and that we all, most likely, originated somewhere in Africa.

What is the longest river in Europe?

The Volga River, which lies entirely within Russia, is Europe's longest river. It flows 2,294 miles (3,692 km) from the Valdai Hills, in northwest-central Russia, near the city

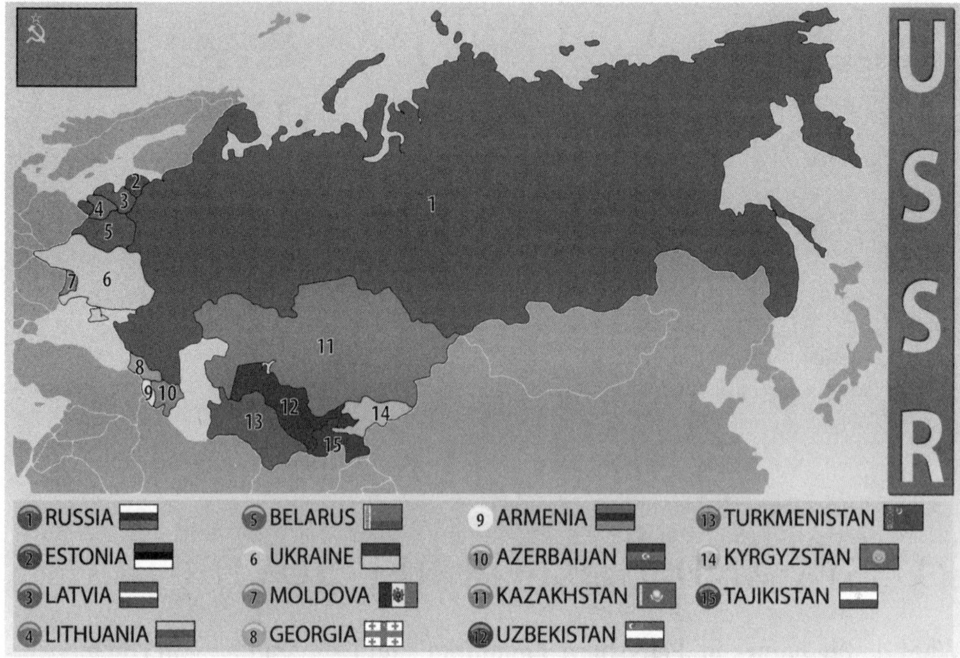

The U.S.S.R. included vast areas of the country that are now independent nations.

of Rzhev, into the Caspian Sea. More than half of the twenty largest cities in Russia are located nearby the Volga. The name is derived from early Slavic language roots meaning "wetness" or "moisture."

What was the U.S.S.R.?

The country called the Union of Soviet Socialist Republics (commonly known as the Soviet Union) was created in 1922, after Vladimir Lenin led a five-year-long revolution against the Tsarist monarchy and its supporters. The Soviet Union consisted of Russia and its neighboring territories, such as Ukraine, Kazakhstan, and the Baltic States. It was governed by a single political party (the Communist Party). By the end of 1991, the former Soviet government created a new commonwealth. Fifteen of the former Soviet republics became independent countries, while others joined the new commonwealth.

How many republics or states comprised the former Soviet Union?

The U.S.S.R. was comprised of the following fifteen Socialist Republics: Armenia, Azerbaijan, Belarus, Estonia, Georgia, Kazakhstan, Kyrgyzstan, Latvia, Lithuania, Moldova, Russia, Tajikistan, Turkmenistan, Ukraine, and Uzbekistan.

How large a part of the Soviet Union was Russia?

While the Soviet Union consisted of fifteen Soviet Socialist Republics, the largest was the Russian Soviet Federated Socialist Republic (RSFSR). The RSFSR, also known as Rus-

What Russian city lost the most people during World War II?

During the German 872-day siege of Leningrad (St. Petersburg) from 1941 to 1943, 1,400,000 residents were evacuated, and nearly 1.5 million people lost their lives.

sia, comprised three-quarters of the Soviet Union's territory and over half of its population.

What role did Russia play in World War II?

Russia was the target of Germany's war machine on its eastern front. More people lost their lives between 1941 and 1945, approximately 30 million people (20 million of whom were civilians), than in all other theaters of World War II combined. Without the active participation of the Russian military against the Nazis, the outcome of the war in Europe may have been drastically different.

How did factories in the U.S.S.R. end up on the east side of the country?

During World War II, the U.S.S.R. enacted its scorched-earth policy as Germany invaded from the west. The scorched-earth policy involved the U.S.S.R. moving everything it could to the east and destroying what it could not move. Factories were disassembled, shipped by train to the region near the Ural Mountains, and reassembled to keep Soviet industry working. The Ural Region is still a major manufacturing area for Russia.

What is the Commonwealth of Independent States?

The Commonwealth of Independent States (CIS), established by Russia in late 1991, just after the fall of the Soviet Union, is an organization with nine member states that serves to keep the resources of the former Soviet Union flowing between the now-independent countries. Nine of the fifteen former Soviet republics currently are members: Armenia, Azerbaijan, Belarus, Kazakhstan, Kyrgyzstan, Moldova, Russia, Tajikistan, and Uzbekistan. The Republic of Georgia withdrew in 2006, and Turkmenistan withdrew in 2005. Two former Soviet Republics are associate members: Ukraine and Turkmenistan.

How big is the Russian Federation?

With 6.6 million square miles (17 million square km) and approximately 144 million people, Russia is Europe's largest country and also its most populous. Additionally, Russia is the largest country in the world in terms of area and the world's ninth most populous. It is situated between 41° and 82° north latitude and 19° east and 169° west longitude and stretches roughly 4,971 miles (8,000 km) from north to south and 4,101 miles (6,600 km) from east to west.

What is Russia's official name?

The official name of Russia is the Russian Federation. It consists of 85 federal subjects, including 46 provinces, 22 republics, 9 territories, 4 autonomous districts, 1 administrative unit, and 3 federal cities (Moscow, St. Petersburg, and Sevastopol).

Is it possible to drive across Russia?

It certainly is. Most of the urban areas of Russia have modern expressways. Even outside these areas, food and fuel are available to travelers. But in the vast expanse of the country east of Moscow, many of the roads are rough gravel or dirt and are only intermittently paved. Driving across Russia is therefore entirely dependent upon the season. During the winter, from November to May, the roads are frozen and can be driven upon; during the summer, many roads may become quagmires and are unusable.

How do most people travel across Russia?

Most people, as well as goods, travel across Russia by airplanes or train. In 1891, Tsar Alexander III launched the building of a railroad that would unify eastern and western Russia. Traveling from Moscow, through Siberia, to Vladivostok on the Pacific Coast, the Trans-Siberian Railroad was opened in 1904 and finished in 1916. It is the longest railroad line in the world, spanning 5,772 miles (9,289 km).

How long does it take to travel by train across Russia?

Depending on the number of stops the train makes, it can take anywhere from five to eight days to travel across Russia. Round-trip ticket prices can be anywhere from $300 to more than $800, depending on class of service. Train lines end in Vladivostok, along the Eastern Pacific Coast.

How much market share does Russia's national airline, Aeroflot, have?

For domestic flights within the country, as well as international flights originating in Russia, Aeroflot has a 37% market share of all passenger traffic. Nearly 10 million people fly Aeroflot each year.

How many tourists visit Russia each year?

Approximately 30 million people travel to Russia each year. The majority of these tourists are from Germany, followed by China, the United States, the United Kingdom, Finland, Italy, Turkey, France, Israel, and Spain.

Which Russian city is among the most expensive in the world?

Moscow is normally ranked between the ninth and twelfth most expensive cities to live in the world, in terms of the cost of food, housing, transportation, products, etc.

Red Square during a Christmas festival in Moscow. You can see the iconic Kremlin building in the background.

How many people live in Moscow?

Approximately twelve million people call Moscow home. An additional six million people live outside the city boundary but within the metropolitan area.

How fast is the Russian economy expanding?

The Russian economy, measured by the gross domestic product, recently grew at approximately 7.9% per year, with most of the growth coming from the energy sector. Its current growth rate is -3.8%. The decline in its growth rate is due to its entanglements in Ukraine and Syria, the decline in oil prices, which is a big contributor to the Russian economy, weak demand for Russian investments, and a decline in household consumption.

How many cell phones are there in Russia?

There are an estimated 256 million cell phones in Russia today, of which 36.2% are classified as "smartphones."

How big is Siberia?

Siberia is an area that covers 5.06 million square miles (13.1 million square km) and makes up approximately 77% of the total area of Russia but only contains 27% of the total Russian population. Siberia is bounded on the west by the Ural Mountains, on the north by the Arctic Ocean, on the east by the Pacific Ocean, and on the south by China, Mongolia, and Kazakhstan. Russia conquered the area now known as Siberia from the late sixteenth century to the seventeenth century.

What is the largest city in Siberia?

Novosibirsk is the largest city in Siberia, and the third-largest city in Russia, with approximately 1.5 million residents. It is located in southwestern Siberia and is a principal stop on the Trans-Siberian Railroad route.

How cold is Siberia?

Siberia holds the record for the world's lowest temperature outside of Antarctica. On February 6, 1933, the temperature reached –90 degrees Fahrenheit (–67.7 degrees Celsius) in Oymyakon, Russia, a village with approximately 500 residents. During the winter, which may last from late September to the middle of May, almost all of Siberia has extremely cold temperatures, often reaching –50 degrees Fahrenheit (-45.56 degrees Celsius).

How big is Siberian Airlines?

Siberian Airlines, now called S7 Airlines, is one of Russia's biggest domestic airline companies. The group of airlines under the S7 umbrella handles more than 10 million passengers each year.

What is a Denisovan?

Denisova Hominins or Denisovans are the third extinct species of humans, discovered in a cave in the Altai Mountains in Siberia. The cave was inhabited by Neanderthals and, later, modern humans for the past 125,000 years. In 2010, DNA was extracted from bone fragments found within the cave complex, allowing scientists to discover and analyze a newly discovered species of humans.

How big is Russia's Lake Baikal?

Lake Baikal, located in eastern Siberia, with a length of 397 miles (640 km), is one of the largest lakes in the world, covering an area of 12,200 square miles (31,500 square km). It is also the deepest lake in the world, with a maximum depth of 5,371 feet (1,637 meters). Lake Baikal holds one-fifth of the world's nonfrozen fresh water and is one of the world's oldest lakes, having been formed 20 to 25 million years ago. The crescent-shaped lake is also famous for its crystal-clear water and bountiful plant and animal life, including the freshwater seal, which has a lifespan of more than fifty years.

What started the fighting in Chechnya?

Chechnya, in the North Caucasus Mountains, was once part of a Soviet republic called Chechen-Ingush. After the fall of the Soviet Union, Chechen-Ingush was divided into two internal republics, Chechnya in the east and Ingushetia in the west. Though the Chechens declared independence in 1992, Russia did not approve and invaded Chechnya in 1994. The Russians crushed the rebellion, killing thousands of Chechens. From 1999 to early 2009, Chechen rebels continued to sporadically fight Russian troops until the Russians withdrew from Chechnya. Lately, Chechen rebels have aligned themselves with radical Islamic groups, hoping to create a new Islamic state. The entire region is still in a state of unrest today.

Where was the Pale of Settlement?

During the eighteenth and nineteenth centuries, the Pale of Settlement was an area in which Tsarist Russia attempted to restrict where Jewish people were permitted to live. It covered an area of 386,100 square miles (1 million square km) and extended from what is now eastern Poland to Ukraine and the country of Belarus. Within the confines of the Pale, Jews were subjected to many anti-Jewish regulations as well as mass killings and deportations to other parts of Russia.

What was the world's worst nuclear disaster?

In April 1986, the Chernobyl nuclear power plant in Ukraine, near the border with Belarus, had a major accident that released radiation into the atmosphere. The explosion was due to flawed reactor design and errors on the part of staff. The protective covering of the nuclear reactor exploded and deadly radiation escaped, immediately killing at least thirty-one people. The radiation exposure that initially occurred is still killing people through related diseases, and this will continue for many more years. An estimated 116,000 people were evacuated from 1,600 square miles (4,140 square km) of contaminated area, and deaths caused by exposure to radiation continue as radioactive isotopes spread across the region.

Protestors against the Ukrainian government march in Kiev in this January 25, 2014, photograph.

How did Ukraine help feed the Soviet Union?

Often called the "breadbasket of the Soviet Union" for decades, Ukraine's rich wheat harvests were used to feed the U.S.S.R. Now, as an independent country, Ukraine produces 3.2% of the world's wheat and exports much of it to Russia.

Why did the Russian government invade Ukraine?

The conflict in Ukraine began in late 2013, when the pro-Russia president, Viktor Yanukovych, was ousted after popular demonstrations against his alleged corrupt regime and his refusal to sign a European Union trade agreement. The country was divided between separatist, ethnic Russians living inside Ukraine, who are oriented toward Moscow, seeking independence and military assistance from Moscow, and ethnic Ukrainians seeking a fully independent country more aligned with Europe. Between 7,000 and 10,000 Russian soldiers, stripped of their insignias, invaded Ukraine in 2014. In the conflict, approximately 8,000 people have lost their lives, mostly civilians.

How many residents of Ukraine have been displaced from their homes since the Russian military invaded their country?

The United Nations does not explicitly track internally displaced people (IDPs), those who stay within the borders of their country yet are forced to leave their homes and towns because of nearby armed conflict. Since the Russian invasion of Ukraine in 2014, it is generally believed that an estimated 1.4 million people have been forced to leave their homes and flee to other, safer parts of their country.

Where is Crimea?

Protruding into the Black Sea, Crimea is a diamond-shaped peninsula attached by the Isthmus of Perekop to southern Ukraine. Though Crimea declared its independence from Ukraine in 1992, it compromised and became an autonomous republic of Ukraine. Later in 2014, Russia's president, Vladimir Putin, annexed Crimea and signed a law incorporating Crimea into the Russian Federation.

Why is there a tiny piece of Russia in the middle of Eastern Europe?

The important seaport of Kaliningrad (formerly named Königsberg), wedged between Poland and Lithuania on the Baltic Sea, was annexed by the U.S.S.R. at the end of World War II. Though once the capital of East Prussia and ethnically German, the Soviets quickly evicted the Germans and replaced them with ethnic Russians. In 1991, when the Russian Federation was formed, many of the autonomous republics within the former U.S.S.R. gained independence. Though Kaliningrad lies west of these new countries, its inhabitants are ethnically Russian and thus remained part of the Russian state. In 2013, Russia moved ballistic nuclear weapons into the enclave. The Russian Baltic Sea Naval Fleet is based in the port of Baltiysk. The city of Kaliningrad has nearly 448,000 residents, while the entire territory has 963,000 residents.

What is the largest landlocked country in the world?

Kazakhstan, part of the Russian Federation, is the ninth-largest country in the world and has no outlet to the ocean. It is over one million square miles (2.725 million square km) in area. While it is located adjacent to the Caspian Sea in Central Asia, the Caspian Sea is a landlocked sea. Approximately 18 million people live in Kazakhstan. The capital was moved from Almaty to Astana in 1997. Other large, landlocked countries include Mongolia in Asia, occupying an area of 604,000 square miles (1.564 million square km), and Chad in Africa, which occupies an area of 496,000 square miles (1.284 million square km).

What are the Baltic States?

The three Baltic States of Estonia, Latvia, and Lithuania are so named because they lie on the Baltic Sea. These three countries became independent after the Soviet Union broke apart in 1990–1991 and joined the European Union and NATO after 2004. The biggest cities by population include: Tallinn, Estonia (439,000 people); Riga, Latvia (641,000 people); and Vilnius, Lithuania (543,000 people).

EASTERN EUROPE

What two countries emerged from Czechoslovakia?

Czechoslovakia was created at the end of World War I (1918) by the Allies as a new country containing the area where the Czechs and Slovaks live. In 1967 and 1968, Czechoslovakia attempted to move away from communism, but the Soviet Union and Warsaw Pact countries invaded, squelching such aspirations. This "Prague Spring," as it was known, was the first military action taken by the countries of the Warsaw Pact. In 1993, the two republics of Czechoslovakia agreed to divide into two independent countries—the Czech Republic and Slovakia. The dissolution of Czechoslovakia was a peaceful one.

How has the word "bohemian" come to mean an unconventional person?

Though Bohemia is a region in the Czech Republic, the term "bohemian" is sometimes used to refer to an artistic or eccentric person. This term comes from the misguided belief that Gypsies (Roma) originated in Bohemia (it is now known that Gypsies originated in India rather than Bohemia). The stereotypes that are often attributed to Gypsies were then attributed to people from Bohemia, or Bohemians. Now, the word "bohemian" with a lowercase "b" is used to refer to an unconventional person, while the same word with a capital "B" refers to a person from Bohemia.

What playwright and author became president of an Eastern European country?

Czech poet and author Václav Havel was both the last president of Czechoslovakia (1989–1992) and the first president of the new Czech Republic (1993–2003). He won numerous literary awards for his dozens of poems, plays, and works of nonfiction. He was also awarded the U.S. Presidential Medal of Freedom, the highest honor bestowed by the president of the United States to a civilian, for his statesmanship. He died in 2011.

What two cities make up Budapest?

Budapest, Hungary, is actually two cities, Buda and Pest. The cities are separated by the Danube River; Buda is on the west bank and Pest is on the east bank. The province in which the twin cities are located is also called Budapest. The cities were joined in November 1873. Budapest is known as one of the most beautiful cities to visit in Europe and attracts more than 4.4 million tourists per year. Approximately 1.74 million people live in the city of Budapest, and approximately 3.3 million people live in the Budapest metropolitan area.

Who are the Magyar?

Magyars are the predominant ethnic group of Hungary. It is thought the ancient Magyar tribe may have originated in Asia thousands of years ago, somewhere east of the Ural Mountains. They share a language much different from any other in Europe.

Who is Lech Walesa?

In 1970, Lech Walesa was one of the leaders of a shipyard workers strike in Gdansk, Poland. Later, he organized workers in noncommunist labor unions, arguing for im-

A panoramic view of Budapest shows the Parliament on the right. Located in Hungary, Budapest is actually two cities (Buda and Pest) on either side of the Danube River.

proved conditions for workers. In 1980, he led the Gdansk shipyard strike, which ended with the government agreeing with his provisions on behalf of workers, including the right to strike and the right to form independent unions. The Catholic Church, which is very influential in Poland, supported his activities, including his formation of the Solidarity Movement. Walesa was later elected president of Poland and served in office from 1990 to 1995.

What are the Balkan States?

The Balkan States is a region in Southeast Europe, named after the Balkan Mountains that run east of Serbia to Bulgaria. The Balkan Peninsula lies between the Adriatic Sea (east of Italy) and the Black Sea. The countries on the peninsula are commonly referred to as the Balkan States (or the Balkans) and include Albania, Bosnia and Herzegovina, Bulgaria, Croatia, Greece, Kosovo, Macedonia, Montenegro, Romania, Serbia, and the portion of Turkey that lies in Europe.

What is balkanization?

Balkanization is taken to mean the fragmentation of a country into ethnic, language, or cultural divisions by territory. It is what happened to Yugoslavia, a former country in the Balkans that disintegrated into different territories and ultimately countries. These new nations engaged in war with each other, practicing forced deportations of ethnic groups and mass killings of unwanted residents. The term was created in the early nineteenth century and has been used to describe the collapse of many countries and empires since then.

Does Yugoslavia still exist?

In 1991, the republics that comprise Yugoslavia fell into disarray and aligned themselves on ethnic and cultural fronts. The old geographic divisions of the country fell apart, and war broke out between various states, including Slovenia, Croatia, Bosnia and Herzegovina, Macedonia, Serbia, Montenegro, and Kosovo. NATO was called in to lead airstrikes in both Bosnia/Herzegovina (1995) and Serbia (1999). Later, in 2006, Montenegro and Serbia also broke apart, and both declared their independence as separate countries in a referendum. Kosovo unilaterally declared its independence from Serbia in 2008. Over the course of just seventeen years, seven new countries were added to the map of the world.

What is ethnic cleansing?

Ethnic cleansing is the policy of forced migration, abuse, or murder of people in certain ethnic or religious groups within a country or region in order to make the geographic area more homogeneous. Ethnic cleansing in practice has been a part of our world history for many millennia. Infamous examples of ethnic cleansing include the mass killings in the former Yugoslavia in the 1990s, the mass killings in Rwanda in 1994, and the Armenian genocide in Turkey in the 1930s.

The Castle of Bran in Transylvania is the setting from which the Dracula myth sprang.

Why did Macedonia's name cause problems between that country and Greece?

When Macedonia declared independence in 1991, Greece felt indignant that a modern country would use what it felt was an historically Greek name. Greece blocked trade to Macedonia until 1995, when the two countries signed an agreement of understanding. The official name that the United Nations recognizes is the Former Yugoslav Republic of Macedonia, although internally it refers to itself as the Republic of Macedonia, its constitutional name.

Why does Romania have so many orphans?

The draconian population policies of Romania, which forbade birth control and abortion and required women to have five children, led to the birth of far more children than could be supported by the country. The population policies were implemented by communist dictator Nicolae Ceausescu, who ruled from 1965 until his capture and execution in 1989. The result of his policies has been thousands of Romanian children living in deplorable conditions. There are still more than 62,000 kids under the care of the state, with nearly 20,000 of them living in orphanages today.

Is Transylvania a country?

The home of Count Dracula, Transylvania, is an historical region located in central Romania. Transylvania is surrounded by the Transylvanian Alps and the Carpathian Mountains. A productive region in the western part of Romania, it contributes more than one-third of the total economic output of that country today. Approximately 7.3 million people live in this region today.

What are steppes?

Common throughout Russia, Asia, and central Europe, a steppe is a dry, short-grass plain that can be flat or hilly. While most steppes were once forested areas, climate, cultivation, and overgrazing by animals have left only short grasses and barren landscapes. The steppe climate is often too dry for forestation.

CHINA, AND EAST ASIA

What three countries have the most people?

China is home to approximately 1.355 billion people, which means that nearly one in every five people on the planet is Chinese. Its rate of growth in population is only .47% per year. India, the second most populous country, has approximately 1.252 billion people, with a growth rate of 1.25% per year, and is expected to surpass China's population sometime before 2050. Both countries have much larger populations than the third-largest country, the United States, which is home to approximately 321 million people.

How many people use cell phones in China?

China's adoption of the widespread use of technology is profound in this world. There are an estimated 1.2 billion cell phone users today, growing at a rate of 12% per year.

What was China's one-child rule?

In the late 1970s, the government of China decided that population control was needed because of a rapidly growing population that would soon outgrow the country's ability to feed itself. The policy mandated that every couple would only be allowed to have one child. The law exempted certain ethnic groups, rural families, and families where neither parent has siblings. Punishment was strict, varied widely, and was primarily economic. The one-child rule helped to stem China's population growth. Unofficially, since 2014, and officially in October 2015, China reversed this policy and now provides economic incentives to families wishing to have a second child to counter the effects of a declining labor force and an aging population.

The Forbidden City in Beijing, China, was home to emperors for centuries and was only opened up to outsiders in 1950.

What is the Forbidden City?

The Forbidden City, located in the center of Beijing, China, was the home of the emperor and the entire imperial court for nearly 500 years. Completed in 1420, the Forbidden City (also known as the Purple Forbidden City or Gugong) was the home of twenty-four emperors from the Ming and Qing dynasties. The last emperor, Puyi, resided there until 1924. For centuries, visitors were prohibited from entering the imperial city without permission from the emperor. In 1950, several decades after the last emperor was expelled, the city was made a museum and opened to the public.

Where can you see the Great Wall of China?

There are many places where tourists may see the northeastern part of the Great Wall of China that are within a one- to two-hour car ride from the capital city of Beijing. One of the most famous and most developed tourist areas is Badaling, where one can see the Great Wall of China, in its restored and untouched splendor, and the nearby Ming Tombs.

Is the Great Wall of China the only man-made object that can be seen from space?

Not very easily, but astronauts can sort of make it out. There are other man-made structures that are more visible from space, such as urban areas, highways, and the Pyramids of Giza in Egypt. Additionally, we can see many natural sites around the world, including the Himalaya Mountains, the Great Barrier Reef, and features of the Amazon River.

How long is China's Great Wall?

Recent archaeological and scientific analysis of both what remains of the Wall and the original foundation concludes that the Great Wall of China stretches approximately 5,500 miles (8,851 km). It averages 25 feet (7.6 meters) tall, is 15 to 30 feet (4.6 to 9.2 meters) wide at the base, and 10 to 15 feet (3 to 4.6 meters) wide at the top. The wall was originally erected as early as 600 B.C.E. to keep northern nomadic tribes from entering the Kingdom of China. The wall was expanded over the course of many succeeding centuries.

What is the Great Wall of China made of?

Walking along sections of the Great Wall that have been restored, it is possible to see that it is made of stones and mortar. In other more remote stretches of the Wall, it was hastily assembled with straw and mud, and in some places, even rice and mud. The trick of the builders was to make it just high enough to prevent people or horses from breaching the walls. In some stretches, the Wall is only a few meters tall. In other places, it is several stories tall.

What is the Terracotta Army?

Located in Xi'an, China, in Shaanxi Province, which was once the capital of the Chinese Empire for thirteen dynasties, the Terracotta Army is one of the great archaeological finds of the twentieth century. It was discovered by two farmers who were drilling for water in 1974. In the third century B.C.E., the first emperor of China, Qin Shi Huang, united all of China, established the longest-running form of government, built the Great Wall, and built his own elaborate tomb. As a symbol of his rule and to guard himself in the afterlife, Qin Shi Huang had an entire clay army replicated—approximately 8,000 soldiers, 130 chariots with 520 horses, and fifteen cavalry horses, all life size, some with remnants indicating a painted surface and each with unique facial expressions, to guard the first emperor's mausoleum at Mt. Li (Lishan).

The Great Wall of China is a structure that still impresses people from all over the world. While it can be seen from space, other man-made objects are more visible.

What is the Three Gorges Dam?

The Yangtze River is the site of the world's largest electricity-generating facility, the Three Gorges Dam. The financing of the project began in 1992 with construction beginning in 1994. The project was completed in 2012, at a cost of approximately $59 billion. The construction of the dam required the relocation of 1.3 million Chinese residents living upstream of the dam, as the rising waters flooded 13 cities, 140 towns, 1,350 villages, and countless archaeological sites. The reservoir is approximately 600 miles (965 km) long, spanning the Yangtze River in Sandouping, Hubei Province, China. The dam is approximately 594 feet (181 meters) high and 1.5 miles (2.3 km) wide. The water level of the dam varies, but typically, it is 557 feet (170 meters) deep. The environmental impacts of the dam are still being felt, with increased drought, less local rainfall, and mudslides that can endanger the millions of people who live nearby.

How much power does the Three Gorges Dam generate?

The Three Gorges Dam is the largest hydroelectric power-generating facility in the world and can produce 22,500 megawatts of power, more than eight times the power produced at the Hoover Dam in the United States.

How much does the United States and China trade with each other?

As a key trading partner of the United States, China exports approximately $467 billion worth of goods and services to the United States each year. The exports are primarily machinery, furniture, toys, and footwear. The United States exports about 26% of that amount or $124 billion to China each year, in the form of agricultural products, aircraft, machinery, and vehicles.

How dependent is the U.S. economy on China?

As the American appetite for Chinese-made products continues to grow, the government of China recirculates the money that we spend back into our economy by buying our debt. The U.S. has sold more than $12.7 trillion in debt in the form of U.S. government bonds in the world. All overseas investors in U.S. Treasury Bonds have invested approximately $6.13 trillion, of which China owns approximately $1.47 trillion or about 12% of the total.

How important is China to the biodiversity in the world?

China's territory covers 3,705,390 square miles (9,596,960 square km) of land. This vast expanse includes a wide diversity of wildlife, as well as 10% of the world's plants with stems, roots, and leaves. It is estimated that 15% to 20% of higher plants in China are endangered, and 44% of wild animals are in a state of decline today. Recently, China has established 2,640 natural reserves, covering nearly 15% of its total land area. China's environmental policies for managing its natural resources are very important for the health of the planet.

How long have communists been in power in China?

China's Communist Party began in 1921. Its communist revolution took place, after World War II, in 1949, after it drove out the Chinese Nationalists (Kuomintang Party) and Mao Zedong became the country's first "chairman." The Communist Party has been ruling China ever since.

What is the world's most commonly spoken language?

Approximately 1.2 billion people around the world speak Mandarin Chinese, the official language of China. This means that approximately 16% of the world's population speak Chinese. Other dialects widely spoken in China include Wu, which is spoken by approximately 77 million people; Min, spoken by more than 71.8 million people; and Yue, which is spoken by an estimated 60 million people. Other minority Chinese dialects include Jin, Xiang, and Hakka.

How much rice does China produce?

China is the world's leading rice producer and is responsible for about 26% of total rice production in the world. The country produces about 228.2 million U.S. tons (207 million metric tons) of rice each year. India, the next biggest producer, produces about 168.7 million U.S. tons (153 million metric tons). Ninety-five percent of the rice in the world is produced in Asia.

How is China changing our map of the world?

Since 2014, China has embarked upon an ambitious plan of filling in and reclaiming archipelago land in a highly disputed territory of tiny islands called the Spratly Islands in the South China Sea. The disputed area, about 500 miles (805 km) southeast of China's Hainan Island and east of Vietnam, is now home to airstrips, helipads, communication centers, and ports, which enable Chinese military air and sea patrols in the area. The islands are made by pumping sand from nearby coastal waters and dumping it on low-lying coral reefs that comprise much of the area in dispute.

What is the highest railroad in the world?

The Qinghai-Tibet Railroad is the highest railroad in the world, climbing through the Tanggula Pass at 16,640 feet (5,072 meters). It connects Lhasa, Tibet, to the rest of China

What is pinyin?

Pinyin is a system for transliterating Chinese characters into the Roman alphabet. It gradually replaced the Wade-Giles system in 1958, when the Chinese government started using pinyin for external press announcements. In the latter half of the twentieth century, it gained common acceptance and is the reason why we now call the Chinese capital Beijing instead of Peking (a Wade-Giles transliteration). The word *Pinyin* means "spelled-out sounds" and is now commonly used to teach Chinese around the world.

on 1,215 miles (1,956 km) of track. Because of the high altitudes of the sections of the track, each rail car includes a supply of oxygen for its passengers. The line also includes the highest railway tunnel in the world called Fenghuoshan Tunnel, which is located at 16,093 feet in elevation (4,905 meters). More than 1.5 million people used the train in its first year of operation in July 2006. The line continues to be expanded today.

Why isn't Tibet on the map?

Tibet is not on the map because it is no longer an independent country. Though Tibet was once a theocratic Buddhist kingdom, China annexed it in 1950. Tibet is now an autonomous region in southwestern China with a communist government installed by China. In addition to the destruction of the Tibetan Buddhist religion in the 1960s, over a period of many decades, China has moved Tibetans out of the area and moved ethnic Chinese into Tibet to help moderate Tibet's secessionist ideas.

What is Hong Kong's official name?

The official name of Hong Kong is Hong Kong Special Administrative Region of the People's Republic of China.

What transition took place within Hong Kong in 1997?

The sovereignty of Hong Kong, a former British territory, was transferred to China on July 1, 1997, following the expiration of a ninety-nine-year lease treaty signed in 1898

The prosperous city of Hong Kong was a British colony for about a century before being turned over, peacefully, to China.

by Imperial China and the United Kingdom. The treaty gave the United Kingdom perpetual use of the trading port of Hong Kong, as well as a section of land on the mainland of China known as the New Territories. Since the treaty for the New Territories was due to expire in the late 1990s, the United Kingdom and China entered into negotiations several years before in order to establish an orderly turnover of both Hong Kong (covered by an in-perpetuity agreement) and the New Territories (covered by the ninety-nine-year lease) back to China. China agreed to the terms of these negotiations, which granted Hong Kong special administrative rights, limited sovereignty, and self-rule, and allowed its economic structure to remain in place until 2047. Hong Kong was and continues to be one of the most developed cities in the world, as well as one of the world's great banking and financial centers. Recent archeological evidence suggests that Hong Kong has been inhabited for 35,000 to 39,000 years.

What is the oldest European settlement in eastern Asia?

In 1557, the Portuguese established the trading colony of Macau on mainland China at the mouth of the Xi (Pearl) River. Macau was a territory of Portugal from 1849 until 1999, when it was also returned to China. Today, it is one of the most densely populated places in the world, with more than 630,000 people living in an 11.6 square-mile (30.3 square km) area that comprises the city. Since 2006, Macau is the largest gambling city in the world and still one of the richest cities in the world.

How was Taiwan created?

In 1949, following the Chinese Civil War and the communist revolution in China, the Chinese Nationalist government, led by Chiang Kai-shek, fled to the island of Taiwan and established a Chinese country there. When the Nationalists arrived, they were met by indigenous Taiwanese and Han Chinese who had been living there for centuries. Although the island has never been controlled by modern China, successive Mainland Chinese and Taiwanese governments have been disputing this territorial issue since 1949. The People's Republic of China considers Taiwan to be a runaway state that will ultimately join the republic again. Until 1971, most governments of the world recognized Taiwan, the Republic of China, as a sovereign country. This changed when a majority of countries at the United Nations adopted a resolution recognizing the People's Republic of China, located on the mainland, as "China."

A view of Taipei, Taiwan. The island state off the coast of China is not recognized as a separate country by the United States.

Does the United States "recognize" Taiwan?

After the implementation of President Richard Nixon's policy to extend relations to the People's Republic of China and recognize the PRC as the legitimate government of all of China in the early 1970s, China has pressured all countries in the world to not recognize Taiwan, which has one of the most developed economies in the world. This is known in China as the "One-China Policy."

The United States extends de facto recognition to Taiwan through unofficial channels, which means they do not exchange ambassadors. The United States still considers Taiwan to be a very strong ally, economically and militarily. China has threatened the use of military action if the Taiwanese ever vote for independence from mainland China and, in effect, officially form their own country. Proposals for degrees of independence by the Taiwanese legislature have been met by many official threats from the Chinese government. Lately, Taiwanese politicians have forged closer ties to the People's Republic of China, although a vast majority of Taiwanese citizens reject any idea of becoming a part of China, wishing to remain independent. Today, approximately twenty-one countries in the world officially recognize Taiwan.

JAPAN AND THE KOREAN PENINSULA

What country will host the 2020 Olympic Games?

Japan will host the 2020 Olympic Games and is planning on featuring advanced robotic technology to surprise visitors in and around the Olympic Village, which is to be located in the capital city of Tokyo, Odaiba District.

How many islands comprise the country of Japan?

There are four main islands that comprise Japan and more than 6,000 smaller islands within its territorial boundaries. The northernmost island, Hokkaido, is home to the city of Sapporo. The largest island, Honshu, is the Japanese core area that includes such large metropolitan areas as Tokyo, Yokohama, Osaka, and Kyoto. Honshu is also the world's seventh-largest island by land area, as well as the second most populous island in the world with an estimated 104 million people living there. It is the "mainland" island of Japan and covers 86,246 square miles (223,656 square km), which makes it larger than the island of Great Britain. In the south are the islands of Shikoku and Kyushu. Kyushu is the southernmost island and was the first island where foreign traders were allowed to enter into Japan in the late nineteenth century. It has a population of over 13 million people with 1.5 million people living in its largest city, Fukuoka. Besides the four main islands, Japan includes more than 6,000 smaller islands. Japan's population is 127 million people and is declining at a rate of approximately 0.7% per year.

What is one of the world's most visited mountains?

Japan's Mt. Fuji, a sacred and important volcanic mountain to the Japanese, is the country's most popular tourist spot and one of the world's most visited mountains. More than 300,000 people visit the mountain each year. While dormant today, Mt. Fuji, which is in the shape of a nearly perfect cone, rises to 12,388 feet (3,776 meters) in height and last erupted in 1708.

Where is the land of the rising sun?

The Japanese name for Japan, Nippon or Nihon, which means "origin of the sun," evolved into "land of the rising sun." The name may be derived from the fact that for centuries, Japan was the easternmost known land and thus where the sun seemed to rise.

How geologically active is Japan?

Both active volcanoes and the potential for severe earthquakes threaten Japan. Japan has 110 active volcanoes, several of which have erupted in the last decade. Forty-seven of these volcanoes are actively monitored by scientists in Japan. Japan's largest volcano, Mt. Aso, which is 79.5 miles (128 km) in circumference and 4,941 feet (1,506 meters) in height, erupted in 2015, disrupting air traffic near the city of Kumamoto.

Earthquakes in Japan are also frequent occurrences, with many very destructive quakes in the last century. In 1923, a major earthquake (approximately 8.3 on the Richter scale) struck Yokohama and killed over 140,000 people. In 1995, an earthquake in Kobe killed 5,500 people. More recently, in 2011, an earthquake centered in the ocean, off the Pacific Coast of Japan, with a magnitude of 9.0 on the Richter scale, caused widespread damage in the prefectures north of Tokyo and killed nearly 16,000 people. It was the most powerful earthquake ever to be reported in Japan.

What is the life expectancy in Japan?

According to experts at the World Health Organization, Japan's overall life expectancy for both women and men is now the highest in the world. People in Japan live, on average, upward of eighty-four years.

How does Japan get its oil?

Having no oil itself, Japan must import all the oil it needs. To accommodate the amount of oil necessary, there is a constant stream of oil tankers spaced approximately 300 miles (483 km) apart that bring oil to Japan twenty-four hours a day, 365 days a year. Behind the United States, China, and India, Japan is the fourth-biggest importer of oil in the world, consuming more than 3.4 million barrels arriving every day.

Where is Iwo Jima?

Iwo Jima, one of the three islands that make up the Bonin Islands, is located southeast of Japan. A tiny island that housed a radar complex and airstrips, it is only 8 square miles (20.7 square km) in area. One of the deadliest battles of World War II was fought on Iwo Jima, with approximately 90,000 combatants. In the end, nearly 20,000 Japanese and 6,000 American soldiers were killed. The Japanese air base on Iwo Jima was captured by the United States on February 23, 1945. The island was returned to Japan in 1968.

Where was the first atomic bomb used on a populated area?

The Japanese city of Hiroshima was leveled by an atomic bomb dropped by the United States on August 6, 1945. Three days later, the United States dropped a second bomb on Nagasaki, Japan. While these events may have hastened the Japanese surrender that ended World War II, over 115,000 residents were killed immediately due to the blasts, and many thousands of people died later due to radiation-related diseases.

What is a bullet train?

Bullet trains, or shinkansen, as they are called in Japan, are similar to traditional passenger trains but have been enhanced to travel at speeds of up to 200 miles per hour (320 km per hour). The Japan Railway Group operates the trains, with approximately 342 daily departures and 1,323 seats per train. They have been used in Japan since 1964. The line that connects just the cities of Tokyo and Osaka carries approximately 151 million passengers per year.

The Atomic Bomb Dome, or Hiroshima Peace Memorial, serves as a reminder of the horrors of war and the frightening power of nuclear weapons.

How have the Kurile Islands kept World War II from ending?

Since 1855, when the Treaty of Shimoda was signed by both Japan and Russia, Japan owned this chain of four islands, located between Russia (south of the Kamchatka Peninsula) and Japan (north of Hokkaido). During World War II, the Soviet Union took control of the islands and hasn't given them back. Japan has been requesting their return but to no avail. Japan claims two of the southernmost islands and two small islets. Because of the Kurile Island controversy, the former Soviet Union never signed the 1951 Treaty of San Francisco, a permanent peace treaty that would effectively end the war between Japan and the Allied Powers, yet Japan and forty-eight other countries did.

How long did Japan occupy and colonize Korea?

Japan began its forced occupation of Korea in 1910 and was removed from the country at the end of World War II in 1945. During the occupation, many Koreans were put into forced labor, slavery, and military conscription. It is estimated that between 200,000 and 800,000 people died, both Korean and Chinese, at the hands of the Japanese in Korea and the Manchuria region of China during the occupation.

How old is the Korean civilization?

Ancient hominid evidence suggests that people have been living in Korea for more than 300,000 years. Pottery found at archaeological sites indicate evidence of a distinct Korean culture as early as 10,000 B.C.E.

How did North and South Korea come to be?

From 1392 to 1910, the Korean Peninsula was the home of the Choson (Joseon) Kingdom. In 1910, Japan took control of the peninsula but lost this territory after its defeat in World War II. At this time, North and South Korea were divided by a line that lies near the latitude of 38 degrees north. This latitude marked the line dividing the Soviet occupation zone in the north and the American occupation zone in the south. From 1950 to 1953, the communist North Koreans fought with the democratic South Koreans. With American forces involved in the Korean War, North Korea's army was eventually forced to retreat into China, though it later recaptured land to the 38th parallel, where the border between the two Koreas remains today.

How does South Korea rank in the world economy?

South Korea ranks twenty-fourth in terms of its per capita GDP (gross domestic product) of $27,970. It is also the United States' sixth-largest trading partner. The economy of South Korea expands at 2.6% per year.

What is South Korea's national slogan?

The English-speaking world sometimes refers to South Korea by using a national slogan, coined in 1885, as the "Land of the Morning Calm."

What are some of the largest South Korean companies?

Daewoo, Hanjin Shipping, Hynix, Hyundai, Kia, LG (Lucky Goldstar), POSCO, and Samsung are South Korean companies with global brand identities. They are some of the largest companies in the world today. For example, POSCO is the fourth-largest maker of steel in the world, and Hynix is the second-largest maker of memory chips used in electronic devices in the world.

What is Truce Village?

The Joint Security Area, or Truce Village, is the nickname given to Panmunjom, a former village destroyed during the Korean War that is 33 miles (53 km) from the capital city of Seoul. It marks the spot along the 151-mile (243-km) border between North and South Korea, and it is the only area between the two countries that the United States still considers a combat zone. The border area strip, which is 2.5 miles (4 km) wide, is patrolled by military forces on both sides. More than 1,000 North and South Koreans have been killed in this area since fighting "ended" in 1953. Approximately 100,000 tourists visit the site each year.

Truce Village is the site of decades of diplomatic negotiations and meetings, numbering in the hundreds, between the totalitarian regime in the north and the democratically elected government of the south. The parties hope to resolve issues ranging from nuclear disarmament of the peninsula and reunification to the blaring of propaganda over loudspeakers across each other's borders.

Who was Kim Il-sung?

He was the supreme leader of the Democratic People's Republic of Korea (North Korea) from its inception in 1948 until his death in 1994. His son, Kim Jong-il, took power after his death and led the republic until his death in 2011. His son, Kim Jong-un, ascended to lead North Korea in early 2012 and is the current leader.

Kim Il-sung was worshipped by his people under a state ideology called Juche (national self-reliance), which sought to establish a unique North Korean political movement independent of Marxist-Leninist ideals and the de-Stalinization of the Soviet Union in the late 1950s.

How many American troops are stationed in South Korea?

Currently, 28,500 U.S. troops still occupy South Korea, but only a small fraction of the forces are actually stationed in the demilitarized zone.

What catastrophe happened in North Korea in 1995?

Because of its isolation from former trading partners in the Soviet Union, coupled with poor relations with neighboring China, failed economic policies that drove up the price of food, excessive military expenditures, and severe floods and droughts, North Korea saw the production of grain fall below what was needed to sustain the population in 1995. This triggered food shortages and caused the deaths of approximately 300,000 people. It is not clear how many people died during the three-year famine because the North Korean government does not report these types of figures.

The United States gave approximately $600 million to North Korea to help prevent mass starvation. The largest donor country in the world, the United States contributed more than 50% of the total aid given to North Korea to help avert the famine. Severe drought conditions occurred again in 2015. With various U.N. and U.S. sanctions against the repressive government, coupled with similar environmental conditions and government mismanagement of agricultural production, famine and starvation are likely to happen again.

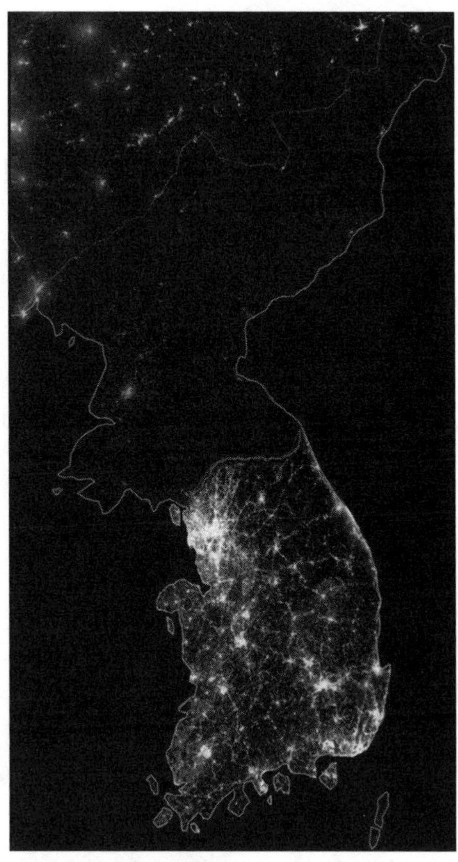

This photograph of the Korean Peninsula taken from space is a graphic depiction of the stark difference between the prosperous South Korea and the communist North Korea.

Who provides food to North Korea today?

The top exporters of food to North Korea are China, Japan, and South Korea.

What is the Arirang Festival?

The Arirang Festival or Arirang Mass Games, held every April 15 and lasting for two months, is a celebration including performances by thousands of students who create gigantic mosaic pictures by holding colored mosaic cards while standing in an enormous stadium. There are also music and dance performances during the ceremony. For this event, the North Korean government has even allowed some American tourists to attend. The event was not held in 2014 and 2015.

What is the least densely populated country in the world?

Mongolia (not to be confused with Inner Mongolia, which is an autonomous region of northern China), with its tiny population of 3 million people spread over 603,902 square miles (1,564,100 square km) of land area, has a population density of fewer than two people per square mile. It is the least densely populated country in the world. Mongolia's population dispersion is limited because only 1% of the land in the country can be used for agriculture; the remainder of the country is dry and used for nomadic herding. Mongolia was originally established in the thirteenth century when Genghis Khan overtook and unified much of mainland Asia.

Where is Ulaanbaatar?

Ulaanbaatar is the largest city and capital of the Republic of Mongolia, with more than 1.3 million residents. The city is situated on the Tuul River, at an elevation of 4,300 feet (1,310 meters) above sea level. It was established in 1639 as a nomadic Buddhist center and was established as a permanent city in 1778. Today, tourists may visit historic Buddhist centers in and near the city.

How big was the Mongol Empire?

The Mongol Empire, under the reign of Genghis Khan (1162–1227 C.E.; also spelled Chingis Khan) and his son Ügödei Khan (1186–1241 C.E.), became one of the greatest contiguous empires in world history. It would eventually stretch from what is now modern-day Korea and China in the east to Poland in the west and to Vietnam and Oman in the south.

Who was Kublai Khan?

Kublai Khan (1215–1294 C.E.) was the grandson of Genghis Khan. Under his rule, the Mongol Empire reached its peak in 1279. He founded the Yuan Dynasty, moved its capital to Beijing in the fourteenth century, and would later move it back to Mongolia, as his empire fell.

Kublai Khan ruled the Mongol Empire in the thirteenth century. His reign included the final years of the Pax Mongolica, a time of peace and prosperity that benefited lands from China to India all the way to Hungary.

SOUTHEAST ASIA

What is Indochina?

Indochina is the name of the peninsula in Southeast Asia composed of Myanmar (Burma), Thailand, Cambodia, Laos, Vietnam, and the mainland portion of Malaysia. During the colonial era, the eastern portion of Indochina was ruled by France, and parts of the west were controlled by the United Kingdom.

Why was the Vietnam War fought, and what were some of the consequences?

After World War II, Vietnamese Nationalists and communists fought the French to gain their freedom from European colonization. Ultimately, the French left Vietnam in 1954 after one hundred years of colonial rule, and the former colonial territory became divided into the primarily communist-controlled North and the pro-West forces in the South. The United States entered Vietnam in 1956 to train South Vietnamese military forces, wanting to help its ally France to regain control of the colony by sending in advisers and aiding in military supplies. Forces in the South saw the chance to gain more American assistance, and they convinced successive U.S. presidential administrations to assist in ridding the country of their enemies. In an effort to stem the spread of communism, the United States continued to support South Vietnam until the fall of Saigon in 1975.

Ironically, during the height of the Vietnam War in 1970, while communist Vietnamese were being portrayed as a threat to the American way of life, President Richard Nixon and Secretary of State Henry Kissinger were courting the largest, most radical communist country, the People's Republic of China, to extend relations with the West. China had been funding and supporting militarily the Vietnamese resistance against both French and American forces for decades.

The American occupation of Vietnam lasted from 1962 to 1975. In 1975, Saigon was overrun by communist forces, and the country was eventually unified under one communist regime. The aftermath and legacy of the war in Vietnam was the destabilization of, and genocide in, neighboring Laos and Cambodia, which cost millions of lives during the upheaval of the 1970s, as well as the displacement of millions of Vietnamese, some of whom emigrated to the United States as "Boat People."

What is Angkor Wat?

Angkor Wat is the location of the archaeological remains of different capitals of Cambodia's Khmer Empire, which lasted from the ninth century to the fifteenth century C.E. The remains cover an area of 154.4 square miles (400 square km) in the northwestern part of the country. It is a UNESCO World Heritage Site and includes the famous Angkor Wat Temple and, at Angkor Thom, the Bayon Temple.

Where is Brunei?

Brunei lies on the northern tip of the island of Borneo, just south of the Philippines. It shares Borneo Island with four provinces of Indonesia (West, Central, South, and East Kalimantan) and two states of Malaysia (Sabah and Sarawak).

Which Southeast Asian country is one of the world's richest?

The tiny nation of Brunei, comprising 2,226 square miles (5,765 square km) in area, located on the island of Borneo, is one of the world's richest countries, with a per capita

When did Burma become Myanmar?

In 1989, the name of the country of Burma changed its name to Myanmar when a military junta took control of the country, following the president's resignation as a result of riots and national turmoil. Although the military has ceded control of the country in recent years to civilian authorities, many of the government leaders are former military officials. The country's official name is the Republic of the Union of Myanmar, with Naypyidaw as its capital. Yangon (formerly Rangoon) is the largest city in the country. The population of the country is estimated to be approximately 51 million people. The country shares borders with Bangladesh, India, China, Laos, and Thailand.

income of $39,355. Brunei's wealth is based on its oil and gas exports. Sultan and Prime Minister His Majesty Paduka Seri Baginda Sultan Haji Hassanal Bolkiah Mu'izzaddin Waddaulah, who is the ruler of Brunei, is one of the richest men in the world.

What is Bandar Seri Begawan?

The capital of Brunei Darussalam and its largest city is Bandar Seri Begawan, or Bandar for short. Approximately 415,000 people live in Brunei. It is said that during the Vietnam War, the airport in Bandar frequently picked up the traces of U.S. and Vietnamese fighter airplanes engaged in missions nearby.

What is an economic tiger?

An "economic tiger" is a term applied to three rapidly developing Asian nations, each with the power and ability to become influential, international economic leaders. South Korea, Taiwan, and Singapore are considered to be the three economic tigers. Hong Kong was once part of this Pacific Rim group, but since its merger with China, it is no longer included in this group.

What part of Southeast Asia is one of the most contested geographic regions in the world?

The Spratly Islands are a group of one hundred small islands, islets, and reefs in the South China Sea. They are located between Vietnam, the Philippines, and East Malaysia. Though only comprised of 2 square miles (5.2 square km) and spread across 158,000 square miles (400,000 square km) of the South China Sea, these islands are of strategic importance. Ownership is contested by Vietnam, China, the Philippines, Malaysia, and Taiwan. Oil fields that yield nearly 15% of the petroleum used by the Philippines have been the center of the dispute. Each country feels that it should be able to negotiate and profit from the lucrative contracts and benefits that oil exploration and production might bring. Of the approximately 200 islands that comprise the Spratly group, Vietnam

What is the only Catholic country in Asia?

The Philippines is the only Catholic country in Asia— approximately 83 percent of its population is Catholic. Catholicism was firmly implanted in the Philippines when the land was under Spanish rule, from the sixteenth through nineteenth centuries.

occupies forty-eight, Taiwan occupies one, the Philippines occupies eight, Malaysia occupies five, and China occupies eight. This is also in the area where China, since 2014, is creating islands for military use today. Reuters reports that more than $5 trillion worth of products traverse this area each year, so it is of great strategic importance to many countries, including the United States, and to the world economy.

How many islands comprise the country of the Philippines?

The Philippines are comprised of 7,107 islands. Only 1,000 of these islands are inhabited, and a further 2,500 still remain unnamed. The islands are divided into three main groups: the Luzon region in the north, the Visayan region in the middle, and the Mindanao and Sulu region in the south. There are approximately 100 million people living in the Philippines today.

When did the United States control the Philippines?

The Philippines were a Spanish colony until 1898, when the United States paid Spain $20 million and took control of the islands at the close of the Spanish–American War. The islands remained under American control through 1946 (except for a two-year period of Japanese occupation during World War II). The Philippines became independent in 1946 and leased land to the United States for military bases until 1992, when the U.S. military presence in the Philippines ended.

How many islands comprise the country of Indonesia?

Indonesia is composed of over 17,508 islands. Of these, only approximately 6,000 of the islands are inhabited. The country is organized into thirty-four provinces. Indonesia is the world's largest archipelago and was formerly known as the Dutch East Indies because of the colonization of the Dutch in the early seventeenth century C.E. The islands had been under the control of the Netherlands since around 1600, but Indonesia declared its independence in 1945 (after being subjected to Japanese invasion and occupation during much of World War II).

What island has the most Muslim inhabitants in the world?

The Indonesian island of Java has the most Muslim inhabitants in the world, with a total population of over 150 million people. Indonesia's capital city, Jakarta, is located on the island of Java.

What is Southeast Asia's largest oil-producing country?

Indonesia produces approximately 800,000 barrels of oil per day, making it the biggest oil producer in Southeast Asia today and putting in the top twenty-five oil-producing countries in the world.

Indonesia, including the island of Java, has more Muslims than any other country in the world.

What is the world's largest Islamic nation?

Indonesia, the world's fourth most populous country, is the world's largest Islamic nation, with a population of approximately 255 million people. About 87% of all Indonesians are Muslims. Islam spread to Indonesia around the thirteenth century C.E.

What was East Timor?

East Timor, the former name of the country, is now the Democratic Republic of Timor-Leste, a country adjacent to the most eastern province of Papua, Indonesia. After being under the control of Portugal for four centuries, the region came under Indonesian rule. Later, in 1975, East Timor declared independence. Indonesia consequently invaded, causing the deaths of about 100,000 people. In 1991, the international community condemned Indonesia for these military actions and the additional massacres and human-rights violations of the residents that followed. Indonesia eventually relinquished control of East Timor, and Timor-Leste became an independent republic in 2002. With a per capita GDP of approximately $1,100 per year, Timor-Leste is one of the poorest countries in the world. Its capital is Dili.

Which country has more languages than any other in the world?

Papua New Guinea is the most linguistically diverse country in the world and is home to more than 850 different languages. The most common language spoken is Tok Pisin. Although the use of English is designated as an official language of the country, it is only spoken by a small fraction of residents. Papua New Guinea is a very poor country where only approximately 57% of the people are literate. The capital of Papua New Guinea is Port Moresby.

What is the smallest country in Southeast Asia?

The Republic of Singapore is the smallest country in Southeast Asia. It covers only 272 square miles (707 square km) of land and lies only 85 miles (137 km) north of the equator.

Where are the largest twin towers in the world?

A twin tower is an architectural design that consists of two adjacent towers, with identical features and height. The highest twin towers in the world are named the Petronas Towers and are located in the capital city of Kuala Lumpur, Malaysia. They stand 1,483 feet (451.9 meters) in height. From 1998 until 2004, it was the tallest building in the world. The steel-and-glass, postmodern-style design is based upon concepts in Islamic art, reflecting Malaysia's culture.

Is Singapore a city or a country?

Singapore, a city-state, is both a city and a country. Born out of the extreme hardship and mass killings by the Japanese during World War II, the country was taken back under United Kingdom control after the Japanese defeat. In 1963, Singapore merged with the states of Malaya, Sabah, and Sarawak to form what is the country of Malaysia today. In 1965, Singapore split from Malaysia and formed its own country. With a population of approximately 5 million people, it is one of the most wealthy and developed countries in the world.

INDIA AND WEST ASIA

What is the Taj Mahal?

Located in Agra, India, the Taj Mahal is a mausoleum for the wife of the Mogul Emperor Shah Jahan. After the death of his wife, Arjumand Banu Begam, in 1631, Shah Jahan began construction of the mausoleum in 1632. Over 300 feet (91.4 meters) tall, the white marble mausoleum is a grandiose and striking memorial to her life and death.

What makes New Delhi so new?

From 1773 to 1912, the capital of India was Calcutta, but it was later moved to Delhi in 1912. Because the British wanted to build a brand-new capital city, they began to construct a new city adjacent to Delhi (now known as Old Delhi). When construction was completed on this new city in 1931, it became the capital of India and was known as New Delhi. The metropolitan area of Old and New Delhi is one of the world's largest urban areas, containing a population of over 24 million people.

Where did Bombay go?

In 1996, India changed the name of the third-largest metropolitan area (population 21 million) from Bombay to Mumbai. One of India's largest corporations, the Tata Group, is headquartered in Mumbai.

Bollywood performances are not only popular in India but also other countries. In this photo, the Indian Bollywood Film Star Song and Dance Troupe perform on stage in Beijing, China.

What is Bollywood?

Known as "Bollywood," Mumbai, India (previously called Bombay), is the world's movie capital. It began producing movies in 1913. The entertainment industry in India produces more than 1,000 films annually, about twice the number of films produced in Hollywood.

Where is Dum Dum Airport?

Each year, over five million passengers pass through Dum Dum International Airport in Kolkata (Calcutta), India. The official name of Kolkata's airport is Netaji Subhas Chandra Bose International Airport.

Who are the biggest exporters and importers of rice in the world?

India currently exports approximately 12.7 million U.S. tons (11.5 million metric tons) of rice per year. The second-biggest exporter of rice in the world, also located in Asia, is Thailand. It exports about 12.1 million U.S. tons (11 million metric tons) of rice per year, about one-quarter of the total rice that is exported in the world.

Does India have a population control program similar to what China implemented?

Though the Indian government does not limit births to one child per family, similar to what was implemented in China, it does have one of the oldest population control programs in the world. The program, begun in the 1950s, encourages the use of birth control and family planning to limit the growth in the country's population. Local Indian governments provide grants to people who undergo sterilization surgery. Approximately 37% of women in India have had the procedure. Although India's population is growing more slowly (1.6%) than in previous years, the world's population growth rate is still lower, at approximately 1.1%.

How does India's caste system work?

The caste system is the extremely rigid, hierarchical social class system of India. The caste system has its origins in approximately 1200 B.C.E. and is based upon the ancient Hindu text *The Laws of Manu*. The caste system consists of four categories: Brahmans (priests and teachers), Kshatryas (warriors and rulers), Vaishyas (farmers, traders, and merchants), and Shudras (servants and laborers). There are many people who are considered to be outside of the caste system, or who have no caste, called the Harijans, or "Untouchables." The Untouchables do the most menial jobs in the society and generally live among each other. The caste system in India dictates not only one's profession, but also whom one can marry, social contacts, and all other aspects of life.

How many Internet users are there in India?

There are approximately 375 million regular users of the Internet in India, and according to an Indian trade group, reported by the *Wall Street Journal*, this number increases to an estimated 400 million users through 2016 because of the growth in the number of smartphones in use. Approximately 276 million people in India access the Internet on a mobile device today.

Where is the world's second-highest mountain?

A mountain named K2, at 28,250 feet (8,611 meters), is the world's second-tallest mountain. It is also known by several other names, including Chhogori, Qogir, Ketu, and Kechu. K2 sits near the disputed Jammu and Kashmir region at the borders of northern India, Pakistan, and China.

Who are the Sherpa?

The Sherpa are an indigenous ethnic group in Tibet and Nepal. They live high among the mountains of the Himalayas and, because of this fact, are often hired as guides for climbing expeditions to such peaks as Mt. Everest and K2. In 1953, Tenzing Norgay (Sherpa) and Edmund Hillary (British) were the first two people to reach the 29,029-foot (8,848-meter) summit of Mt. Everest. There are an estimated 113,000 Sherpa living in Nepal. The capital of Nepal is Kathmandu.

Does a cashmere sweater come from Kashmir?

The Kashmir goats, which make the fine wool known as cashmere, do come from the Kashmir region in India (a disputed area not controlled by any one nation). This region, which is located in northern India bordering on Pakistan, is plagued by violence and unrest. In 1947 the British decided to split the colony of India into two separate countries—one Hindu (India) and one Islamic (Pakistan). The state of Jammu and Kashmir has a mixed population of Hindus and Muslims, which has led to conflict within the region. India and Pakistan have waged three wars over this region, and sporadic violence continues today. China and Mongolia produce approximately 87% of the cashmere wool in the world today.

What is Gross National Happiness?

The term "Gross National Happiness" is a phrase that was coined by the former king of Bhutan, Jigme Singye Wangchuck, in the early 1970s, to refer to a measurement of development that utilizes Buddhist principles, encapsulated in several pillars of social-economic development that contribute to overall happiness of the people in a country. These categories include sustainable development, the preservation and promotion of cultural values, respect for the natural environment, and good governance of the people. The king said that "Gross National Happiness is more important than Gross National Product." In simple terms, the king asserted that it is important to have social-economic development, while attending to values such as "kindness, equality, and humanity." Bhutan's capital is Thimphu.

What are the principal industries in the Maldives?

Tourism and fishing contribute the most to the Maldives's GDP, which is approximately $2.854 billion per year.

How many islands make up the Maldives?

There are approximately 1,192 islands and islets, spread across 35,000 square miles (90,000 square km) of area within the territory of the Maldives, of which only 250 are inhabited. The capital of the country is Malé. Of the inhabited islands, some are actually hotel resorts only, as many of the islands have only enough space for one or two hotels. The Maldives lie about 600 miles (965 km) southwest of India.

Which country has the lowest maximum elevation in the world?

The Maldives has a natural maximum elevation of only 7.9 feet (2.4 meters) above sea level. The problem is that sea levels over the past twenty-five years are rising at a rate of approximately .14 inches (3.5 millimeters) per year, which may make the country, or parts of it, disappear entirely. The 2004 Asian tsunami caused the ocean to completely cover parts of the Maldives. Only nine of the more than 1,192 islands were not impacted.

What is a dhoni?

A dhoni or doni is a traditional, wooden, motorized Maldivian fishing boat, sometimes used with sails.

What language is spoken in the Maldives?

Dhivehi, an Indo-Aryan language, is spoken by the 341,000 inhabitants of the Maldives.

Why does Bangladesh flood so often?

Given that most of Bangladesh's elevation is near sea level and that it is located within the delta of the Ganges and Brahmaputra rivers, it is not remarkable that the country is easily flooded by regular monsoons (periodic, seasonal heavy rains often accompanied by strong winds) and hurricanes. Unfortunately, Bangladesh also suffers from a poor emergency warning system, so people in Bangladesh are not adequately warned of impending disasters, causing many fatalities for residents each year.

Where was East Pakistan?

When the British left South Asia, they divided the region into India and Pakistan. Muslim regions in Pakistan were located to the east and west of Hindu India. These two separate territories became East and West Pakistan, and they were separated by approximately 1,000 miles (1,600 km). Having been extremely geographically separated from West Pakistan for over two decades, East Pakistan declared independence and became Bangladesh at the end of 1971.

What does the suffix "-stan" mean?

The suffix "-stan" (as in Afghanistan) is from a Persian word meaning "place of" or "country." Therefore, Afghanistan literally means "land of the Afghans."

How many countries end in the suffix "-stan"?

There are eight. Six are republics of the former Soviet Union: Kazakhstan, Turkestan, Turkmenistan, Kyrgyzstan, Tajikistan, and Uzbekistan. Afghanistan and Pakistan are the two others.

Where did Osama bin Laden spend the last days of his notorious life?

Osama bin Laden, the mastermind behind the destruction of New York City's World Trade Center, was hiding in plain sight, with assistance and protection of elements high in the Pakistan government, in a house in the city of Abbottabad, Pakistan, in the northeast part of the country, only minutes away from Pakistan's capital, Islamabad. In May

2011, he was killed by U.S. Navy Seals, who stormed his residential compound in a neighborhood in the city.

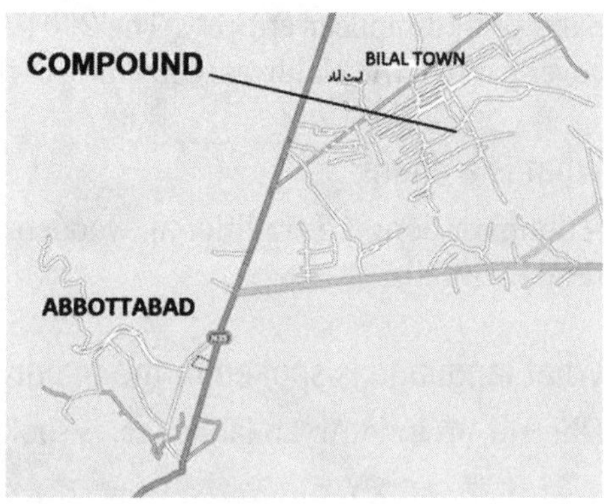

The compound where terrorist leader Osama bin Laden was found and killed was located near Abbottabad, Pakistan.

Why did the Soviet Union invade Afghanistan in 1979?

The Soviet Union sent troops to Afghanistan in 1979 because it wanted to come to the aid of its ally, the Parcham faction of the People's Democratic Party of Afghanistan, which was more moderate in its outlook toward moving the country toward communism. The Parcham faction had signed a treaty of friendship and cooperation with the U.S.S.R. the previous year. The United States began to fund insurgent groups, the Mujahideen resistance, who were opposed to both the current Afghan government and the Soviet occupation of their country. These same Mujahideen would later form the ideological corps of the people who masterminded the attack on the World Trade Center in 2001.

How many people lost their lives during the Soviet occupation of Afghanistan?

An estimated 850,000 to 1.5 million Afghani civilians were killed during the war, which lasted until 1989. Of the 600,000 Soviet troops who served in Afghanistan, 14,453 lost their lives, and more than 469,000 became sick or were wounded. More than five million Afghan people were displaced and fled to Pakistan and Iran as refugees. Another two million Afghan civilians were displaced within their own country, seeking shelter from the violence of the war and the factional fighting that ensued.

What were the Buddhas of Bamiyan?

The Buddhas of Bamiyan were two gigantic sculptures—one measuring 180 feet (55 meters) high and the other 121 feet (37 meters) high—that were built in the sixth century C.E. Located 143 miles (230 km) northwest of Kabul, Afghanistan, they were designated as a UNESCO World Heritage Site and were one of the great archaeological and religious sites in the world. In 2001, the Taliban regime in Afghanistan ordered their destruction, and they were destroyed.

Why has Afghanistan been contested and invaded so many times?

Afghanistan sits at a crossroads linking Asia with the Middle East. For millennia, it was the major transit point along the Silk Road, which brought goods from Asia to and from the Middle East. In recent times, it has been treated as a geographically strategic location because of its potential for influencing the policies of many countries in both re-

Why is the Aral Sea shrinking?

The area of the Aral Sea, located on the border of Kazakhstan and Uzbekistan, has been reduced in size by one-half since 1960. Though it was once the world's fourth-largest lake, diversion of its feeder rivers for nearby agricultural irrigation has severely shrunken the lake. This shrinkage has exposed soil saturated with salt, which now destroys plants and vegetation across the nearby plains. A heavily polluted lake, it represents one the world's great ecological disasters today.

gions. It is an ethnically diverse region that is home to many tribes and cultures, all of whom have been vying for some form of control or voice in the way in which their country is governed. Therefore, in the past several hundred years, major geopolitical players have sought to control, occupy, or colonize this country. Much of what we see happening in Afghanistan today has roots in conflicts dating back many centuries.

THE LAND AND ITS HISTORY

What is the difference between the Near East, Middle East, and Far East?

At one time, it was common to refer to the Near, Middle, and Far East to describe broad parts of the world. Though two of the terms have fallen into disuse, the term "Middle East" is still commonly used. The Near East, at its greatest extent in the sixteenth century, once referred to the Ottoman Empire, which included Eastern Europe, Western Asia, and part of Northern Africa. The Middle East referred to the area from Iran to Myanmar (formerly Burma). The Far East used to refer to Southeast Asia, China, Japan, and Korea.

Where exactly is the Middle East?

It is generally agreed that the Middle East includes Egypt, Israel, Palestine, Syria, Lebanon, Jordan, the countries of the Arabian Peninsula (Saudi Arabia, Yemen, Oman, the United Arab Emirates, Bahrain, and Qatar), Iraq, Kuwait, Turkey, Cyprus, and Iran. Most regional specialists also include the countries of northern Africa (Morocco, Algeria, Tunisia, and Libya). The new countries of Azerbaijan, Georgia, and Armenia (former Soviet republics) are sometimes included in the region.

Is the Middle East a desert?

Actually, very little of the region is filled with sand dunes and sand storms. Coastal areas are temperate, and several areas boast very pleasant and moist Mediterranean climates. A lack of water is a problem for agriculture throughout the region, so some wealthier countries are implementing technological solutions such as desalinization plants and drip agriculture for water conservation.

What is the Fertile Crescent?

The Fertile Crescent is an area located in a crescent-shaped region between the Persian Gulf in the east and northeastern Egypt in the west. The development of this region, mostly located along the Tigris and Euphrates rivers, was begun in approximately 10,000 B.C.E. and was based on the availability of water from these rivers and the fertile soil they deposited. This area was a center of human civilization, sometimes referred to as the "Cradle of Civilization," and was the location of the ancient Mesopotamian Empire. The region is widely believed to be the location where urbanization, agriculture, written language, trade, scientific inquiry, historical record, and religion first began. The term was first used and widely publicized in the early twentieth century C.E.

How far did the Islamic Empire spread?

Although not a unified empire controlled by one person, at its widest extent, the Islamic empire and its cultural/economic influence included most of northern Africa, Spain and Portugal, the Balkan Peninsula, India, Indonesia, Kazakhstan, and the southern reaches of Russia. Islamic influence also spread east into western China. The areas that were under Islamic rule from the eighth through the sixteenth century C.E. still keep the Islamic religion as a major aspect of their culture today.

What was Mesopotamia?

This ancient region is situated between the Tigris and Euphrates rivers, from contemporary southern Turkey to the Persian Gulf, and includes modern-day Turkey, Syria, and Kuwait. Mesopotamia was the center, at one time or another, of such civilizations as Babylonia, Assyria, and Sumer. The area is often credited with such developments as the invention of the wheel, cereal crop cultivation, mathematics, and astronomy.

What was the Byzantine Empire?

After the fall of Rome in 476 C.E., the eastern portion of the Roman Empire became the Byzantine Empire. The capital city of the Byzantine Empire was moved far east, to Constantinople (now Istanbul, Turkey). At one point, the empire included most of the eastern and southern coast of the Mediterranean Sea. The empire shrank in size until 1453, when Constantinople was conquered by the Ottoman Empire.

What was the Ottoman Empire?

The Ottoman Empire began as a tiny state in the fourteenth century. It was centered in the city of Bursa in what is now northeast Turkey, and it rapidly expanded through the conquest of neighboring states. Its greatest expansion was in the sixteenth century, when it included southeastern Europe, the Middle East, and North Africa. Due to wars

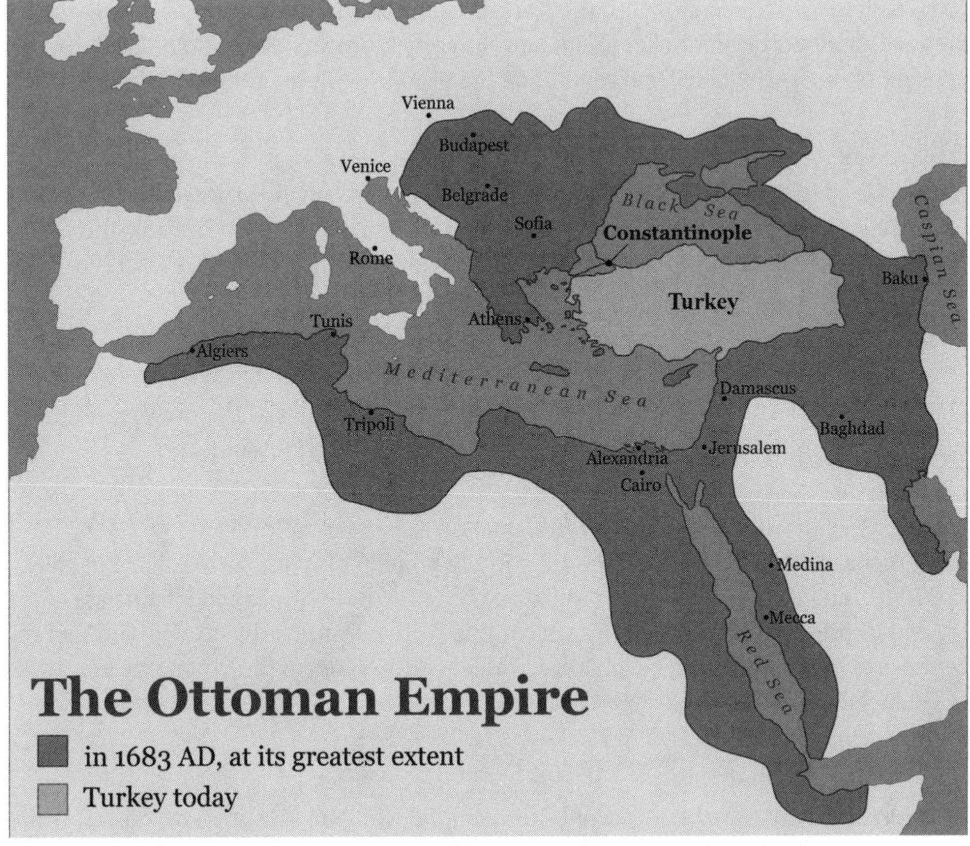

The extent of the Ottoman Empire in the seventeenth century rivaled that of the earlier Roman and Byzantine empires.

with other European countries in the seventeenth and eighteenth centuries, the Ottoman Empire began to decline and became known as the "sick man of Europe." The Empire's successor was Turkey, which became an independent country in 1922.

RELIGION

What religions have their roots in the Middle East?

Judaism, Zoroastrianism, Christianity, and Islam all have their roots in the Middle East. The region is consequently filled with sites that are considered holy by all four religions.

What is Zoroastrianism?

Zoroastrianism is a religion that began in present-day Iran around 1200–1500 B.C.E. It is based upon the teachings of the prophet Zoroaster and is an ancient religion that some believe may have influenced the later religions of Christianity and Islam. Zoroaster believed in monotheism (belief in one omnipresent, compassionate, and just God), the concept of dualism (good versus evil), and the free will of humankind.

Which religions came to be practiced first?

Before the advent of Judaism, many earlier religions were practiced, including animism, which is the worship of nature and natural phenomena like fire, wind, water, and earth. Judaism came into practice around 5000 B.C.E., followed by Zoroastrianism in 1200–1500 B.C.E., Christianity in 30 C.E., and Islam in 600 C.E.

What is the difference between Islam and a Muslim?

Islam is the name of the religion that adheres to the teachings of the prophet Muhammad, and a Muslim is a person who is a follower of the teachings of Islam.

Who was the prophet Muhammad?

The prophet Muhammad was the founder of the religion of Islam. He was born in Mecca in 570 C.E. and fled to Medina later in the year 622 C.E. Both Mecca and Medina are cities located in present-day Saudi Arabia. According to the Islamic religion, Muhammad received prophecies from God, which were subsequently written in the Quran, the holy book of Islam. Muhammad's death in 632 C.E. led to the expansion of Islam around Eurasia.

Who are the Sunni and Shi'a Muslims?

The Sunni and Shi'a are two sects of Islam. About 87–90% of Muslims are Sunnis, while Shi'ism has 10–13% of Muslim adherents and is especially prevalent in Iran. After Muhammad's death, two relatives claimed to succeed him to lead the religious movement. Their followers developed into the two sects.

PEOPLE, COUNTRIES, AND CITIES

Can you ski in the Middle East?

Yes, you can ski in the Middle East. Since many parts of the region are mountainous, thousands of people ski in such popular places as Dizin and Shemshak, Iran; Mt. Hermon, Israel; and Mzaar, Lebanon. People can even ski in Dubai, along the coast of the Persian Gulf, where each year, thousands enjoy skiing in one of the largest indoor ski facilities in the world, called Ski Dubai, located at the Mall of the Emirates.

What are some of the largest urban areas in the Middle East?

Istanbul, Turkey, is an urban area with approximately 13.9 million residents, and Tehran, Iran, and its surrounding area is home to 13.3 million people.

What countries are located on the Persian Gulf?

The countries that have a coastline adjacent to the Persian Gulf are Iran, a small enclave of the country of Oman, the United Arab Emirates, Saudi Arabia, Qatar, Bahrain, Kuwait, and Iraq.

What commodity makes the Persian Gulf so strategically important?

The Persian Gulf is strategically very important because 21% of the world's petroleum is transported through the area by sea in very large oil tankers. At the southeastern end of the Gulf is the Strait of Hormuz, a narrow chokepoint that can be (and has been) controlled to prevent ships from sailing in or out of the Gulf.

How many people speak Arabic in the world?

Approximately 220 million people are native speakers of Arabic, and an additional 200 million are non-native speakers, including people who speak many dialects of Arabic. Arabic is also the language of the Islamic religion, practiced by 1.6 billion Muslims throughout the world.

Are all Israelis Jewish?

No, they are not. Approximately 80% of Israelis are Jewish, and the remaining 20.7% are predominantly Muslim Arabs.

How was Israel founded?

European Jews, who had been persecuted for centuries in Europe, began to emigrate to the Palestine area in the late nineteenth century. These new immigrants purchased land and set up communities in the desert and coastal areas. This area, which was under British administration, began to see a new influx of European Jews in the early part of the twentieth century. After World War I, the United Kingdom issued the Balfour Declaration, which accepted the establishment of a Jewish state in the area that we call Israel and Palestine today. At this time, approximately 750,000 people lived in the disputed area, comprised of 78% Muslims, 11% Jews, and 10% Christians. The general area was merely a collection of territories ruled by colonial powers such as the United Kingdom and France.

After World War II, hundreds of thousands of European Jews emigrated to the region, enough to cause the Arab community to revolt and riot against them. At this time, nearly 33% of all people in Palestine were Jewish. Jewish paramilitary groups increasingly fought against British occupying forces. The United Nations recognized Jerusalem as an international city for both Arabs and Jews, and it approved an interim plan to divide Palestine into Arab and Jewish states. The Jewish authorities accepted the plan, but new Arab countries that were formed after World War I and World War II did not. One day before the British Mandate of control over the Palestine region expired in May 1948, the State of Israel was proclaimed.

Which member of the United Nations could not serve on the Security Council?

In order to sit on one of the four rotating spots on the nine-member Security Council, a member of the United Nations must also be a member of one of five regional groups. Israel did not gain enough support from other U.N. nations to be allowed entry into a regional group until 2000, when it joined the Western Europe and Others group on a temporary basis. In 2004, Israel was made a permanent member of this group, making it eligible to be elected to the U.N. Security Council in later years. Since 1946, sixty-eight U.N. member countries have not served on the Security Council.

What is the Gaza Strip?

The area of land along the Mediterranean Sea at the border of Israel and Egypt is named the Gaza Strip. It is only approximately 141 square miles (365 square km) in area, and with nearly 1.8 million residents squeezed inside of its narrow confines, it is one of the most densely populated areas in the world. The Gaza Strip was once a part of Egypt until it was captured by Israel for a brief period in the years 1956 and 1957 and then permanently after the 1967 war. The Palestinian Liberation Organization (PLO) and Israel agreed to Palestinian limited self-rule in the Gaza Strip in 1994.

How has the West Bank caused conflict?

The West Bank, which refers to the western bank of the Jordan River, was supposed to become part of an independent Palestine at the same time Israel became a state. However, Arab attacks following the United Nation's 1947 proclamation of Israeli statehood led Israel to take over the West Bank militarily, when Israel became an independent state in 1948. Following a 1950 truce, Jordan occupied the West Bank, but Israel retook the territory during a 1967 war against its Arab neighbors. Peace talks in the late 1980s led to an agreement between Israel and the Palestine Liberation Organization (PLO) for limited Palestinian self-rule in the area. The disputed area includes much of the western shoreline of the Dead Sea and East Jerusalem. Approximately 2.7 million people live in this area, including approximately 500,000 Israeli Jews. The remaining residents are Palestinian Arabs, comprising approximately 80% of the total population. The global community does not support Israeli settlements in this area and considers it to be illegal under international law. This is a major source of the conflict, with all parties seeking a negotiated solution in this area.

How many countries supported the Palestinian bid to become an officially recognized state within the United Nations?

In November 2012, the U.N. General Assembly considered a resolution to change the status of the State of Palestine from "observer" to "nonmember state." There were 138 votes in favor of the resolution, forty-one countries abstained from voting, and nine countries opposed the resolution.

The West Bank and Gaza Strip have long been regions of contention between Israel and its Muslim neighbors.

Which two countries will build a canal connecting the Red Sea and the Dead Sea?

The countries of Jordan and Israel are planning a five-year, $800 million project to construct a canal that will link the Red Sea with the Dead Sea in order to provide drinking water to communities of people located in Israel, Jordan, and Palestine and stabilize the Dead Sea. A large desalination plant located near the city of Aqaba, Jordan, will process seawater from the new canal.

> ### What was the Persian Gulf War?
>
> In late 1990, Iraq attacked Kuwait, claiming that it was Iraq's long-lost nineteenth province. A coalition of countries, led by the United States, fought against Iraq. The land war lasted a mere one hundred hours (the air war began January 17, 1991, and ended, along with the land war, on February 28), and Kuwait was liberated from Iraq in February 1991. Most of Kuwait was destroyed, especially oil facilities, by Iraq's "scorched earth" policy.

Where is Asia Minor?

Asia Minor is the term used for the larger part of Turkey that lies in Asia, east of the Strait of Bosporus. This part of Turkey is also known as Anatolia.

Which Middle Eastern country has the most tourism?

Turkey has the most tourist arrivals, with more than 20 million people coming to visit each year. Other countries in the Middle East have significant numbers of tourists. Over eight million people visit Egypt each year, and because of the annual Hajj or pilgrimage to Saudi Arabia's religious sites at the cities of Mecca and Medina, more than two million people visit there each year.

Why is Cyprus divided?

The island of Cyprus, with its capital at Nicosia, is part of the European Union and is located in the eastern Mediterranean Sea. It was a colony of the United Kingdom until its independence in 1960. At that time, approximately 77% of the population of the island were ethnically Greek, and 18% of the population were ethnically Turkish. In 1974, a coup d'etat organized by Greek Cypriots occurred that overthrew the president of the country. Turkey then invaded the island and succeeded in taking control of its northern half. This became the Turkish Republic of Northern Cyprus, which is not internationally recognized as an independent country. Southern Cyprus remained an independent country called the Republic of Cyprus. Since 1974, the U.N. Peacekeeping Force in Cyprus (UNFICYP) maintains a force of 920 soldiers in order to monitor the ceasefire line between the two parts of the country.

What was Babylonia?

Babylonia was an ancient country along the Euphrates River in what is now Iraq. It began in the twenty-first century B.C.E. and was led by King Hammurabi in the eighteenth century B.C.E. Babylonia dissolved into chaos in approximately 1000 B.C.E.

When did the Iraq War begin?

The war, which has been called the Second Persian Gulf War and Operation Iraqi Freedom, began on March 20, 2003. Although President George W. Bush proclaimed the end of the war aboard the aircraft carrier U.S.S. *Abraham Lincoln* on May 1, 2003, declaring the "mission accomplished," the war was still being waged five years later, with no end in sight. It was not until 2011 that the United States, under the Obama administration, officially withdrew its military forces from Iraq. Since 2014, the United States has found itself again militarily entangled in Iraq, fighting against elements who seek to destroy the nascent state.

Many hold that the United States invasion of Iraq was based upon false pretenses that the Iraqis possessed weapons of mass destruction.

Why did the United States invade Iraq?

The administration under President George W. Bush claimed that Iraq, led by Saddam Hussein, was a threat to the region because it believed that Iraq was developing weapons of mass destruction, including nuclear and biological weapons, and that Iraq posed a threat to the United States and its neighbors. This was later proven false by many U.S. and United Nations investigators. The weapons programs had stopped at the time of the first Gulf War in 1991. The Bush administration also claimed that Iraq was somehow responsible for the 9/11 terrorist attacks, and the president repeated these claims in the news media for many years. These allegations have also been proven false. There was never a connection between the perpetrators of the 9/11 attacks, who were primarily Saudi Arabian nationals, and the country of Iraq.

How many people have been killed during the U.S. war with Iraq?

Numbers and methods used to compute the statistics of dead and wounded in Iraq vary considerably, from a high of over 1.2 million people (Opinion Research Business) to a low of 654,000 people (Lancet Survey). From 2003 to 2011, an estimated 4,452 Americans had lost their lives in Iraq.

How much money have U.S. taxpayers spent on the war in Iraq?

Although it is very difficult to estimate the costs of the war, including disability expenses for decades of care of returning veterans, estimates of the costs of the war are in the $3–6 trillion range.

What is ISIL (or ISIS)?

ISIL is an acronym for the Islamic State of Iraq and the Levant, a terrorist group that seeks to establish an Islamic state across the region, especially in Iraq and Syria. It was formed in 1999 and supported Al-Qaeda in 2004 and grew out of the chaos created by the Iraq War, from a minority of armed and disaffected Sunni Muslims who were dissatisfied with the democratic ideals of the elected and predominantly Shi'a Iraqi government and military leadership. ISIS, which stands for the Islamic State of Iraq and Syria, is another acronym used for the group.

While participating in civil warfare, along with many other insurgent groups, the group currently controls large portions of the territories of both Iraq and Syria, affecting the lives of more than 10 million residents. They are also responsible for committing terrorist acts in other countries in the region (like Egypt) and in Europe.

How do Syria and Russia cooperate in the Syrian Civil War?

Russia, an historic ally of Syria, joined the military conflict in Syria in 2015 in order to fight against the Islamic State groups as well as to support the military groups of Syria's embattled leader and ally, President Bashar al-Assad. Syria has been in a state of civil war since 2011, which has caused the deaths of approximately 250,000 people and has caused more than 11 million people to leave their homes and towns.

What is Greater Syria?

Some Syrians, Lebanese, Jordanians, and Palestinians see the boundaries of their countries as artificially imposed decades ago by colonial rulers and hope to see a united Greater Syria composed of the four areas.

What was the United Arab Republic?

In 1958, non-neighbors Egypt and Syria united to form the country known as the United Arab Republic, with its capital at Cairo. This union lasted until 1961, after a Syrian coup d'etat that overthrew the country's leader. Syria then decided to secede from the union and become an independent country. Even after the dissolution, Egypt kept the name "United Arab Republic" for a decade.

Which Middle Eastern country has the largest population?

If you consider Egypt to be part of the Middle East, it is the most populous country, with over 89 million people. Just behind Egypt is Iran, with over 78 million people, followed by Turkey, with over 77.7 million people.

Who are the Kurds?

The Kurds are an indigenous ethnic group who live in southeastern Turkey, northwestern Iran, northeastern Iraq, Syria, and Armenia. Though they are one of the largest ethnic groups living in the Middle East, they have no country of their own. It is estimated that there are 25–35 million Kurds living in the region. Since the Kurds have been persecuted in Iraq, the United Nations established a security zone for them in an area above 36 degrees north latitude. The Kurds hope that one day they will have their own nation-state in the region where they now live.

Which country has the most Azeri-speaking people?

Surprisingly, between 11.2 and 20 million Azerbaijanis (Azeri is their native language) live in the northeastern provinces of Iran, considerably more than the approximately eight million who live in the Republic of Azerbaijan, Iran's neighbor to the north. It is difficult to know the exact number of Azeri speakers, since the government of Iran doesn't release such figures.

The Ayatollah Ali Khamenei is shown casting his vote in the 2016 Iranian general elections.

Who is an ayatollah?

The title of ayatollah is used for a high-ranking Shi'a religious scholar, religious leader, or cleric. In 1979, Iran went through a revolution and a period of religious fundamentalism that replaced the pro-West and secular Shah (the former leader) with an ayatollah as the religious and political leader of the nation. Under the ayatollah's rule, traditional social values are mandated.

What is a theocracy?

A theocracy is a country where the civil leader is believed to receive direct guidance from his or her God. Iran has the world's largest theocracy, ruled by people who are national leaders as well as religious leaders. While there is a secular president, the division of power between mullahs (a Muslim trained in religious law) and the president is loosely defined.

Which Middle Eastern country currently has the longest-ruling leader?

Ali Khamenei has been the head of state of Iran (as president and, later, supreme leader) since October of 1981.

Who are the Bedouins?

The Bedouins are a seminomadic group that have inhabited areas of the Middle East, from the Arabian and Syrian deserts, stretching all the way to Northern Africa, for centuries. Grazing their goat and camel herds, the Bedouins travel wide distances and frequently cross international borders. Many Middle Eastern countries have effectively halted the crossings. As a result, many descendants of nomadic tribes have become urbanized. The largest population concentration of Bedouin people is in Rahat, Israel, where approximately 62,500 live.

Where does the name "Saudi Arabia" come from?

The name "Saudi Arabia" comes from the last name of the ruling Al-Saud family, who have been in control of the country since 1932. Saudi Arabia is one of a few countries in the world that are named after a ruling family.

Is the Empty Quarter empty?

The Empty Quarter, also known as the Rub' al Khali, is Saudi Arabia's vast, open desert. The Empty Quarter is devoid of any significant population and stretches for 620 miles (1,000 km). It is the largest, contiguous sand desert in the world, covering 250,000 square miles of area (650,000 square km), extending through Saudi Arabia and into Oman, the United Arab Emirates, and Yemen. Parts of the Rub' al Khali contain major oil production operations.

The Kaaba in Masjid al-Haram in Mecca, Saudi Arabia.

Why do people travel to Mecca?

The city of Mecca is 43 miles (70 km) from Jeddah, Saudi Arabia. The holy book of Islam, the Quran, requires that all Muslims make a journey, known as a Hajj, to the city of Mecca at some time in their lives. Mecca is the holiest city of Islam, as it was the birthplace of Muhammad and the place where the Prophet and founder of Islam in the late sixth century C.E. received the first of a series of revelations of the Quran. Non-Muslims are forbidden to enter Mecca, which contains many religious sites for Muslims.

What is the holiest site in Mecca?

The most important site in Mecca is the Great Mosque (called the Masjid al-Haram), the largest mosque in the world. It is located in the center of the city and houses the sacred Black Stone (Kaaba). Muslims around the world always pray in the direction of the Black Stone during daily prayers.

How many Muslims make a pilgrimage to Mecca each year?

Each year, approximately two million Muslims from around the world make a pilgrimage to Mecca during the last month of the Islamic calendar. While hundreds of years ago, the trip took weeks or months to complete from the vast reaches of the Islamic Empire, today people use the modern conveniences of travel and fly to Saudi Arabia.

What nationalities were the 9/11 terrorists?

Out of the nineteen terrorists who participated in the 9/11 attacks in 2001, fifteen of them were citizens of Saudi Arabia, two were citizens of the United Arab Emirates, one was from Yemen, and one was from Lebanon.

Who was Osama bin Laden?

Osama bin Laden was a Saudi Arabian militant and the leader of the terrorist organization Al-Qaeda ("the Base" in Arabic). Al-Qaeda is a Sunni Muslim organization that was created from veterans of the struggle against the Soviet occupation of Afghanistan. Bin Laden was the son in a wealthy Saudi Arabian family that made its fortune in the Saudi civil construction business. He was expelled from Saudi Arabia, moved to the Sudan, and spent his final years in Abbottabad, Pakistan, before being killed by American military specialists in May 2011. He was the mastermind behind the attacks on the World Trade Center in New York and the Pentagon in Virginia.

Bin Laden used his personal wealth to fund mujahideen insurgents in Afghanistan after the Soviet withdrawal in 1989, and he organized bombings against various American interests, including the American embassies in Tanzania and Kenya in 1998. He demanded the end of foreign influence in Muslim countries and the creation of a new Muslim caliphate. He had been waging a jihad ("holy war," although many Muslims say this is a misuse of the term) to achieve his aims through violence and the killing of both civilians and military personnel.

Which country is composed of seven sheikdoms?

The country of the United Arab Emirates, located on the Arabian Peninsula, is a federation composed of seven autonomous sheikdoms (also known as emirates). The seven emirates were defended by the United Kingdom in the nineteenth century but merged to become an independent country when the U.K. withdrew from the area in 1971. Two of the emirates and a few islands offshore were invaded by Iran shortly afterward. The seven emirates are Abu Dhabi, Ajman, Fujairah, Sharjah, Dubai, Ra's al-Khaimah, and Umm al-Qaiwain.

How do you drive to the Kingdom of Bahrain?

The Kingdom of Bahrain, with its capital at Manama, is located 15 miles (24 km) off the coast of Saudi Arabia and has been connected by the four-lane King Fahd Causeway (bridge) to Saudi Arabia since 1986. Saudi Arabia paid for the construction of the causeway.

PHYSICAL FEATURES AND RESOURCES

How big is the Sahara Desert?

The Sahara is the world's largest hot desert. It covers approximately 3.6 million square miles (9.4 million square km) in northern Africa. The Sahara receives less than 3.93 inches (100 mm) of rain each year but contains hundreds of individual oases. The elevation in the Sahara Desert ranges from 436 feet (133 meters) below sea level to 11,204 feet (3,415 meters) above sea level. There are people who live in the Sahara Desert, mostly at or near approximately ninety major oases.

Where is the Maghreb?

From the Arabic word for "place of sunset" or "western," the Maghreb is a region of Northern Africa composed of the Atlas Mountains and coast of Morocco, Algeria, Tunisia, and Libya farther east. The name "Maghreb" originated from the term used by the Islamic Empire for this region of the world.

Where is the Barbary Coast?

The Barbary Coast, or "Berber Coast," refers to the countries that are situated in the middle and western part of the Northern African coast along the Mediterranean Sea: Morocco, Algeria, Tunisia, and Libya. Though the name Barbary Coast comes from the Berber people who inhabit the region, it is best known for its association with pirates from the sixteenth through the nineteenth centuries. In 1805, in its first foreign military action, the United States invaded Tripoli, Libya, in an attempt to put an end to piracy in the region.

Where is the Horn of Africa?

The Horn of Africa is the name used to describe the eastern peninsula of continental Africa that includes the countries of Djibouti, Eritrea, Ethiopia, and Somalia. The easternmost tip of the Horn, which is located in Somalia, is called Cape Guardafui or Ras Asir.

Which African country is the world's leading producer of cocoa beans?

Côte d'Ivoire, also known as the Ivory Coast (named for the large amount of ivory collected from elephant herds that once roamed the country), produces 34.2% of the world's cocoa beans, the beans used to make chocolate. It produces more than 1.82 million tons (1.65 million metric tons) each year. Ghana produces another 18.2% of the world's total, and both Cameroon and Nigeria contribute 13.3% to the total production of cocoa in the world. This means that nearly 65.7% of the world's cocoa beans are grown in Africa, mostly in the western part of Africa.

How did a lake kill more than 2,000 people?

In August 1986, Cameroon's Lake Nyos, a crater lake that sits upon a volcanic vent, produced an eruption of carbon dioxide and hydrogen sulfide gasses. The cloud of acidic gas blew into nearby villages, killing more than 1,700 people while they slept.

What is the largest lake on the continent of Africa?

The largest lake on the continent of Africa is Lake Victoria, which touches the African nations of Uganda, Kenya, and Tanzania. It is the third-biggest lake in the world and covers an area of 26,828 square miles (69,485 square km).

Is there any permanent ice in Africa?

Though located within 198.8 miles (320 km) of the equator, there is ice and snow year-round at the top of Mt. Kilimanjaro.

How was Mt. Kilimanjaro formed?

Mt. Kilimanjaro, the tallest mountain in Africa at 19,341 feet (5,895 meters), was formed as a volcano, but it is now dormant. The mountain is located in northeastern Tanzania

What is the Bight of Bonny?

Also called the Bight of Biafra, the Bight of Bonny is a bay located in the Gulf of Guinea. It shares a coastline with several west African countries, including Nigeria, Cameroon, Equatorial Guinea, and Gabon. The word "bight" is from the Old English word for "bay."

The tallest mountain in Africa, at 19,341 feet (5,895 meters), is Mt. Kilimanjaro.

and was first successfully summited in 1889 by German geology professor Hans Meyer and Austrian climber Ludwig Purtscheller.

Has global warming affected the ice on Mt. Kilimanjaro?

Global warming has affected the glacial ice on top of Mt. Kilimanjaro and is thought by many scientists to have caused as much as 80% of the ice to permanently disappear. Scientists believe that by the year 2040, there will be no more snow on top of the mountain.

What is the world's longest freshwater lake?

Lake Tanganyika is 420 miles (676 km) long from north to south and is the longest freshwater lake in the world, but it averages only 31 miles (50 km) in width. The lake borders several African countries, including the Democratic Republic of the Congo, Tanzania, Burundi, and Zambia. It is also the world's second-deepest lake, with a maximum depth of 4,823 feet (1,470 meters).

Which country produces the most gold on the African continent?

South Africa's gold mines produce 5.2% of the world's gold annually. In 1884, gold was first discovered in South Africa at the mines near Witwatersrand, which is now South Africa's largest gold-producing area. In 2007, China overtook South Africa by producing 304.2 tons (276 metric tons) of gold, beating South Africa for the title of biggest

What is the Great Rift Valley?

In eastern Africa there lies a long, narrow, and deep valley known as the Great Rift Valley. Over 3,700 miles (6,000 km) in length and 20 to 60 miles (32 to 97 kilometers) wide, the rift spans nearly the length of Africa and was formed by tectonic plates sliding apart. In 10 million years, eastern Africa will eventually detach from the rest of Africa along this rift and form its own subcontinent.

gold producer by just four tons (3.63 metric tons). This was the first time that South Africa has not been first in gold production since 1905. Today, South Africa is the seventh-biggest producer of gold in the world, producing approximately 165 tons (150 metric tons) per year.

What is the Kalahari Desert?

The Kalahari Desert covers much of Botswana and Namibia. It is one of the largest deserts in the world, covering over 100,000 square miles (259,000 square km) of area and lies on a high plateau at 3,000 feet (914 meters) above sea level.

What country in Africa has the world's highest minimum elevation?

Lesotho, with its capital at Maseru, surrounded by the mountains of southeastern South Africa, has an absolute minimum elevation of 4,593 feet (1,400 meters) above sea level in the valley of the Orange River, but approximately 80% of the land in the country lies above 5,906 feet (1,800 meters).

What two countries owned Walvis Bay?

When Namibia gained independence from South Africa in 1990, South Africa kept control of Walvis Bay, located approximately 400 miles (644 km) north of the South African border. The excellent harbor and deep-water port of Walvis Bay was retained for four years until it was returned to Namibia on March 1, 1994.

What is the Caprivi Strip?

The Caprivi Strip is the name of the narrow strip of land that protrudes from the northeast corner of Namibia, surrounded on three sides by the countries of Angola, Zambia, and Botswana. The land was acquired by German chancellor Leo von Caprivi from the British in 1890 in order for Namibia, known then as German Southwest Africa, to have access to the Zambezi River. This strip of land is approximately 300 miles (480 km) long but no more than 65 miles (105 km) at its widest.

HISTORY

How did a meeting in Berlin change the map of Africa?

In 1884, at a time when most of Africa had yet to be colonized, thirteen European countries and the United States met at the Berlin Conference to divide Africa up among themselves. There was no African representation at the meeting. Geographic borders were drawn across Africa, ignoring the continent's cultural and historical differences. Though the borders were established over one hundred years ago, many of these delineations still exist today and are responsible for some of the turmoil and conflict among the now independent African countries that we see today.

How many African countries were independent in 1950?

In 1950, there were only four independent countries on the continent: Egypt, South Africa, Ethiopia, and Liberia. All other countries gained their independence in the decades that followed. Most recently, Namibia became an independent country in 1990, Eritrea became independent of Ethiopia in 1993, and South Sudan became independent of the Sudan in 2011.

Which countries colonized Africa?

Seven European countries colonized Africa: Belgium, Germany, France, Italy, Portugal, Spain, and the United Kingdom. Private interests in the United States maintained a colony in Liberia during the nineteenth century C.E.

How many African countries were colonized by Italy?

Italy established colonies in what are now Libya, Eritrea, and Somalia.

Of all fifty-four countries in Africa, which ones were never colonized?

Only one country in Africa escaped the onslaught of colonial occupation: Ethiopia. Two other countries, Liberia and the Sudan, are commonly thought to have escaped the colonialism of the nineteenth century C.E. In fact, Liberia was colonized by private U.S. organizations for a brief period in the nineteenth century, and the Sudan fell under British colonial control in 1899.

What was the Organisation of African Unity?

Founded in 1963, the Organisation of African Unity helped strengthen and defend African unity across the continent. The thirty-two member-countries that were independent at that time sought to increase development and economic unity within and between member-countries. Twenty-one more African countries joined the organization before it was disbanded in 2002. A new organization, the African Union, was created with the same fifty-three member states at that time. With the addition of South Sudan in 2011, there are fifty-four members today. The African Union seeks to integrate political and socioeconomic interests of the member states, speaking with a common voice on issues involving Africa. Its headquarters is in Addis Ababa, Ethiopia.

Who was Haile Selassie?

Haile Selassie was not only emperor of Ethiopia; many considered him to literally be a god.

Haile Selassie was emperor of Ethiopia and was one of the great leaders of the twentieth century, as well as one of the most prominent figures in modern African history. He served as head of Ethiopia from 1930 to 1974 and was responsible for modernizing Ethiopia, forming the African Union, and uniting Africans both in Ethiopia and throughout the continent. He was a committed internationalist, and because of this, the United Nations made Ethiopia a charter member when the U.N. was formed in 1945. His leadership has been criticized by some human rights groups for its autocratic methods of implementing reforms. Haile Selassie is revered by the Rastafarian religion of Jamaica as a living god.

PEOPLE, COUNTRIES, AND CITIES

How many countries are there on the continent of Africa?

Africa is home to more than a quarter of the world's countries. There are forty-eight independent countries on the continent itself and fifty-four if you include the nearby island countries of Cape Verde, Comoros, Madagascar, Mauritius, São Tomé and Príncipe, and Seychelles.

How many African countries are landlocked?

Only sixteen of Africa's forty-eight continental countries have no access to seas or oceans. These include Botswana, Burkina Faso, Burundi, Central African Republic, Chad, Ethiopia, Lesotho, Malawi, Mali, Niger, Rwanda, Swaziland, South Sudan, Uganda, Zambia, and Zimbabwe.

How prevalent is AIDS in Africa?

Approximately 26.04 million Africans have AIDS, which makes up more than 70% of the world's AIDS cases. Almost all AIDS transmission in Africa is through heterosexual contact. Approximately 800,000 people die of AIDS-related illnesses in Africa each year. It is the number-one cause of death of teenagers aged ten to nineteen in Africa today because only one-third of the estimated 2.6 million children who are infected with the virus are able to obtain proper treatment.

Which country in Africa is the only country that is not a member of the African Union?

Morocco withdrew from the Organisation of African Unity (later named the African Union) in 1984 over a dispute with members on the recognition of the Sahrawi Arab Democratic Republic by the organization. It did not become a member of the newly formed African Union and is the only country in Africa that is not a member. Lately, the country of the Central African Republic has had its membership in the Union suspended because of internal civil and political unrest.

What is the Sahrawi Arab Democratic Republic (SADR)?

Although not a fully recognized country because of an international border dispute with Morocco, Sahrawi claims sovereignty over all of the territory of the former Spanish colony of Western Sahara. A part of the territory of Western Sahara is considered by Morocco to be a province of Morocco. Sahrawi is currently a member of the African Union. Approximately eighty-five countries recognize the SADR as having a legitimate right to self-determination.

When did the Suez Canal begin to operate?

After a decade of construction, the Suez Canal, built by the French, opened on November 17, 1869. The 101-mile-long (162.5-km-long) canal cuts through northeastern Egypt, making a passageway for ships to sail between the Mediterranean Sea and the Red Sea, which eventually leads to the Indian Ocean.

Where are Egypt's pyramids?

The most famous group of pyramids in Egypt is located near the city of Giza, which is just outside of Cairo. This group includes the Great Pyramid, which is known as one of the Seven Wonders of the Ancient World. The pyramids in Egypt served as tombs for pharaohs and were built from the twenty-seventh to the tenth century B.C.E. Other pyramids are located along the Nile River in southern Egypt and the northern Sudan. Approximately seventy pyramids remain in the region.

Where in Egypt do most Egyptians live?

About 95% of Egypt's approximately 89 million residents live within 12 miles (19 km) of the Nile River. Since the rest of Egypt mainly consists of desert, the remaining residents live scattered across the country, primarily near oases or along the coast.

The pyramids of Egypt have long been objects of wonder and speculation.

What is the largest metropolitan area by population in continental Africa?

Cairo, Egypt, is by far the largest metropolitan area in continental Africa, with over 16.3 million people.

Who lives in the City of the Dead?

There lies on the outskirts of Cairo, Egypt, a 700-year-old cemetery filled with mausoleums, memorials, and mosque-shaped tombs and shrines. Because of severe overcrowding in Cairo, squatters, some of whom have resided there for decades, now live inside these memorials to the dead in what is called by many el-Arafa necropolis or the City of the Dead. Recently, the Egyptian government provided the area with electricity and water. Built as a home for the dead, this cemetery is now becoming a home for the living.

Where is Timbuktu?

Though the name is commonly used to refer to an extremely distant place, Timbuktu (or Tombouctou) is actually a town near the Niger River, in the west African country of Mali. It has a population estimated to be 54,000 people and is a major salt-trading post for Saharan Desert camel caravan routes.

Why are there so many starving people in Ethiopia?

Droughts in the 1970s and 1980s, which were especially severe from 1984 to 1986, destroyed Ethiopian agriculture. Approximately 90% of Ethiopian agriculture relies solely on rainwater. Though relief food was shipped to Ethiopia, internal political corruption kept the food from reaching the starving victims. During the 1980s, approximately one million people died of starvation in Ethiopia.

Where is Ouagadougou?

Ouagadougou (pronounced wah-gah-doo-goo) is the capital of the west African country of Burkina Faso (population 17.3 million). The city, with an estimated 2.7 million residents, is the largest city in the country and is home to the University of Ouagadougou.

How did American slavery help found Liberia?

In 1822 the American Colonization Society succeeded in founding a colony on the western coast of Africa for freed American slaves. The colony was named Liberia from the Latin word *liber,* which means "free." This colony became independent in July 1847, although it was not recognized by the United States until 1862. It was Africa's first independent country and is home to 4.5 million people.

Why are so many ships registered in Liberia?

Though most are owned by foreign corporations and spend little time in Liberian territorial waters, many shipping companies choose to register their ocean vessels in Liberia because of its low fees. Over 3,900 ships are registered there, giving Liberia one of the world's largest merchant fleets.

What is Africa's most populous country?

Nigeria, with a population of 182.2 million people, is Africa's most populous country. Nigeria is also the world's seventh most populous country. If Nigeria's population continues to increase at its present 2.7% growth rate, the population will increase to over 208 million by the year 2020. The average Nigerian woman gives birth to six children in her lifetime.

Where is the largest church in Africa?

The Basilica of Our Lady of Peace in Yamoussoukro, Ivory Coast, is Africa's largest church, with an interior that encompasses 85,993 square feet (7,989 square meters). Completed

Where was Kunta Kinte from?

Kunta Kinte, the protagonist of Alex Haley's novel *Roots,* was from the Republic of the Gambia. Though the Gambia follows 200 miles (322 kilometers) of the Gambia River, it is a very thin country, averaging only 12 miles (19 kilometers) in width. The Gambia is bordered in the north, south, and east by the African country of Senegal and to the west by the Atlantic Ocean.

The largest church in the world is the Basilica of Our Lady of Peace in Yamoussoukro, Ivory Coast.

in 1989 by President Félix Houphouët-Boigny, the church has seating for 18,000 people. In 1983, Houphouët-Boigny relocated the capital of the Ivory Coast from Abidjan to his hometown of Yamoussoukro.

Is Equatorial Guinea on the equator?

No, Equatorial Guinea's southernmost point is still one degree north of the equator. Though most of Equatorial Guinea is located only a few degrees north of the equator, its southern neighbor, Gabon, is truly equatorial.

Which African country has Spanish as an official language?

The west African country of Equatorial Guinea, which is composed of five islands in the Gulf of Guinea and a tiny sliver of land between Cameroon and Gabon, was a colony of Spain until 1968 and has kept Spanish as its official language. The capital city of Malabo is located on the island of Bioko, previously named Bubis.

How many countries named the Congo are there?

There are two countries named the Congo, and just to make things a little more confusing, they're neighbors. The similarity between the names of the two—the Democratic Republic of the Congo and its western neighbor, the Republic of the Congo—makes distinguishing them a little difficult. In 1908 the Democratic Republic of the Congo was named the Belgian Congo; in 1960 it was called Republic of the Congo; in 1964 it became People's Republic of the Congo; in 1966 it was called Democratic Republic of the Congo;

Who speaks Swahili?

Approximately 5–15 million people speak the Bantu language Swahili as a first language, and approximately 60–150 million East Africans speak it as a second language. It is widely used in such places as Tanzania, Kenya, Uganda, Rwanda, Burundi, Mozambique, the Democratic Republic of the Congo, and Comoros. Swahili is a mixture of Arabic and African languages that gradually developed through trading between Africans and Arabs. Though there are over 1,000 different languages in Africa, Swahili is the second-most popular language (Arabic is the first).

in 1971 it was called Zaire; and finally, in 1997, again became the Democratic Republic of the Congo.

What was the Rwandan genocide?

The Rwandan genocide was the systematic killing of an estimated 500,000–1 million people belonging to the Tutsi minority ethnic group and moderate Hutus by the majority Hutus. Most of the killing occurred within a span of one hundred days in the summer of 1994. It was carried out by two Hutu extremist military groups during the east African country of Rwanda's Civil War. The genocide had its roots in the Civil War, which lasted from 1990 to 1994 and pitted the majority Hutus against the Uganda-supported Tutsi minority.

Why is Cabinda separate from Angola?

Cabinda, though a province of Angola, is separated from the bulk of Angola by approximately 25 miles (40.2 km) of land that is part of the Democratic Republic of the Congo. In 1975, a coup in Portugal caused a change in Portuguese foreign policy, and a move was made to free all Portuguese colonial territories, including Angola, which controlled Cabinda. For decades, Cabindans have used armed resistance against Angola in the hope of obtaining their independence.

What was Zimbabwe's previous name?

In April 1980, after decades of armed conflict, the former British colony of Rhodesia was granted independence and renamed itself Zimbabwe. The capital city of Salisbury was renamed Harare. Rhodesia had been originally named for British imperialist and businessman Cecil Rhodes in 1895.

What is the name of the currency of Botswana?

Since 1976, the southern African country of Botswana, consisting primarily of the arid Kalahari Desert, uses the pula, which means "rain," as its currency. The capital of Botswana is Gabarone (pronounced GA-ba-ROH-nee).

How many provinces are there in South Africa?

In 1994, several new provinces (states) were created. South Africa is now divided into nine provinces: Eastern Cape, Free State, Gauteng, KwaZulu-Natal, Limpopo, Mpumalanga, North West, Northern Cape, and Western Cape. The nine provinces incorporated existing homelands (Bantustans) of indigenous peoples.

Nelson Mandela brought an end to apartheid in South Africa and became the country's first black president.

How many capital cities does South Africa have?

South Africa has three capital cities: Pretoria is the administrative capital, Bloemfontein is the capital of the judiciary, and Cape Town is the legislative capital.

Who was Nelson Mandela?

Nelson Mandela was a South African leader, who was the head of the African National Congress and its armed resistance group, Umkhonto we Sizwe. For many years, this group fought against the apartheid policies of South Africa. Mandela spent twenty-seven years imprisoned on Robben Island for his political activities. Upon his release in 1990, Mandela asserted a policy of reconciliation and negotiation to end the racist policies of apartheid. He was the first fully representative, democratically elected president of South Africa, serving as head of state for one term, from 1994 to 1999. He was awarded the Nobel Peace Prize for his work in 1993. Nelson Mandela died in 2013.

What was apartheid?

South Africa's legalized form of systemic, government-mandated racial discrimination was known as apartheid (meaning "separateness" in Afrikaans). It classified all individuals into one of four ethnic groups: white, black, coloured (mixed race), and Asian. Apartheid laws limited where people of different racial groups could live, work, and gather. Apartheid was finally repealed in 1991.

What is one of the only countries in the world to provide constitutional protection to the gay, lesbian, bisexual, and transgendered community?

In 1996, South Africa became the first country in the world to officially protect the rights of its LGBT residents. South Africa's 1996 constitution protects the LGBT community against discrimination in both the public and private sectors. The "Equality Clause" in the South African Bill of Rights protects people from discrimination based on race, gender, pregnancy, marital status, ethnic or social origin, color, sexual orientation, age, disability, religion, conscience, belief, culture, language, and birth.

How many official languages are there in South Africa?

The Republic of South Africa has eleven official languages: Afrikaans, English, Ndebele, Northern Sotho, Sotho, Swazi, Tsonga, Tswana, Venda, Xhosa, and Zulu.

Which country lies completely within South Africa?

The tiny country of the Kingdom of Lesotho is completely surrounded by South Africa. Its population is approximately 2 million people, with 43,000 (40%) male workers migrating to South Africa to work in various mines for employment. The country gained independence from the British in 1966 and today is one of the biggest exporters of clothing to the United States in Sub-Saharan Africa.

What was the Nazis' plan for Madagascar?

In 1940, after the French defeat at the hands of the Germans, the Nazis developed a plan to forcibly relocate European Jews to the French colony of Madagascar. This plan never materialized, however, and the Nazis continued the extermination of Jews throughout Europe.

What language do people in Madagascar speak?

The people of Madagascar speak Malagasy. Malagasy is more closely related to the languages spoken in Indonesia and Polynesia than to any African language. The people of Madagascar are of Indonesian and Malaysian descent, having migrated there nearly 2,000 years ago. The French language is widely spoken among the country's educated population.

What fish, once thought to be extinct, suddenly appeared near the Comoros?

In December 1938, a fisherman discovered a very strange-looking fish near the Comoros islands. Scientists discovered that this highly endangered, lobe-finned fish was a coela-

What is significant about the country of Comoros?

The Union of the Comoros, an island nation off the coast of Southeast Africa and just northwest of the island of Madagascar, consists of four main islands. In 1974 and 1976, when referendums were held to become independent from France, one of the four islands, Mahori (Mayotte), decided to remain a colony of France. Since 2011, Mahori (Mayotte) is an overseas department of France.

canth, thought to have been extinct for over 70 million years. One of two species that exist continues to live in the waters off of the Comoros Islands, located between Madagascar and continental Africa. The country is officially known as the Union of the Comoros, with its capital in Moroni.

How many people lived on the islands of Seychelles before 1770?

None. The country, which is composed of 115 islands northeast of Madagascar, was first inhabited by the colonizing French in the 1770s. The British later gained control of the area and brought Africans to the islands. The islands gained independence from the U.K. in 1976. With only an estimated 92,000 residents, it is the least populated country in Africa.

OCEANIA

What is Oceania?

Oceania is the region in the central and southern Pacific that includes Australia, New Zealand, Papua New Guinea, and the islands that compose Polynesia, Melanesia, and Micronesia.

Who owns all of those islands in the Pacific Ocean?

There are hundreds of islands in the Pacific Ocean, most of which are uninhabited. While many of these islands are part of the territory of nearby independent countries, others are managed by the governments of remote countries as vestiges of past colonial empires.

What is Polynesia?

Polynesia consists of more than 1,000 islands in a great, triangular region bounded by Hawaii to the north, New Zealand to the southwest, and Easter Island to the southeast. The region includes the countries of Samoa, Tonga, and Tuvalu. Also located in this area is Tokelau, a territory of New Zealand, and the colony of French Polynesia, which includes Tahiti and 117 other islands and atolls.

What is Micronesia?

The region known as Micronesia consists of islands east of the Philippines, west of the International Date Line, north of the equator, and south of the Tropic of Cancer. In 1986, approximately 600 of the islands in this area united to form an independent country, the Federated States of Micronesia. Though 600 islands joined the federation, there are approximately 1,500 other islands in the region, including the independent countries of the Marshall Islands, Kiribati, Nauru, and Palau, as well as three U.S. territories (Guam Island, Northern Mariana Islands, and Wake Island). Although people have inhabited Micronesia for millennia, the first contact with Europeans was when explorer Ferdinand Magellan arrived in 1521.

What is Melanesia?

Located northeast of Australia, Melanesia is a small region that lies south of the equator and west of 180 degrees longitude but excludes New Zealand. Melanesia includes the countries of Vanuatu, Fiji, Papua New Guinea, and the Solomon Islands, as well as the French overseas territory of New Caledonia.

How did the Bikini Atoll get its name?

In 1946, the United States began to test atomic weapons on the Bikini Atoll, which consists of twenty-three islands, in the Marshall Islands. It was also in the late 1940s when the two-piece bathing suit made its debut and took its name from the intensely publicized Bikini Atoll nuclear tests. The original name of the atoll is derived from the Marshallese language name of the islands, Pikinni, which means "surface of coconuts."

How do the people on the tiny islands in the Southern Pacific Ocean go shopping?

Most basic goods can be bought on each populated island, and larger items may be shipped via airplane. When residents need to travel between islands, they often take to the air. Each island has an airstrip of adequate length for its own transportation needs. In the past, inhabitants used boats as their primary means of transportation between islands as well as a means to obtain goods from abroad.

Where did the mutineers of the *Bounty* land?

In 1789, members of the crew of the HMS *Bounty* mutinied. After having dropped off nineteen other members of the crew, including Captain William Bligh, the mutineers landed on the uninhabited island of Pitcairn. While the captain and his loyal crew suc-

How did the Guano Islands Act help fertilize America?

In 1856, the U.S. Congress passed the Guano Islands Act, which allowed citizens of the United States to take possession of any unclaimed island, not under the jurisdiction of another government, that contained guano. Guano, the excrement of sea birds, was mined for use as fertilizer before the widespread use of chemical fertilizers. Beginning in 1857, the Baker and Howland islands, located southwest of Hawaii, were mined by the United States until their guano was depleted in 1891.

A 1790 illustration depicts Lieutenant Bligh and other officers being cast off the *Bounty* during the famous 1789 mutiny.

cessfully returned to England, the mutineers established a community composed of nine male mutineers, six male Polynesians, and twelve female Polynesians who had also been on board the *Bounty*. In 1856, approximately 200 of the mutineers' descendants voluntarily moved from Pitcairn Island to Norfolk Island because of overpopulation. Descendants of the original mutineers continue to live on Pitcairn and other islands nearby into the early twenty-first century.

Where did Gauguin live?

The French painter Paul Gauguin first visited Tahiti in 1891 and later moved there from 1895 to 1901 in order to escape European civilization.

What two Pacific Ocean island neighbors have the most difference in time?

Although the cities of Pago Pago, American Samoa, and Apia, Samoa, are only 77.79 miles (125.19 km) apart, because of

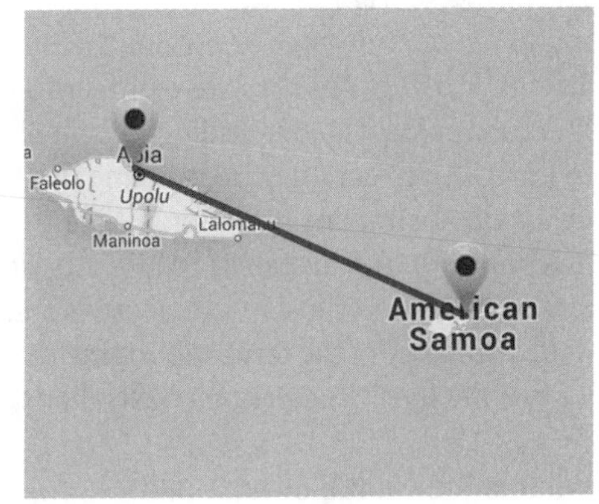

Pago Pago is 77.79 miles (125.19 kilometers) from Apia, but the official difference in time between the two is twenty-five hours because of the International Date Line.

Daylight Saving Time observed during Apia's spring, pushing the clocks ahead by one hour, the time difference between them is twenty-five hours. So in November, when it is 9:42 A.M. on Thursday morning in Pago Pago, it is 10:42 A.M. on Friday in Apia. Pago Pago does not observe Daylight Saving Time, but Apia does.

AUSTRALIA

How was Australia discovered?

Long throughout history, there remained an assumed, yet completely undiscovered, land called Terra Australis Incognita, or ""Unknown Southern Land." As early as the fourth century B.C.E., the Greek philosopher Aristotle believed that an extremely large continent, located in the Southern Hemisphere, lay undiscovered and would complete the symmetry of the landmasses. For centuries, this unknown landmass remained a treasured legend and often appeared on maps in varied sizes and shapes. When the territory now known as Australia was discovered, but not colonized, by the Dutch navigator Willem Janszoon in the early seventeenth century, no one believed that this was the famed Terra Australis Incognita. During the early seventeenth century, the western coast of this territory was named New Holland and claimed for the Netherlands. In 1770, James Cook claimed the east coast of this territory for England and called it New South Wales. It wasn't until 1803 that Matthew Flinders circumnavigated this territory and proved that it was a continent and was the long sought-after Terra Australis Incognita. Finally, in the early nineteenth century, nearly two centuries after having been discovered, this land was finally named Australia.

Who were the Aborigines?

The term "Aborigines" was a name used to describe the indigenous inhabitants of Australia, having migrated from Southeast Asia to the continent approximately 42,000–48,000 years ago. In the late eighteenth century, when European colonization began, there were over 300,000 Indigenous Australians in Australia. Many died from exposure to European infectious diseases, and by 1920, there were only approximately 60,000 Indigenous Australians remaining. Like the Maori people of New Zealand, Australia's indigenous population rebounded in the late twentieth century and now number over 200,000. Most now live in urban areas and have considerable political and cultural support. Today, use of the term "Aborigine" has a negative connotation, and, since the 1980s, we use the term "Indigenous Australians" instead.

What is the capital of Australia?

The capital of Australia is Canberra, which is located in a federal territory (similar to Washington, D.C.) within the Australian state of New South Wales. When Australia was founded in 1901, both Sydney and Melbourne wanted to become the capital city. In 1908,

Was Australia really used as a penal colony?

Yes, approximately two-thirds of Australia's initial settlers were convicts from Great Britain. From 1788 to 1868, when Australia was used as a penal colony, approximately 162,000 prisoners were sent to the continent until 1868, when Britain stopped sending prisoners to Australia. Free colonists who were settling the new territory began arriving with the first ship of convicts.

it was ultimately decided that a brand-new capital city would be built and located away from the coast. Canberra is the largest city in Australia that is not on the coastline (population 381,488).

What's the big, red rock in the middle of Australia?

The big, red, sandstone monolith in the center of Australia is called Uluru (previously named Ayers Rock). It is found within the Uluru-Kaṭa Tjuta National Park and is the world's largest monolith, measuring approximately 1.5 miles (2.4 km) wide and 1,100 feet (335 meters) high.

Are kangaroos native to Australia?

Yes, kangaroos are native to Australia. Kangaroos range in size from giant kangaroos, which are five feet (1.5 meters) tall, to tiny, rat-sized marsupials, named potoroos. The potoroo is highly endangered.

Is the Tasmanian devil a real animal?

The Tasmanian devil is a real animal, though it resembles few characteristics of its cartoon counterpart. The real Tasmanian devil is a carnivorous marsupial that lives in the wild, only on the island of Tasmania, just southeast of the Australian mainland.

Is Australia the smallest continent?

Despite being the sixth-largest country in the world by land area, Australia is the smallest continent in the world. Australia is approximately 2.969 million square miles (7.692 million square km) in area, slightly smaller than Brazil. Australia is also the largest country in Oceania, the largest country surrounded entirely by water, and the largest country entirely lying in the Southern Hemisphere.

Which country is the world's leading bauxite producer?

Bauxite ore is a necessary component in the production of aluminum. Australia mines process more bauxite ore than any other country in the world, producing approximately

35% of the world's supply annually. Bauxite ore is especially prevalent in Australia's Darling Range, located in the southwestern part of the country.

Which country is one of the world's leading lead producers?

Australia produces approximately 13% of the world's lead annually. Most of the lead is mined near Mt. Isa in the northeast area of the country and near Broken Hill in the southeast of the country. China is the leader in lead production, producing approximately 55% of the world's lead.

How much beef does Australia export?

Australia is the third-biggest exporter of beef in the world, behind India and Brazil, and is responsible for exporting 1,851 metric tons per year.

What is a boomerang?

The boomerang was thought to have been developed as a hunting tool by the indigenous people of Australia. There are two types of boomerangs—those that return and those that do not return. The returning boomerang is used to kill small animals; the nonreturning boomerang is used to kill large game. Although scientists have dated Australian boomerangs to be approximately 10,000 years old, other boomerang-like tools have been unearthed in such unlikely places as Europe and Egypt and have been dated to be approximately 30,000 years old.

What is the Royal Flying Doctor Service?

Created in 1928, the Royal Flying Doctor Service (RFDS) is a nonprofit organization established to provide health care and emergency services to the sparse population of Australia's outback. With sixty-six aircraft, the RFDS flies approximately 16.7 million miles (26.9 million km) each year and serves approximately 300,000 people through the use of such services as clinics, flights, and telephone consultations.

Where is the largest coral reef in the world?

The Great Barrier Reef is the largest coral reef system in the world. Located just off the northeastern coast of Queensland, Australia, it is comprised of approximately 2,900 reefs

Where is the outback?

The outback is the general term used to describe one of seven regions of Australia, encompassing all areas north and west of the population centers along the east coast of the country. Most of Australia's population is concentrated along the coast, since the interior is extremely dry and barren.

and 900 islands and extends for over 1,400 miles (2,300 km). Much of the area is now protected as a marine park. Google created a street view of the reef system, in its Google Maps application in 2014, free for users to peruse.

What is ANZUS?

In 1951, Australia, New Zealand, and the United States signed the Australia–New Zealand–United States (ANZUS) Treaty to protect each other militarily against any regional instability. In 1985, New Zealand banned nuclear weapons from its country and thereafter refused to allow U.S. nuclear-powered or nuclear-armed ships to dock in its harbors. New Zealand was summarily suspended from the treaty, although it still participates in many ways to collectively defend the region.

NEW ZEALAND

Who are the Maori?

The Maori are the indigenous inhabitants of New Zealand. Between 1250 and 1300 C.E., the Maori arrived in New Zealand from other Pacific islands. In 1769 there were over 100,000 Maori, but their population decreased significantly (to 40,000) by the end of the nineteenth century due to the effects of European colonization. In the twenty-first century, the Maori population has expanded to approximately 600,000 in New Zealand, with another 120,000 living in Australia. The Maori comprise 15% of New Zealand's population.

What is Aotearoa?

Aotearoa is the Maori name for New Zealand. It means "The Land of the Long White Cloud." Originally, it was used to describe just the North Island of the country.

Are there more people or sheep in New Zealand?

There are approximately 4.63 million people living in New Zealand but approximately 29.8 million sheep. The number of sheep peaked in 1982, at approximately 70.3 million. New Zealand has long been a leading exporter of wool.

An illustration of a Maori chief, who is sporting traditional face tattoos.

Can you ski in New Zealand?

New Zealand's South Island is famous for having some of the best skiing in the Southern Hemisphere. The center of New Zealand skiing is located in the mountain resort community of Queenstown, which is only a few hours by plane from the capital city of Auckland. The ski season is typically July through September, when winter hits the countries of the southern latitudes.

Who are Kiwis?

One nickname for a New Zealander is "Kiwi," but kiwis are also a flightless bird and a type of fruit found in New Zealand. Kiwi birds, a national icon of New Zealand, have long, thin beaks and lay eggs larger in proportion to their body size than any other bird. Most of the world's supply of kiwi fruit is also grown in New Zealand.

What was the *Rainbow Warrior*?

In 1985, the Greenpeace ship *Rainbow Warrior* was in the Auckland, New Zealand, harbor when it exploded and sank, killing one staff member. It was later discovered that French secret agents planted bombs on board the *Rainbow Warrior* in order to stop the organization from protesting French nuclear weapon tests in the Pacific. Following the incident, the French minister of defense and head of the Secret Service were forced to resign. As a result of Greenpeace activities, nuclear testing came to an end in the Pacific. New Zealand, a country very much opposed to nuclear weapons, maintained a poor relationship with France for many years following the bombing.

Greenpeace's *Rainbow Warrior*.

Which country was one of the world's earliest welfare states?

A welfare state is a government that enacts policies, either universal or selective, that promote economic and social well-being to its people. In 1898, while still a colony of the United Kingdom, New Zealand created the first publicly funded pension to elderly people. Later, in the 1930s, New Zealand began to offer its citizens full social security and health benefits.

Which country first granted women the right to vote?

By the late nineteenth century, democracies throughout the world did not extend voting rights to women. This changed in September of 1893, when New Zealand became the first country to give women, universally, the right to vote. Women voted in the next election, held in November of that same year. Also in 1893, a woman was elected mayor of the small town of Onehunga. It was the first time a woman had been elected mayor of a town anywhere in the British Empire.

ANTARCTICA

How big is the continent of Antarctica?

The continent of Antarctica is the fifth-biggest continent on Earth, covering approximately 5,400,000 square miles (14,000,000 square kilometers) in area. It is nearly twice the size of the continent of Australia and 1.3 times bigger than Europe.

How many research stations does the United States maintain in Antarctica?

The United States maintains three research stations in Antarctica: McMurdo Station (the main station), South Pole Station, and Palmer Station.

How much of the world's fresh water is located in Antarctica?

The ice in Antarctica is referred to as the Antarctic ice sheet (East and West). Most of this ice is approximately one mile (1.6 km) thick. Approximately 90% of all ice on the planet is in Antarctica, which means that Antarctica contains an estimated 70% of the world's fresh water. Some have suggested that large chunks of ice could be cut off from Antarctica and shipped to dry regions of the world, but this has yet to be done.

How much ice is being added to the Antarctic ice sheet each year?

From 1992 to 2001, snowfall added 112 billion tons (101.6 billion metric tons) of ice to the Antarctic ice sheet each year. But from 2003 to 2008, the rate of snowfall fell to 82 billion tons (74.39 billion metric tons) per year, a significant decrease.

Only about a thousand people live in Antarctica. This photo shows a whaling' station on Deception Island.

What time is it in Antarctica?

Because of the many research stations in Antarctica, most researchers who visit the continent generally observe Greenwich Mean Time, the same time zone as London, England, or the time zone of their home countries.

How many people are living on Antarctica?

The number of researchers who visit and live on Antarctica varies each year, but normally there are approximately 1,000 people in the winter months and 4,500 people during the summer months.

Which continent has the highest average elevation?

The average elevation of Antarctica, approximately 6,500 feet (2,000 meters), is higher than that of any other continent. The highest point in Antarctica is Vinson Massif, with a peak elevation of 16,050 feet (4,892 meters).

How dry is Antarctica?

Though Antarctica is covered with ice, it is the driest continent on the planet. Some of the ice in Antarctica has been there for thousands of years, and the continent receives less than an average of eight inches (203 mm) of precipitation along the coastal areas annually and much less farther inland. By contrast, the Sahara Desert receives approximately 10 inches (254 mm) each year.

Who owns Antarctica?

Though Antarctica is a cold, icy, barren territory, seven countries claimed portions of it in the early twentieth century. All of these claims were defined by lines of longitude, and problems arose as many of these claims overlapped. In 1959, the Antarctic Treaty was established, proclaiming that no additional claims could be made upon Antarctica and that the continent would be used solely for scientific purposes. Furthermore, the Antarctic Treaty provides the legal framework for the resolution of disputes, reserves the entire region for peace, promotes scientific investigations and international cooperation, requires an annual exchange of information about activities, and encourages environmental stewardship. Representatives of twenty-nine voting nations and twenty-one nonvoting nations meet on a regular basis to discuss the provisions of the treaty.

Countries of the World

There are currently 195 countries in the world, and in this appendix you'll find key statistics on each one. You can use this chapter to find important geographic, political, and cultural information on countries mentioned in this book. The source of this information is the CIA's *World Factbook.*

Afghanistan

Long Name: Islamic Republic of Afghanistan

Location: Southern Asia, north and west of Pakistan, east of Iran

Area: 251,827 sq. mi. (652,230 sq. km)

Climate: arid to semiarid; cold winters and hot summers

Terrain: mostly rugged mountains; plains in north and southwest

Population: 32,564,342

Population Growth Rate: 2.32%

Birth Rate (per 1,000): 38.57

Death Rate (per 1,000): 13.89

Life Expectancy: 50.87 years

Ethnic Groups: Pashtun, Tajik, Hazara, Uzbek, other (includes smaller numbers of Baloch, Turkmen, Nuristani, Pamiri, Arab, Gujar, Brahui, Qizilbash, Aimaq, Pashai, and Kyrghyz)

Religion: Muslim 99.7% (Sunni 84.7-89.7%, Shia 10-15%), other 0.3% (2009 est.)

Languages: Afghan Persian or Dari (official) 50%, Pashto (official) 35%, Turkic languages (primarily Uzbek and Turkmen) 11%, 30 minor languages (primarily Balochi and Pashai) 4%, much bilingualism, but Dari functions as the lingua franca

Literacy: 38.2%

Government Type: Islamic republic

Capital: Kabul

Independence: 19 August 1919 (from United Kingdom control over Afghan foreign affairs)

GDP Per Capita: $2,000

Occupations: agriculture 78.6%, industry 5.7%, services 15.7%

Currency: afghani (AFA)

Albania

Long Name: Republic of Albania

Location: Southeastern Europe, bordering the Adriatic Sea and Ionian Sea, between Greece in the south and Montenegro and Kosovo to the north

Area: 11,099 sq. mi. (28,748 sq. km)

Climate: mild temperate; cool, cloudy, wet winters; hot, clear, dry summers; interior is cooler and wetter

Terrain: mostly mountains and hills; small plains along coast

Population: 3,029,278

Population Growth Rate: 0.3%

Birth Rate (per 1,000): 12.92

Death Rate (per 1,000): 6.58

Life Expectancy: 78.13 years

Ethnic Groups: Albanian 82.6%, Greek 0.9%, other 1% (including Vlach, Roma (Gypsy), Macedonian, Montenegrin, and Egyptian), unspecified 15.5%; note: in 1989, other estimates of the Greek population ranged from 1% (official Albanian statistics) to 12% (from a Greek organization)

Religion: Muslim 56.7%, Roman Catholic 10%, Orthodox 6.8%, atheist 2.5%, Bektashi (a Sufi order) 2.1%, other 5.7%, unspecified 16.2%

Languages: Albanian 98.8% (official-derived from Tosk dialect), Greek 0.5%, other 0.6% (including Macedonian, Roma, Vlach, Turkish, Italian, and Serbo-Croatian), unspecified 0.1%

Literacy: 97.6%

Government Type: parliamentary democracy

Capital: Tirana (Tirane)

Independence: 28 November 1912 (from the Ottoman Empire)

GDP Per Capita: $11,900

Occupations: agriculture 41.8%, industry 11.4%, services 46.8%

Currency: lek (ALL)

Algeria

Long Name: People's Democratic Republic of Algeria

Location: Northern Africa, bordering the Mediterranean Sea, between Morocco and Tunisia

Area: 919,590 sq. mi. (2,381,741 sq. km)

Climate: arid to semiarid; mild, wet winters with hot, dry summers along coast; drier with cold winters and hot summers on high plateau; sirocco is a hot, dust/sand-laden wind especially common in summer

Terrain: mostly high plateau and desert; some mountains; narrow, discontinuous coastal plain

Population: 39,542,166

Population Growth Rate: 1.84%

Birth Rate (per 1,000): 23.67

Death Rate (per 1,000): 4.31

Life Expectancy: 76.59 years

Ethnic Groups: Arab-Berber 99%, European less than 1%

Religion: Muslim (official; predominantly Sunni) 99%, other (includes Christian and Jewish) less than 1%

Languages: Arabic (official), French (lingua franca), Berber or Tamazight (official); dialects include Kabyle Berber (Taqbaylit), Shawiya Berber (Tacawit), Mzab Berber, Tuareg Berber (Tamahaq)

Literacy: 80.2%

Government Type: republic

Capital: Algiers

Independence: 5 July 1962 (from France)

GDP Per Capita: $14,400

Occupations: agriculture 14%, industry 13.4%, construction and public works 10%, trade 14.6%, government 32%, other 16%

Currency: Algerian dinar (DZD)

Andorra

Long Name: Principality of Andorra

Location: Southwestern Europe, between France and Spain

Area: 181 sq. mi. (468 sq. km)

Climate: temperate; snowy, cold winters and warm, dry summers

Terrain: rugged mountains dissected by narrow valleys

Population: 85,580

Population Growth Rate: 0.12%

Birth Rate (per 1,000): 8.13

Death Rate (per 1,000): 6.96

Life Expectancy: 82.72 years

Ethnic Groups: Andorran 49%, Spanish 24.6%, Portuguese 14.3%, French 3.9%, other 8.2%

Religion: Roman Catholic (predominant)

Languages: Catalan (official), French, Castilian, Portuguese

Literacy: 100%

Government Type: parliamentary democracy (since March 1993) that retains its chiefs of state in the form of a co-principality; the two princes are the President of France and Bishop of Urgell, whose diocese is located in neighboring Spain; both co-princes maintain offices and representatives in Andorra

Capital: Andorra la Vella

Independence: 1278 (formed under the joint suzerainty of the French Count of Foix and the Spanish Bishop of Urgel)

GDP Per Capita: $37,200

Occupations: agriculture 0.4%, industry 4.7%, services 94.9%

Currency: euro (EUR)

Angola

Long Name: Republic of Angola

Location: Southern Africa, bordering the South Atlantic Ocean, between Namibia and Democratic Republic of the Congo

Area: 481,351.35 sq. mi. (1,246,700 sq. km)

Climate: semiarid in south and along coast to Luanda; north has cool, dry season (May to October) and hot, rainy season (November to April)

Terrain: narrow coastal plain rises abruptly to vast interior plateau

Population: 19,625,353

Population Growth Rate: 2.78%

Birth Rate (per 1,000): 38.78

Death Rate (per 1,000): 11.49

Life Expectancy: 55.63 years

Ethnic Groups: Ovimbundu 37%, Kimbundu 25%, Bakongo 13%, mestico (mixed European and native African) 2%, European 1%, other 22%

Religion: indigenous beliefs 47%, Roman Catholic 38%, Protestant 15%

Languages: Portuguese (official), Bantu and other African languages

Literacy: 71.1%

Government Type: republic; multiparty presidential regime

Capital: Luanda

Independence: 11 November 1975 (from Portugal)

GDP Per Capita: $7,600

Occupations: agriculture 85%, industry and services 15%

Currency: kwanza (AOA)

Antigua and Barbuda

Long Name: Antigua and Barbuda

Location: Caribbean, islands between the Caribbean Sea and the North Atlantic Ocean, east-southeast of Puerto Rico

Area: 171 sq. mi. (442.6 sq. km)

Climate: tropical maritime; little seasonal temperature variation

Terrain: mostly low-lying limestone and coral islands, with some higher volcanic areas

Population: 92,436

Population Growth Rate: 1.24%

Birth Rate (per 1,000): 15.85

Death Rate (per 1,000): 5.69

Life Expectancy: 76.33 years

Ethnic Groups: black 87.3%, mixed 4.7%, hispanic 2.7%, white 1.6%, other 2.7%, unspecified 0.9

Religion: Protestant 68.3% (Anglican 17.6%, Seventh Day Adventist 12.4%, Pentecostal 12.2%, Moravian 8.3%, Methodist 5.6%, Wesleyan Holiness 4.5%, Church of God 4.1%, Baptist 3.6%), Roman Catholic 8.2%, other 12.2%, unspecified 5.5%, none 5.9%

Languages: English (official), Antiguan creole

Literacy: 99%

Government Type: constitutional monarchy with a parliamentary system of government and a Commonwealth realm

Capital: Saint John's

Independence: 1 November 1981 (from United Kingdom)

GDP Per Capita: $23,700

Occupations: agriculture 7%, industry 11%, services 82%

Currency: East Caribbean dollar (XCD)

Argentina

Long Name: Argentine Republic

Location: Southern South America, bordering the South Atlantic Ocean, between Chile and Uruguay

Area: 1,073,518 sq. mi. (2,780,400 sq. km)

Climate: mostly temperate; arid in southeast; subantarctic in southwest

Terrain: rich plains of the Pampas in northern half, flat to rolling plateau of Patagonia in south, rugged Andes along western border

Population: 43,431,886

Population Growth Rate: 0.93%

Birth Rate (per 1,000): 16.64

Death Rate (per 1,000): 7.33

Life Expectancy: 77.69 years

Ethnic Groups: white (mostly Spanish and Italian) 97%, mestizo (mixed white and Amerindian ancestry), Amerindian, or other non-white groups 3%

Religion: nominally Roman Catholic 92% (less than 20% practicing), Protestant 2%, Jewish 2%, other 4%

Languages: Spanish (official), Italian, English, German, French

Literacy: 98.1%

Government Type: republic

Capital: Buenos Aires

Independence: 9 July 1816 (from Spain)

GDP Per Capita: $22,400

Occupations: agriculture 5%, industry 23%, services 72%

Currency: Argentine peso (ARS)

Armenia

Long Name: Republic of Armenia

Location: Southwestern Asia, east of Turkey

Area: 11,484 sq. mi. (29,743 sq. km)

Climate: highland continental, hot summers, cold winters

Terrain: Armenian Highland with mountains; little forest land; fast flowing rivers; good soil in Aras River valley

Population: 3,056,382

Population Growth Rate: -0.15%

Birth Rate (per 1,000): 13.61

Death Rate (per 1,000): 9.34

Life Expectancy: 74.37 years

Ethnic Groups: Armenian 98.1%, Yezidi (Kurd) 1.1%, other 0.7%

Religion: Armenian Apostolic 92.6%, Evangelical 1%, other 2.4%, none 1.1%, unspecified 2.9%

Languages: Armenian (official) 97.9%, Kurdish (spoken by Yezidi minority) 1%, other 1%

Literacy: 99.7%

Government Type: republic

Capital: Yerevan

Independence: 21 September 1991 (from Soviet Union)

GDP Per Capita: $8,400

Occupations: agriculture 39%, industry 17%, services 44%

Currency: dram (AMD)

Australia

Long Name: Commonwealth of Australia

Location: Oceania, continent between the Indian Ocean and the South Pacific Ocean

Area: 2,988,901 sq. mi. (7,741,220 sq. km)

Climate: generally arid to semiarid; temperate in south and east; tropical in north

Terrain: mostly low plateau with deserts; fertile plain in southeast

Population: 22,751,014

Population Growth Rate: 1.07%

Birth Rate (per 1,000): 12.15

Death Rate (per 1,000): 7.14

Life Expectancy: 82.15 years

Ethnic Groups: English 25.9%, Australian 25.4%, Irish 7.5%, Scottish 6.4%, Italian 3.3%, German 3.2%, Chinese 3.1%, Indian 1.4%, Greek 1.4%, Dutch 1.2%, other 15.8% (includes Australian aboriginal 0.5%), unspecified 5.4%

Religion: Protestant 30.1% (Anglican 17.1%, Uniting Church 5.0%, Presbyterian and Reformed 2.8%, Baptist, 1.6%, Lutheran 1.2%, Pentecostal 1.1%, other Protestant 1.3%), Catholic 25.3% (Roman Catholic 25.1%, other Catholic 0.2%), other Christian 2.9%, Orthodox 2.8%, Buddhist 2.5%, Muslim 2.2%, Hindu 1.3%, other 1.3%, none 22.3%, unspecified 9.3%

Languages: English 76.8%, Mandarin 1.6%, Italian 1.4%, Arabic 1.3%, Greek 1.2%, Cantonese 1.2%, Vietnamese 1.1%, other 10.4%, unspecified 5%

Literacy: 99%

Government Type: federal parliamentary democracy

Capital: Canberra

Independence: 1 January 1901 (federation of United Kingdom colonies)

GDP Per Capita: $65,400

Occupations: agriculture 3.6%, industry 21.1%, services 75.3%

Currency: Australian dollar (AUD)

Austria

Long Name: Republic of Austria

Location: Central Europe, north of Italy and Slovenia

Area: 32,382 sq. mi. (83,871 sq. km)

Climate: temperate; continental, cloudy; cold winters with frequent rain and some snow in lowlands and snow in mountains; moderate summers with occasional showers

Terrain: in the west and south mostly mountains (Alps); along the eastern and northern margins mostly flat or gently sloping

Population: 8,665,550

Population Growth Rate: 0.55%

Birth Rate (per 1,000): 9.41

Death Rate (per 1,000): 9.42

Life Expectancy: 81.39 years

Ethnic Groups: Austrians 91.1%, former Yugoslavs 4% (includes Croatians, Slovenes, Serbs, and Bosniaks), Turks 1.6%, German 0.9%, other or unspecified 2.4%

Religion: Catholic 73.8% (includes Roman Catholic 73.6%, other Catholic 0.2%), Protestant 4.9%, Muslim 4.2%, Orthodox 2.2%, other 0.8% (includes other Christian), none 12%, unspecified 2%

Languages: German (official nationwide) 88.6%, Turkish 2.3%, Serbian 2.2%, Croatian (official in Burgenland) 1.6%, other (includes Slovene, official in South Carinthia, and Hungarian, official in Burgenland) 5.3%

Literacy: 98%

Government Type: federal republic

Capital: Vienna

Independence: 976 (Margravate of Austria established); 17 September 1156 (Duchy of Austria founded); 11 August 1804 (Austrian Empire proclaimed); 12 November 1918 (republic proclaimed)

GDP Per Capita: $47,500

Occupations: agriculture 5.5%, industry 26%, services 68.5%

Currency: euro (EUR)

Azerbaijan

Long Name: Republic of Azerbaijan

Location: Southwestern Asia, bordering the Caspian Sea, between Iran and Russia, with a small European portion north of the Caucasus range

Area: 33,436 sq. mi. (86,600 sq. km)

Climate: dry, semiarid steppe

Terrain: large, flat Kur-Araz Ovaligi (Kura-Araks Lowland) (much of it below sea level) with Great Caucasus Mountains to the north, Qarabag Yaylasi (Karabakh Upland) in west; Baku lies on Abseron Yasaqligi (Apsheron Peninsula) that juts into Caspian Sea

Population: 9,780,780

Population Growth Rate: 0.96%

Birth Rate (per 1,000): 16.64

Death Rate (per 1,000): 7.07

Life Expectancy: 72.2 years

Ethnic Groups: Azerbaijani 91.6%, Lezgian 2%, Russian 1.3%, Armenian 1.3%, Talysh 1.3%, other 2.4%

Religion: Muslim 96.9% (predominantly Shia), Christian 3%, other <0.1

Languages: Azerbaijani (Azeri) (official) 92.5%, Russian 1.4%, Armenian 1.4%, other 4.7%

Literacy: 99.8%

Government Type: republic

Capital: Baku

Independence: 30 August 1991 (from Soviet Union)

GDP Per Capita: $18,700

Occupations: agriculture 6%, industry 58%, services 36.1%

Currency: Azerbaijani manat (AZN)

The Bahamas

Long Name: Commonwealth of The Bahamas

Location: Caribbean, chain of islands in the North Atlantic Ocean, southeast of Florida, northeast of Cuba

Area: 5,359 sq. mi. (13,880 sq. km)

Climate: tropical marine; moderated by warm waters of Gulf Stream

Terrain: long, flat coral formations with some low rounded hills

Population: 324,597

Population Growth Rate: 0.85%

Birth Rate (per 1,000): 15.5

Death Rate (per 1,000): 7.05

Life Expectancy: 72.2 years

Ethnic Groups: black 90.6%, white 4.7%, black and white 2.1%, other 1.9%, unspecified 0.7%

Religion: Protestant 69.9% (includes Baptist 34.9%, Anglican 13.7%, Pentecostal 8.9% Seventh Day Adventist 4.4%, Methodist 3.6%, Church of God 1.9%, Brethren 1.6%), Roman Catholic 12%, other Christian 13% (includes Jehovah's Witness 1.1%), other 0.6%, none 1.9%, unspecified 2.6%

Languages: English (official), Creole (among Haitian immigrants)

Literacy: 95.6%

Government Type: constitutional parliamentary democracy

Capital: Nassau

Independence: 10 July 1973 (from United Kingdom)

GDP Per Capita: $25,600

Occupations: agriculture 3%, industry 11%, tourism 49%, other services 37%

Currency: Bahamian dollar (BSD)

Bahrain

Long Name: Kingdom of Bahrain

Location: Middle East, archipelago in the Persian Gulf, east of Saudi Arabia

Area: 293 sq. mi. (760 sq. km)

Climate: arid; mild, pleasant winters; very hot, humid summers

Terrain: mostly low desert plain rising gently to low central escarpment

Population: 1,346,613

Population Growth Rate: 2.41%

Birth Rate (per 1,000): 13.66

Death Rate (per 1,000): 2.69

Life Expectancy: 78.73 years

Ethnic Groups: Bahraini 46%, Asian 45.5%, other Arabs 4.7%, African 1.6%, European 1%, other 1.2%

Religion: Muslim 70.3%, Christian 14.5%, Hindu 9.8%, Buddhist 2.5%, Jewish 0.6%, folk religion <0.1, unaffiliated 1.9%, other 0.2%

Languages: Arabic, English, Farsi, Urdu

Literacy: 95.7%

Government Type: constitutional monarchy

Capital: Manama

Independence: 15 August 1971 (from United Kingdom)

GDP Per Capita: $51,200

Occupations: agriculture 1%, industry 32%, services 67%

Currency: Bahraini dinar (BHD)

Bangladesh

Long Name: People's Republic of Bangladesh

Location: Southern Asia, bordering the Bay of Bengal, between Burma and India

Area: 57,320 sq. mi. (148,460 sq. km)

Climate: tropical; mild winter (October to March); hot, humid summer (March to June); humid, warm rainy monsoon (June to October)

Terrain: mostly flat alluvial plain; hilly in southeast

Population: 168,957,745

Population Growth Rate: 1.6%

Birth Rate (per 1,000): 21.14

Death Rate (per 1,000): 5.61

Life Expectancy: 70.94 years

Ethnic Groups: Bengali at least 98%, ethnic groups 1.1%

Religion: Muslim 89.1%, Hindu 10%, other 0.9% (includes Buddhist, Christian)

Languages: Bangla 98.8% (official, also known as Bengali), other 1.2%

Literacy: 61.5%

Government Type: parliamentary democracy

Capital: Dhaka

Independence: 16 December 1971 (from West Pakistan)

GDP Per Capita: $3,600

Occupations: agriculture 16%, industry 30.4%, services 53.6%

Currency: taka (BDT)

Barbados

Long Name: Barbados

Location: Caribbean, island in the North Atlantic Ocean, northeast of Venezuela

Area: 166 sq. mi. (430 sq. km)

Climate: tropical; rainy season (June to October)

Terrain: relatively flat; rises gently to central highland region

Population: 290,604

Population Growth Rate: 0.31%

Birth Rate (per 1,000): 11.87

Death Rate (per 1,000): 8.44

Life Expectancy: 75.18 years

Ethnic Groups: black 92.4%, white 2.7%, mixed 3.1%, East Indian 1.3%, other 0.2%, unspecified 0.2%

Religion: Protestant 66.3% (includes Anglican 23.9%, other Pentecostal 19.5%, Adventist 5.9%, Methodist 4.2%, Wesleyan 3.4%, Nazarene 3.2%, Church of God 2.4%, Baptist 1.8%, Moravian 1.2%, other Protestant 0.8%), Roman Catholic 3.8%, other Christian 5.4% (includes Jehovah's Witness 2.0%, other 3.4%), Rastafarian 1%, other 1.5%, none 20.6%, unspecified 1.2%

Languages: English

Literacy: 99.7%

Government Type: parliamentary democracy

Capital: Bridgetown

Independence: 30 November 1966 (from United Kingdom)

GDP Per Capita: $16,700

Occupations: agriculture 10%, industry 15%, services 75%

Currency: Barbadian dollar (BBD)

Belarus

Long Name: Republic of Belarus

Location: Eastern Europe, east of Poland

Area: 80,154 sq. mi. (207,600 sq. km)

Climate: cold winters, cool and moist summers; transitional between continental and maritime

Terrain: generally flat and contains much marshland

Population: 9,589,689

Population Growth Rate: -0.2%

Birth Rate (per 1,000): 10.7

Death Rate (per 1,000): 13.36

Life Expectancy: 72.48 years

Ethnic Groups: Belarusian 83.7%, Russian 8.3%, Polish 3.1%, Ukrainian 1.7%, other 2.4%, unspecified 0.9%

Religion: Eastern Orthodox 80%, other (including Roman Catholic, Protestant, Jewish, and Muslim) 20%

Languages: Russian (official) 70.2%, Belarusian (official) 23.4%, other 3.1% (includes small Polish- and Ukrainian-speaking minorities), unspecified 3.3%

Literacy: 99.7%

Government Type: republic in name, although in fact a dictatorship

Capital: Minsk

Independence: 25 August 1991 (from Soviet Union)

GDP Per Capita: $17,800

Occupations: agriculture 9.3%, industry 32.7%, services 58%

Currency: Belarusian ruble (BYB/BYR)

Belgium

Long Name: Kingdom of Belgium

Location: Western Europe, bordering the North Sea, between France and the Netherlands

Area: 11,787 sq. mi. (30,528 sq. km)

Climate: temperate; mild winters, cool summers; rainy, humid, cloudy

Terrain: flat coastal plains in northwest, central rolling hills, rugged mountains of Ardennes Forest in southeast

Population: 11,323,973

Population Growth Rate: 0.76%

Birth Rate (per 1,000): 11.41

Death Rate (per 1,000): 9.63

Life Expectancy: 80.88 years

Ethnic Groups: Flemish 58%, Walloon 31%, mixed or other 11%

Religion: Roman Catholic 75%, other (includes Protestant) 25%

Languages: Dutch (official) 60%, French (official) 40%, German (official) less than 1%

Literacy: 99%

Government Type: federal parliamentary democracy under a constitutional monarchy

Capital: Brussels

Independence: 4 October 1830 (a provisional government declared independence from the Netherlands); 21 July 1831 (King LEOPOLD I ascended to the throne)

GDP Per Capita: $44,100

Occupations: agriculture 1.3%, industry 18.6%, services 80.1%

Currency: euro (EUR)

Belize

Long Name: Belize

Location: Central America, bordering the Caribbean Sea, between Guatemala and Mexico

Area: 8,867 sq. mi. (22,966 sq. km)

Climate: tropical; very hot and humid; rainy season (May to November); dry season (February to May)

Terrain: flat, swampy coastal plain; low mountains in south

Population: 347,369

Population Growth Rate: 1.87%

Birth Rate (per 1,000): 24.68

Death Rate (per 1,000): 5.97

Life Expectancy: 68.59 years

Ethnic Groups: mestizo 52.9%, Creole 25.9%, Maya 11.3%, Garifuna 6.1%, East Indian 3.9%, Mennonite 3.6%, white 1.2%, Asian 1%, other 1.2%, unknown 0.3%

Religion: Roman Catholic 40.1%, Protestant 31.5% (includes Pentecostal 8.4%, Seventh Day Adventist 5.4%, Anglican 4.7%, Mennonite 3.7%, Baptist 3.6%, Methodist 2.9%, Nazarene 2.8%), Jehovah's Witness 1.7%, other 10.5% (includes Baha'i, Buddhist, Hindu, Morman, Muslim, Rastafarian), unknown 0.6%, none 15.5%

Languages: English 62.9% (official), Spanish 56.6%, Creole 44.6%, Maya 10.5%, German 3.2%, Garifuna 2.9%, other 1.8%, unknown 0.3%, none 0.2% (cannot speak)

Literacy: 76.9%

Government Type: parliamentary democracy

Capital: Belmopan

Independence: 21 September 1981 (from United Kingdom)

GDP Per Capita: $8,600

Occupations: agriculture 10.2%, industry 18.1%, services 71.7%

Currency: Belizean dollar (BZD)

Bénin

Long Name: Republic of Benin

Location: Western Africa, bordering the Bight of Benin, between Nigeria and Togo

Area: 43,484 sq. mi. (112,622 sq. km)

Climate: tropical; hot, humid in south; semiarid in north

Terrain: mostly flat to undulating plain; some hills and low mountains

Population: 10,448,647

Population Growth Rate: 2.78%

Birth Rate (per 1,000): 36.02

Death Rate (per 1,000): 8.21

Life Expectancy: 61.47 years

Ethnic Groups: Fon and related 39.2%, Adja and related 15.2%, Yoruba and related 12.3%, Bariba and related 9.2%, Fulani and related 7%, Ottamari and related 6.1%, Yoa-Lokpa and related 4%, Dendi and related 2.5%, other 1.6%, unspecified 2.9%

Religion: Catholic 27.1%, Muslim 24.4%, Vodoun 17.3%, Protestant 10.4% (Celestial 5%, Methodist 3.2%, other Protestant 2.2%), other traditional religions 6%, other Christian 5.3%, other 1.9%, none 6.5%, unspecified 1.1%

Languages: French (official), Fon and Yoruba (most common vernaculars in south), tribal languages (at least six major ones in north)

Literacy: 38.4%

Government Type: republic

Capital: Porto-Novo

Independence: 1 August 1960 (from France)

GDP Per Capita: $2,000

Occupations: N/A

Currency: Communaute Financiere Africaine franc (XOF)

Bhutan

Long Name: Kingdom of Bhutan

Location: Southern Asia, between China and India

Area: 14,824 sq. mi. (38,394 sq. km)

Climate: varies; tropical in southern plains; cool winters and hot summers in central valleys; severe winters and cool summers in Himalayas

Terrain: mostly mountainous with some fertile valleys and savanna

Population: 741,919

Population Growth Rate: 1.11%

Birth Rate (per 1,000): 17.78

Death Rate (per 1,000): 6.69

Life Expectancy: 69.51 years

Ethnic Groups: Bhote 50%, ethnic Nepalese 35% (includes Lhotsampas-one of several Nepalese ethnic groups), indigenous or migrant tribes 15%

Religion: Lamaistic Buddhist 75.3%, Indian- and Nepalese-influenced Hinduism 22.1%, other 2.6%

Languages: Sharchhopka 28%, Dzongkha (official) 24%, Lhotshamkha 22%, other 26% (includes foreign languages)

Literacy: 64.9%

Government Type: in transition to constitutional monarchy; special treaty relationship with India

Capital: Thimphu

Independence: 1907 (became a unified kingdom under its first hereditary king)

GDP Per Capita: $8,200

Occupations: agriculture 57%, industry 21%, services 22%

Currency: ngultrum (BTN); Indian rupee (INR)

Bolivia

Long Name: Republic of Bolivia

Location: Central South America, southwest of Brazil

Area: 424,162 sq. mi. (1,098,581 sq. km)

Climate: varies with altitude; humid and tropical to cold and semiarid

Terrain: rugged Andes Mountains with a highland plateau (Altiplano), hills, lowland plains of the Amazon Basin

Population: 10,800,882

Population Growth Rate: 1.56%

Birth Rate (per 1,000): 22.76

Death Rate (per 1,000): 6.52

Life Expectancy: 68.86 years

Ethnic Groups: mestizo (mixed white and Amerindian ancestry) 68%, indigenous 20%, white 5%, cholo/chola 2%, black 1%, other 1%, unspecified 3%

Religion: Roman Catholic 76.8%, Evangelical and Pentecostal 8.1%, Protestant 7.9%, other 1.7%, none 5.5%

Languages: Spanish (official) 60.7%, Quechua (official) 21.2%, Aymara (official) 14.6%, foreign languages 2.4%, Guarani (official) 0.6%, other native languages 0.4%, none 0.1%

Literacy: 95.7%

Government Type: republic

Capital: La Paz

Independence: 6 August 1825 (from Spain)

GDP Per Capita: $6,500

Occupations: agriculture 32%, industry 20%, services 47.9%

Currency: boliviano (BOB)

Bosnia and Herzegovina

Long Name: Bosnia and Herzegovina

Location: Southeastern Europe, bordering the Adriatic Sea and Croatia

Area: 19,767 sq. mi. (51,197 sq. km)

Climate: hot summers and cold winters; areas of high elevation have short, cool summers and long, severe winters; mild, rainy winters along coast

Terrain: mountains and valleys

Population: 3,867,055

Population Growth Rate: –0.13%

Birth Rate (per 1,000): 8.87

Death Rate (per 1,000): 9.75

Life Expectancy: 76.55 years

Ethnic Groups: Bosniak 48.4%, Serb 32.7%, Croat 14.6%, other 4.3%

Religion: Muslim 40%, Orthodox 31%, Roman Catholic 15%, other 14%

Languages: Bosnian (official), Croatian (official), Serbian (official)

Literacy: 98.5%

Government Type: emerging federal democratic republic

Capital: Sarajevo

Independence: 1 March 1992 (from Yugoslavia; referendum for independence completed 1 March 1992; independence declared 3 March 1992)

GDP Per Capita: $10,200

Occupations: N/A

Currency: konvertibilna marka (convertible mark) (BAM)

Botswana

Long Name: Republic of Botswana

Location: Southern Africa, north of South Africa

Area: 224,607 sq. mi. (581,730 sq. km)

Climate: semiarid; warm winters and hot summers

Terrain: predominantly flat to gently rolling tableland; Kalahari Desert in southwest

Population: 2,182,719

Population Growth Rate: 1.21%

Birth Rate (per 1,000): 20.96

Death Rate (per 1,000): 13.39

Life Expectancy: 54.18 years

Ethnic Groups: Tswana (or Setswana) 79%, Kalanga 11%, Basarwa 3%, other, including Kgalagadi and white 7%

Religion: Christian 71.6%, Badimo 6%, other 1.4% (includes Baha'i, Hindu, Muslim), unspecified 0.4%, none 20.6%

Languages: Setswana 78.2%, Kalanga 7.9%, Sekgalagadi 2.8%, English (official) 2.1%, Sesarwa 1.9%, Sempukushu 1.7%, other 5.1%, unspecified 0.2%

Literacy: 88.5%

Government Type: parliamentary republic

Capital: Gaborone

Independence: 30 September 1966 (from United Kingdom)

GDP Per Capita: $17,700

Occupations: N/A

Currency: pula (BWP)

Brazil

Long Name: Federative Republic of Brazil

Location: Eastern South America, bordering the Atlantic Ocean

Area: 3,287,957 sq. mi. (8,515,770 sq. km)

Climate: mostly tropical, but temperate in south

Terrain: mostly flat to rolling lowlands in north; some plains, hills, mountains, and narrow coastal belt

Population: 204,259,812

Population Growth Rate: 0.77%

Birth Rate (per 1,000): 14.46

Death Rate (per 1,000): 6.58

Life Expectancy: 73.53 years

Ethnic Groups: white 47.7%, mulatto (mixed white and black) 43.1%, black 7.6%, Asian 1.1%, indigenous 0.4%

Religion: Roman Catholic 64.6%, other Catholic 0.4%, Protestant 22.2% (includes Adventist 6.5%, Assembly of God 2.0%, Christian Congregation of Brazil 1.2%, Universal Kingdom of God 1.0%, other Protestant 11.5%), other Christian 0.7%, Spiritist 2.2%, other 1.4%, none 8%, unspecified 0.4%

Languages: Portuguese (official and most widely spoken language)

Literacy: 92.6%

Government Type: federal republic

Capital: Brasilia

Independence: 7 September 1822 (from Portugal)

GDP Per Capita: $15,800

Occupations: agriculture 15.7%, industry 13.3%, services 71%

Currency: real (BRL)

Brunei

Long Name: Brunei Darussalam

Location: Southeastern Asia, bordering the South China Sea and Malaysia

Area: 2,226 sq. mi. (5,765 sq. km)

Climate: tropical; hot, humid, rainy

Terrain: flat coastal plain rises to mountains in east; hilly lowland in west

Population: 429,646

Population Growth Rate: 1.62%

Birth Rate (per 1,000): 17.32

Death Rate (per 1,000): 3.52

Life Expectancy: 76.97 years

Ethnic Groups: Malay 65.7%, Chinese 10.3%, other indigenous 3.4%, other 20.6%

Religion: Muslim (official) 78.8%, Christian 8.7%, Buddhist 7.8%, other (includes indigenous beliefs) 4.7%

Languages: Malay (official), English, Chinese dialects

Literacy: 96%

Government Type: constitutional sultanate

Capital: Bandar Seri Begawan

Independence: 1 January 1984 (from United Kingdom)

GDP Per Capita: $79,700

Occupations: agriculture 4.2%, industry 62.8%, services 33%

Currency: Bruneian dollar (BND)

Bulgaria

Long Name: Republic of Bulgaria

Location: Southeastern Europe, bordering the Black Sea, between Romania and Turkey

Area: 42,811 sq. mi. (110,879 sq. km)

Climate: temperate; cold, damp winters; hot, dry summers

Terrain: mostly mountains with lowlands in north and southeast

Population: 7,186,893

Population Growth Rate: -0.58%

Birth Rate (per 1,000): 8.92

Death Rate (per 1,000): 14.44

Life Expectancy: 74.39 years

Ethnic Groups: Bulgarian 76.9%, Turkish 8%, Roma 4.4%, other 0.7% (including Russian, Armenian, and Vlach), other (unknown) 10%

Religion: Eastern Orthodox 59.4%, Muslim 7.8%, other (including Catholic, Protestant, Armenian Apostolic Orthodox, and Jewish) 1.7%, none 3.7%, unspecified 27.4%

Languages: Bulgarian (official) 76.8%, Turkish 8.2%, Roma 3.8%, other 0.7%, unspecified 10.5%

Literacy: 98.4%

Government Type: parliamentary democracy

Capital: Sofia

Independence: 3 March 1878 (as an autonomous principality within the Ottoman Empire); 22 September 1908 (complete independence from the Ottoman Empire)

GDP Per Capita: $18,400

Occupations: agriculture 6.7%, industry 30.2%, services 63.1%

Currency: lev (BGN)

Burkina Faso

Long Name: Burkina Faso

Location: Western Africa, north of Ghana

Area: 105,869 sq. mi. (274,200 sq. km)

Climate: tropical; warm, dry winters; hot, wet summers

Terrain: mostly flat to dissected, undulating plains; hills in west and southeast

Population: 18,931,686

Population Growth Rate: 3.03%

Birth Rate (per 1,000): 42.03

Death Rate (per 1,000): 11.72

Life Expectancy: 55.12 years

Ethnic Groups: Mossi 52.5%, Fulani 8.4%, Gurma 6.8%, Bobo 4.8%, Gurunsi 4.5%, Senufo 4.4%, Bissa 3.9%, Lobi 2.5%, Dagara 2.4%, Tuareg/Bella 1.9%, Dioula 0.8%, unspecified/no answer 0.1%, other 7%

Religion: Muslim 61.6%, Catholic 23.2%, traditional/animist 7.3%, Protestant 6.7%, other/no answer 0.2%, none 0.9%

Languages: French (official), native African languages belonging to Sudanic family spoken by 90% of the population

Literacy: 36%

Government Type: parliamentary republic

Capital: Ouagadougou

Independence: 5 August 1960 (from France)

GDP Per Capita: $1,800

Occupations: agriculture 90%, industry and services 10%

Currency: Communaute Financiere Africaine franc (XOF)

Burundi

Long Name: Republic of Burundi

Location: Central Africa, east of Democratic Republic of the Congo

Area: 10,745 sq. mi. (27,830 sq. km)

Climate: equatorial; high plateau with considerable altitude variation (772 m to 2,670 m above sea level); average annual temperature varies with altitude from 23 to 17 degrees centigrade but is generally moderate as the average altitude is about 1,700 m; average annual rainfall is about 150 cm; two wet seasons (February to May and September to November), and two dry seasons (June to August and December to January)

Terrain: hilly and mountainous, dropping to a plateau in east, some plains

Population: 10,742,276

Population Growth Rate: 3.28%

Birth Rate (per 1,000): 42.01

Death Rate (per 1,000): 8.27

Life Expectancy: 60.09 years

Ethnic Groups: Hutu (Bantu) 85%, Tutsi (Hamitic) 14%, Twa (Pygmy) 1%, Europeans 3,000, South Asians 2,000

Religion: Catholic 62.1%, Protestant 23.9% (includes Adventist 2.3% and other Protestant 21.6%), Muslim 2.5%, other 3.6%, unspecified 7.9%

Languages: Kirundi 29.7% (official), Kirundi and other language 9.1%, French (official) and French and other language 0.3%, Swahili and Swahili and other language 0.2% (along Lake Tanganyika and in the Bujumbura area), English and English and other language 0.06%, more than 2 languages 3.7%, unspecified 56.9%

Literacy: 85.6%

Government Type: republic

Capital: Bujumbura

Independence: 1 July 1962 (from UN trusteeship under Belgian administration)

GDP Per Capita: $900

Occupations: agriculture 93.6%, industry 2.3%, services 4.1%

Currency: Burundi franc (BIF)

Cambodia

Long Name: Kingdom of Cambodia

Location: Southeastern Asia, bordering the Gulf of Thailand, between Thailand, Vietnam, and Laos

Area: 69,898 sq. mi. (181,035 sq. km)

Climate: tropical; rainy, monsoon season (May to November); dry season (December to April); little seasonal temperature variation

Terrain: mostly low, flat plains; mountains in southwest and north

Population: 15,708,756

Population Growth Rate: 1.58%

Birth Rate (per 1,000): 23.83

Death Rate (per 1,000): 7.68

Life Expectancy: 64.14 years

Ethnic Groups: Khmer 90%, Vietnamese 5%, Chinese 1%, other 4%

Religion: Buddhist (official) 96.9%, Muslim 1.9%, Christian 0.4%, other 0.8%

Languages: Khmer (official) 96.3%, other 3.7%

Literacy: 77.2%

Government Type: multiparty democracy under a constitutional monarchy

Capital: Phnom Penh

Independence: 9 November 1953 (from France)

GDP Per Capita: $3,500

Occupations: agriculture 48.7%, industry 19.9%, services 31.5%

Currency: riel (KHR)

Cameroon

Long Name: Republic of Cameroon

Location: Western Africa, bordering the Bight of Biafra, between Equatorial Guinea and Nigeria

Area: 183,568 sq. mi. (475,440 sq. km)

Climate: varies with terrain, from tropical along coast to semiarid and hot in north

Terrain: diverse, with coastal plain in southwest, dissected plateau in center, mountains in west, plains in north

Population: 23,739,218

Population Growth Rate: 2.59%

Birth Rate (per 1,000): 36.17

Death Rate (per 1,000): 10.11

Life Expectancy: 57.93 years

Ethnic Groups: Cameroon Highlanders 31%, Equatorial Bantu 19%, Kirdi 11%, Fulani 10%, Northwestern Bantu 8%, Eastern Nigritic 7%, other African 13%, non-African less than 1%

Religion: indigenous beliefs 40%, Christian 40%, Muslim 20%

Languages: 24 major African language groups, English (official), French (official)

Literacy: 75%

Government Type: republic; multiparty presidential regime

Capital: Yaounde

Independence: 1 January 1960 (from French-administered UN trusteeship)

GDP Per Capita: $3,200

Occupations: agriculture 70%, industry 13%, services 17%

Currency: Communaute Financiere Africaine franc (XAF)

Canada

Long Name: Canada

Location: Northern North America, bordering the North Atlantic Ocean on the east, North Pacific Ocean on the west, and the Arctic Ocean on the north, north of the continental US

Area: 3,855,085 sq. mi. (9,984,670 sq km)

Climate: varies from temperate in south to subarctic and arctic in north

Terrain: mostly plains with mountains in west and lowlands in southeast

Population: 35,099,836

Population Growth Rate: 0.75%

Birth Rate (per 1,000): 10.28

Death Rate (per 1,000): 8.42

Life Expectancy: 81.76 years

Ethnic Groups: Canadian 32.2%, English 19.8%, French 15.5%, Scottish 14.4%, Irish 13.8%, German 9.8%, Italian 4.5%, Chinese 4.5%, North American Indian 4.2%, other 50.9%

Religion: Catholic 40.6% (includes Roman Catholic 38.8%, Orthodox 1.6%, other Catholic 0.2%), Protestant 20.3% (includes United Church 6.1%, Anglican 5%, Baptist 1.9%, Lutheran 1.5%, Pentecostal 1.5%, Presbyterian 1.4%, other Protestant 2.9%), other Christian 6.3%, Muslim 3.2%, Hindu 1.5%, Sikh 1.4%, Buddhist 1.1%, Jewish 1%, other 0.6%, none 23.9%

Languages: English (official) 58.7%, French (official) 22%, Punjabi 1.4%, Italian 1.3%, Spanish 1.3%, German 1.3%, Cantonese 1.2%, Tagalog 1.2%, Arabic 1.1%, other 10.5%

Literacy: 99%

Government Type: constitutional monarchy that is also a parliamentary democracy and a federation

Capital: Ottawa

Independence: 1 July 1867 (union of British North American colonies); 11 December 1931 (recognized by United Kingdom)

GDP Per Capita: $45,900

Occupations: agriculture 2%, manufacturing 13%, construction 6%, services 76%, other 3%

Currency: Canadian dollar (CAD)

Cape Verde

Long Name: Republic of Cape Verde

Location: Western Africa, group of islands in the North Atlantic Ocean, west of Senegal

Area: 1,557 sq. mi. (4,033 sq. km)

Climate: temperate; warm, dry summer; precipitation meager and very erratic

Terrain: steep, rugged, rocky, volcanic

Population: 545,993

Population Growth Rate: 1.36%

Birth Rate (per 1,000): 20.33

Death Rate (per 1,000): 6.11

Life Expectancy: 71.85 years

Ethnic Groups: Creole (mulatto) 71%, African 28%, European 1%

Religion: Roman Catholic 77.3%, Protestant 3.7% (includes Church of the Nazarene 1.7%, Adventist 1.5%, Universal Kingdom of God 0.4%, and God and Love 0.1%), other Christian 4.3% (includes Christian Rationalism 1.9%, Jehovah's Witness 1%, Assembly of God 0.9%, and New Apostolic 0.5%), Muslim 1.8%, other 1.3%, none 10.8%, unspecified 0.7%

Languages: Portuguese, Crioulo (a blend of Portuguese and West African words)

Literacy: 87.6%

Government Type: republic

Capital: Praia

Independence: 5 July 1975 (from Portugal)

GDP Per Capita: $6,700

Occupations: N/A

Currency: Cape Verdean escudo (CVE)

Central African Republic

Long Name: Central African Republic

Location: Central Africa, north of Democratic Republic of the Congo

Area: 240,534 sq. mi. (622,984) sq. km

Climate: tropical; hot, dry winters; mild to hot, wet summers

Terrain: vast, flat to rolling, monotonous plateau; scattered hills in northeast and southwest

Population: 5,391,539

Population Growth Rate: 2.3%

Birth Rate (per 1,000): 35.08

Death Rate (per 1,000): 13.8

Life Expectancy: 51.81 years

Ethnic Groups: Baya 33%, Banda 27%, Mandjia 13%, Sara 10%, Mboum 7%, M'Baka 4%, Yakoma 4%, other 2%

Religion: indigenous beliefs 35%, Protestant 25%, Roman Catholic 25%, Muslim 15%

Languages: French (official), Sangho (lingua franca and national language), tribal languages

Literacy: 36.8%

Government Type: republic

Capital: Bangui

Independence: 13 August 1960 (from France)

GDP Per Capita: $600

Occupations: N/A

Currency: Communaute Financiere Africaine franc (XAF)

Chad

Long Name: Republic of Chad

Location: Central Africa, south of Libya

Area: 495,753 sq. mi. (1,284,000 sq. km)

Climate: tropical in south, desert in north

Terrain: broad, arid plains in center, desert in north, mountains in northwest, lowlands in south

Population: 11,631,456

Population Growth Rate: 1.89%

Birth Rate (per 1,000): 36.6

Death Rate (per 1,000): 14.28

Life Expectancy: 49.81 years

Ethnic Groups: Sara 27.7%, Arab 12.3%, Mayo-Kebbi 11.5%, Kanem-Bornou 9%, Ouaddai 8.7%, Hadjarai 6.7%, Tandjile 6.5%, Gorane 6.3%, Fitri-Batha 4.7%, other 6.4%, unknown 0.3%

Religion: Muslim 53.1%, Catholic 20.1%, Protestant 14.2%, animist 7.3%, other 0.5%, unknown 1.7%, atheist 3.1%

Languages: French (official), Arabic (official), Sara (in south), more than 120 different languages and dialects

Literacy: 40.2%

Government Type: republic

Capital: N'Djamena

Independence: 11 August 1960 (from France)

GDP Per Capita: $2,800

Occupations: agriculture 80%, industry 20%

Currency: Communaute Financiere Africaine franc (XAF)

Chile

Long Name: Republic of Chile

Location: Southern South America, bordering the South Pacific Ocean, between Argentina and Peru

Area: 291,933 sq. mi. (756,102 sq. km)

Climate: temperate; desert in north; Mediterranean in central region; cool and damp in south

Terrain: low coastal mountains; fertile central valley; rugged Andes in east

Population: 17,508,260

Population Growth Rate: 0.82%

Birth Rate (per 1,000): 13.83

Death Rate (per 1,000): 6

Life Expectancy: 78.61 years

Ethnic Groups: white and non-indigenous 88.9%, Mapuche 9.1%, Aymara 0.7%, other indigenous groups 1% (includes Rapa Nui, Likan Antai, Quechua, Colla, Diaguita, Kawesqar, Yagan or Yamana), unspecified 0.3%

Religion: Roman Catholic 66.7%, Evangelical or Protestant 16.4%, Jehovah's Witnesses 1%, other 3.4%, none 11.5%, unspecified 1.1%

Languages: Spanish 99.5% (official), English 10.2%, indigenous 1% (includes Mapudungun, Aymara, Quechua, Rapa Nui), other 2.3%, unspecified 0.2%

Literacy: 97.5%

Government Type: republic

Capital: Santiago

Independence: 18 September 1810 (from Spain)

GDP Per Capita: $23,800

Occupations: agriculture 13.2%, industry 23%, services 63.9%

Currency: Chilean peso (CLP)

China

Long Name: People's Republic of China

Location: Eastern Asia, bordering the East China Sea, Korea Bay, Yellow Sea, and South China Sea, between North Korea and Vietnam

Area: 3,705,390 sq. mi. (9,596,960 sq. km)

Climate: extremely diverse; tropical in south to subarctic in north

Terrain: mostly mountains, high plateaus, deserts in west; plains, deltas, and hills in east

Population: 1,367,485,388

Population Growth Rate: 0.45%

Birth Rate (per 1,000): 12.49

Death Rate (per 1,000): 7.53

Life Expectancy: 75.41 years

Ethnic Groups: Han Chinese 91.6%, Zhuang 1.3%, other (includes Hui, Manchu, Uighur, Miao, Yi, Tujia, Tibetan, Mongol, Dong, Buyei, Yao, Bai, Korean, Hani, Li, Kazakh, Dai and other nationalities) 7.1%

Religion: Buddhist 18.2%, Christian 5.1%, Muslim 1.8%, folk religion 21.9%, Hindu <0.1%, Jewish <0.1%, other 0.7% (includes Daoist [Taoist]), unaffiliated 52.2%

Languages: Standard Chinese or Mandarin (official; Putonghua, based on the Beijing dialect), Yue (Cantonese), Wu (Shanghainese), Minbei (Fuzhou), Minnan (Hokkien-Taiwanese), Xiang, Gan, Hakka dialects, minority languages

Literacy: 96.4%

Government Type: Communist state

Capital: Beijing

Independence: 221 BC (unification under the Qin or Ch'in Dynasty); 1 January 1912 (Manchu Dynasty replaced by a Republic); 1 October 1949 (People's Republic established)

GDP Per Capita: $14,300

Occupations: agriculture 8.9%, industry 42.7%, services 48.4%

Currency: Renminbi (RMB), also referred by the unit yuan (CNY)

Colombia

Long Name: Republic of Colombia

Location: Northern South America, bordering the Caribbean Sea, between Panama and Venezuela, and bordering the North Pacific Ocean, between Ecuador and Panama

Area: 439,734 sq. mi. (1,138,910 sq. km)

Climate: tropical along coast and eastern plains; cooler in highlands

Terrain: flat coastal lowlands, central highlands, high Andes Mountains, eastern lowland plains

Population: 46,736,728

Population Growth Rate: 1.04%

Birth Rate (per 1,000): 16.47

Death Rate (per 1,000): 5.4

Life Expectancy: 75.48 years

Ethnic Groups: mestizo and white 84.2%, Afro-Colombian (includes multatto, Raizal, and Palenquero) 10.4%, Amerindian 3.4%, Roma <0.01, unspecified 2.1%

Religion: Roman Catholic 90%, other 10%

Languages: Spanish

Literacy: 94.7%

Government Type: republic; executive branch dominates government structure

Capital: Bogotá

Independence: 20 July 1810 (from Spain)

GDP Per Capita: $14,000

Occupations: agriculture 17%, industry 21%, services 62%

Currency: Colombian peso (COP)

Comoros

Long Name: Union of the Comoros

Location: Southern Africa, group of islands at the northern mouth of the Mozambique Channel, about two-thirds of the way between northern Madagascar and northern Mozambique

Area: 863 sq. mi. (2,235 sq. km)

Climate: tropical marine; rainy season (November to May)

Terrain: volcanic islands, interiors vary from steep mountains to low hills

Population: 780,971

Population Growth Rate: 1.77%

Birth Rate (per 1,000): 27.84

Death Rate (per 1,000): 7.57

Life Expectancy: 63.85 years

Ethnic Groups: Antalote, Cafre, Makoa, Oimatsaha, Sakalava

Religion: Sunni Muslim 98%, Roman Catholic 2%

Languages: Arabic (official), French (official), Shikomoro (a blend of Swahili and Arabic)

Literacy: 77.8%

Government Type: republic

Capital: Moroni

Independence: 6 July 1975 (from France)

GDP Per Capita: $1,600

Occupations: agriculture 80%, industry and services 20%

Currency: Comoran franc (KMF)

The Congo

See Democratic Republic of the Congo or Republic of the Congo

Costa Rica

Long Name: Republic of Costa Rica

Location: Central America, bordering both the Caribbean Sea and the North Pacific Ocean, between Nicaragua and Panama

Area: 19,730 sq. mi. (51,100 sq. km)

Climate: tropical and subtropical; dry season (December to April); rainy season (May to November); cooler in highlands

Terrain: coastal plains separated by rugged mountains including over 100 volcanic cones, of which several are major volcanoes

Population: 4,814,144

Population Growth Rate: 1.22%

Birth Rate (per 1,000): 15.91

Death Rate (per 1,000): 4.55

Life Expectancy: 78.4 years

Ethnic Groups: white or mestizo 83.6%, mulato 6.7%, indigenous 2.4%, black of African descent 1.1%, other 1.1%, none 2.9%, unspecified 2.2%

Religion: Roman Catholic 76.3%, Evangelical 13.7%, Jehovah's Witnesses 1.3%, other Protestant 0.7%, other 4.8%, none 3.2%

Languages: Spanish (official), English

Literacy: 97.8%

Government Type: democratic republic

Capital: San Jose

Independence: 15 September 1821 (from Spain)

GDP Per Capita: $15,500

Occupations: agriculture 6%, industry 19.7%, services 74.3%

Currency: Costa Rican colon (CRC)

Côte d'Ivoire

Long Name: Republic of Cote d'Ivoire

Location: Western Africa, bordering the North Atlantic Ocean, between Ghana and Liberia

Area: 124,503 sq. mi. (322,463 sq. km)

Climate: tropical along coast, semiarid in far north; three seasons-warm and dry (November to March), hot and dry (March to May), hot and wet (June to October)

Terrain: mostly flat to undulating plains; mountains in northwest

Population: 23,295,302

Population Growth Rate: 1.91%

Birth Rate (per 1,000): 28.67

Death Rate (per 1,000): 8.55

Life Expectancy: 58.34 years

Ethnic Groups: Akan 32.1%, Voltaique or Gur 15%, Northern Mande 12.4%, Krou 9.8%, Southern Mande 9%, other 21.2% (includes European and Lebanese descent), unspecified 0.5%

Religion: Muslim 40.2%, Catholic 19.4%, Evangelical 19.3%, Methodist 2.5%, other Christian 4.5%, animist or no religion 12.8%, other religion/ unspecified 1.4%

Languages: French (official), 60 native dialects with Dioula the most widely spoken

Literacy: 43.1%

Government Type: republic; multiparty presidential regime established 1960

Capital: Yamoussoukro

Independence: 7 August 1960 (from France)

GDP Per Capita: $3,400

Occupations: N/A

Currency: Communaute Financiere Africaine franc (XOF)

Croatia

Long Name: Republic of Croatia

Location: Southeastern Europe, bordering the Adriatic Sea, between Bosnia and Herzegovina and Slovenia

Area: 21,851 sq. mi. (56,594 sq. km)

Climate: Mediterranean and continental; continental climate predominant with hot summers and cold winters; mild winters, dry summers along coast

Terrain: geographically diverse; flat plains along Hungarian border, low mountains and highlands near Adriatic coastline and islands

Population: 4,464,844

Population Growth Rate: -0.13%

Birth Rate (per 1,000): 9.45

Death Rate (per 1,000): 12.18

Life Expectancy: 76.61 years

Ethnic Groups: Croat 90.4%, Serb 4.4%, other 4.4% (including Bosniak, Hungarian, Slovene, Czech, and Roma), unspecified 0.8%

Religion: Roman Catholic 86.3%, Orthodox 4.4%, Muslim 1.5%, other 1.5%, unspecified 2.5%, not religious or atheist 3.8%

Literacy: 99.3%

Government Type: presidential/parliamentary democracy

Capital: Zagreb

Independence: 25 June 1991 (from Yugoslavia)

GDP Per Capita: $21,300

Occupations: agriculture 1.9%, industry 27.6%, services 70.4%

Currency: kuna (HRK)

Cuba

Long Name: Republic of Cuba

Location: Caribbean, island between the Caribbean Sea and the North Atlantic Ocean, 150 km south of Key West, Florida

Area: 42,803 sq. mi. (110,860 sq. km)

Climate: tropical; moderated by trade winds; dry season (November to April); rainy season (May to October)

Terrain: mostly flat to rolling plains, with rugged hills and mountains in the southeast

Population: 11,031,433

Population Growth Rate: –0.15%

Birth Rate (per 1,000): 9.9

Death Rate (per 1,000): 7.72

Life Expectancy: 78.39 years

Ethnic Groups: white 64.1%, mestizo 26.6%, black 9.3%

Religion: nominally Roman Catholic 85%, Protestant, Jehovah's Witnesses, Jewish, Santeria

Languages: Spanish

Literacy: 99.8%

Government Type: Communist state

Capital: Havana

Independence: 20 May 1902 (from Spain 10 December 1898; administered by the US from 1898 to 1902)

GDP Per Capita: $10,200

Occupations: agriculture 18%, industry 10%, services 72%

Currency: Cuban peso (CUP) and Convertible peso (CUC)

Cyprus

Long Name: Republic of Cyprus

Location: Middle East, island in the Mediterranean Sea, south of Turkey

Area: 3,572 sq. mi. (9,251 sq. km)

Climate: temperate; Mediterranean with hot, dry summers and cool winters

Terrain: central plain with mountains to north and south; scattered but significant plains along southern coast

Population: 1,189,197

Population Growth Rate: 1.43%

Birth Rate (per 1,000): 11.41

Death Rate (per 1,000): 6.62

Life Expectancy: 78.51 years

Ethnic Groups: Greek 98.8%, other 1% (includes Maronite, Armenian, Turkish-Cypriot), unspecified 0.2%

Religion: Orthodox Christian 89.1%, Roman Catholic 2.9%, Protestant/Anglican 2%, Muslim 1.8%, Buddhist 1%, other (includes Maronite, Armenian Church, Hindu) 1.4%, unknown 1.1%, none/atheist 0.6%

Languages: Greek (official) 80.9%, Turkish (official) 0.2%, English 4.1%, Romanian 2.9%, Russian 2.5%, Bulgarian 2.2%, Arabic 1.2%, Filipino 1.1%, other 4.3%, unspecified 0.6%

Literacy: 99.1%

Government Type: republic

Capital: Nicosia

Independence: 16 August 1960 (from United Kingdom); note: Turkish Cypriots proclaimed self-rule on 13 February 1975 and independence in 1983, but these proclamations are only recognized by Turkey

GDP Per Capita: $31,000

Occupations: agriculture 14.5%, industry 29%, services 56.5%

Currency: Cypriot pound (CYP); euro (EUR) after 1 January 2008

Czech Republic

Long Name: Czech Republic

Location: Central Europe, southeast of Germany

Area: 30,451 sq. mi. (78,867 sq. km)

Climate: temperate; cool summers; cold, cloudy, humid winters

Terrain: Bohemia in the west consists of rolling plains, hills, and plateaus surrounded by low mountains; Moravia in the east consists of very hilly country

Population: 10,644,842

Population Growth Rate: 0.16%

Birth Rate (per 1,000): 9.63

Death Rate (per 1,000): 10.34

Life Expectancy: 78.48 years

Ethnic Groups: Czech 64.3%, Moravian 5%, Slovak 1.4%, other 1.8%, unspecified 27.5%

Religion: Roman Catholic 10.4%, Protestant (includes Czech Brethren and Hussite) 1.1%, other and unspecified 54%, none 34.5%

Languages: Czech (official) 95.4%, Slovak 1.6%, other 3%

Literacy: 99%

Government Type: parliamentary democracy

Capital: Prague

Independence: 1 January 1993 (Czechoslovakia split into the Czech Republic and Slovakia)

GDP Per Capita: $31,500

Occupations: agriculture 2.6%, industry 37.4%, services 60%

Currency: Czech koruna (CZK)

Democratic Republic of the Congo

Long Name: Democratic Republic of the Congo

Location: Central Africa, northeast of Angola

Area: 905,355 sq. mi. (2,344,858 sq. km)

Climate: tropical; hot and humid in equatorial river basin; cooler and drier in southern highlands; cooler and wetter in eastern highlands; north of Equator-wet season (April to October), dry season (December to February); south of Equator-wet season (November to March), dry season (April to October)

Terrain: vast central basin is a low-lying plateau; mountains in east

Population: 79,375,136

Population Growth Rate: 2.45%

Birth Rate (per 1,000): 34.88

Death Rate (per 1,000): 10.07

Life Expectancy: 56.93 years

Ethnic Groups: over 200 African ethnic groups of which the majority are Bantu; the four largest tribes—Mongo, Luba, Kongo (all Bantu), and the Mangbetu-Azande (Hamitic)—make up about 45% of the population

Religion: Roman Catholic 50%, Protestant 20%, Kimbanguist 10%, Muslim 10%, other (includes syncretic sects and indigenous beliefs) 10%

Languages: French (official), Lingala (a lingua franca trade language), Kingwana (a dialect of Kiswahili or Swahili), Kikongo, Tshiluba

Literacy: 63.8%

Government Type: republic

Capital: Kinshasa

Independence: 30 June 1960 (from Belgium)

GDP Per Capita: $800

Occupations: N/A

Currency: Congolese franc (CDF)

Denmark

Long Name: Kingdom of Denmark

Location: Northern Europe, bordering the Baltic Sea and the North Sea, on a peninsula north of Germany (Jutland); also includes two major islands (Sjaelland and Fyn)

Area: 16,639 sq. mi. (43,094 sq. km)

Climate: temperate; humid and overcast; mild, windy winters and cool summers

Terrain: low and flat to gently rolling plains

Population: 5,581,503

Population Growth Rate: 0.22%

Birth Rate (per 1,000): 10.27

Death Rate (per 1,000): 10.25

Life Expectancy: 79.25 years

Ethnic Groups: Scandinavian, Inuit, Faroese, German, Turkish, Iranian, Somali

Religion: Evangelical Lutheran (official) 80%, Muslim 4%, other (denominations of less than 1% each, includes Roman Catholic, Jehovah's Witness, Serbian Orthodox Christian, Jewish, Baptist, and Buddhist) 16%

Languages: Danish, Faroese, Greenlandic (an Inuit dialect), German (small minority), English (as the predominant second language)

Literacy: 99%

Government Type: constitutional monarchy

Capital: Copenhagen

Independence: first organized as a unified state in 10th century; in 1849 became a constitutional monarchy

GDP Per Capita: $45,800

Occupations: agriculture 2.6%, industry 20.3%, services 77.1%

Currency: Danish krone (DKK)

Djibouti

Long Name: Republic of Djibouti

Location: Eastern Africa, bordering the Gulf of Aden and the Red Sea, between Eritrea and Somalia

Area: 8,858 sq. mi. (23,200 sq. km)

Climate: desert; torrid, dry

Terrain: coastal plain and plateau separated by central mountains

Population: 828,324

Population Growth Rate: 2.2%

Birth Rate (per 1,000): 23.65

Death Rate (per 1,000): 7.73

Life Expectancy: 62.79 years

Ethnic Groups: Somali 60%, Afar 35%, other 5% (includes French, Arab, Ethiopian, and Italian)

Religion: Muslim 94%, Christian 6%

Languages: French (official), Arabic (official), Somali, Afar

Literacy: 67.9%

Government Type: republic

Capital: Djibouti

Independence: 27 June 1977 (from France)

GDP Per Capita: $3,300

Occupations: N/A

Currency: Djiboutian franc (DJF)

Dominica

Long Name: Commonwealth of Dominica

Location: Caribbean, island between the Caribbean Sea and the North Atlantic Ocean, about half way between Puerto Rico and Trinidad and Tobago

Area: 290 sq. mi. (751 sq. km)

Climate: tropical; moderated by northeast trade winds; heavy rainfall

Terrain: rugged mountains of volcanic origin

Population: 73,607

Population Growth Rate: 0.21%

Birth Rate (per 1,000): 15.41

Death Rate (per 1,000): 7.91

Life Expectancy: 76.79 years

Ethnic Groups: black 86.6%, mixed 9.1%, indigenous 2.9%, other 1.3%, unspecified 0.2%

Religion: Roman Catholic 61.4%, Protestant 20.6% (inclues Evangelical 6.7%, Seventh Day Adventist 6.1%, Pentecostal 5.6%, Baptist 4.1%, Methodist 3.7%, Church of God 1.2%, other 1.3%), Rastafarian 1.3%, Jehovah's Witnesses 1.2%, other 0.3%, none 6.1%, unspecified 1.1%

Languages: English (official), French patois

Literacy: 94%

Government Type: parliamentary democracy

Capital: Roseau

Independence: 3 November 1978 (from United Kingdom)

GDP Per Capita: $11,500

Occupations: agriculture 40%, industry 32%, services 28%

Currency: East Caribbean dollar (XCD)

Dominican Republic

Long Name: Dominican Republic

Location: Caribbean, eastern two-thirds of the island of Hispaniola, between the Caribbean Sea and the North Atlantic Ocean, east of Haiti

Area: 18,792 sq. mi. (48,670 sq. km)

Climate: tropical maritime; little seasonal temperature variation; seasonal variation in rainfall

Terrain: rugged highlands and mountains with fertile valleys interspersed

Population: 10,478,756

Population Growth Rate: 1.23%

Birth Rate (per 1,000): 18.73

Death Rate (per 1,000): 4.55

Life Expectancy: 77.97 years

Ethnic Groups: mixed 73%, white 16%, black 11%

Religion: Roman Catholic 95%, other 5%

Languages: Spanish

Literacy: 91.8%

Government Type: democratic republic

Capital: Santo Domingo

Independence: 27 February 1844 (from Haiti)

GDP Per Capita: $14,900

Occupations: agriculture 14.4%, industry 20.8%, services 64.7%

Currency: Dominican peso (DOP)

Ecuador

Long Name: Republic of Ecuador

Location: Western South America, bordering the Pacific Ocean at the Equator, between Colombia and Peru

Area: 109,484 sq. mi. (283,561 sq. km)

Climate: tropical along coast, becoming cooler inland at higher elevations; tropical in Amazonian jungle lowlands

Terrain: coastal plain (costa), inter-Andean central highlands (sierra), and flat to rolling eastern jungle (oriente)

Population: 15,868,396

Population Growth Rate: 1.35%

Birth Rate (per 1,000): 18.51

Death Rate (per 1,000): 5.06

Life Expectancy: 76.56 years

Ethnic Groups: mestizo (mixed Amerindian and white) 71.9%, Montubio 7.4%, Amerindian 7%, white 6.1%, Afroecuadorian 4.3%, mulato 1.9%, black 1%, other 0.4%

Religion: Roman Catholic 74%, Evangelical 10.4%, Jehovah's Witness 1.2%, other 6.4% (includes Mormon Buddhist, Jewish, Spiritualist, Muslim, Hindu, indigenous religions, African American religions, Pentecostal), atheist 7.9%, agnostic 0.1%

Languages: Spanish (Castilian) 93% (official), Quechua 4.1%, other indigenous 0.7%, foreign 2.2%

Literacy: 94.5%

Government Type: republic

Capital: Quito

Independence: 24 May 1822 (from Spain)

GDP Per Capita: $11,300

Occupations: agriculture 27.8%, industry 17.8%, services 54.4%

Currency: U.S. dollar (USD)

Egypt

Long Name: Arab Republic of Egypt

Location: Northern Africa, bordering the Mediterranean Sea, between Libya and the Gaza Strip, and the Red Sea north of Sudan, and includes the Asian Sinai Peninsula

Area: 386,660 sq. mi. (1,001,450 sq. km)

Climate: desert; hot, dry summers with moderate winters

Terrain: vast desert plateau interrupted by Nile valley and delta

Population: 88,487,396

Population Growth Rate: 1.79%

Birth Rate (per 1,000): 22.9

Death Rate (per 1,000): 4.77

Life Expectancy: 73.7 years

Ethnic Groups: Egyptian 99.6%, other 0.4% (2006 census)

Religion: Muslim (predominantly Sunni) 90%, Christian (majority Coptic Orthodox, other Christians include Armenian Apostolic, Catholic, Maronite, Orthodox, and Anglican) 10%

Languages: Arabic (official), English and French widely understood by educated classes

Literacy: 73.8%

Government Type: republic

Capital: Cairo

Independence: 28 February 1922 (from United Kingdom)

GDP Per Capita: $11,500

Occupations: agriculture 29.2%, industry 23.5%, services 47.3%

Currency: Egyptian pound (EGP)

El Salvador

Long Name: Republic of El Salvador

Location: Central America, bordering the North Pacific Ocean, between Guatemala and Honduras

Area: 8,125 sq. mi. (21,041 sq. km)

Climate: tropical; rainy season (May to October); dry season (November to April); tropical on coast; temperate in uplands

Terrain: mostly mountains with narrow coastal belt and central plateau

Population: 6,141,350

Population Growth Rate: 0.25%

Birth Rate (per 1,000): 16.46

Death Rate (per 1,000): 5.69

Life Expectancy: 74.42 years

Ethnic Groups: mestizo 86.3%, white 12.7%, Amerindian 0.2% (includes Lenca, Kakawira, Nahua-Pipil), black 0.1%, other 0.6% (2007 est.)

Religion: Roman Catholic 57.1%, Protestant 21.2%, Jehovah's Witnesses 1.9%, Mormon 0.7%, other religions 2.3%, none 16.8% (2003 est.)

Languages: Spanish, Nahua (among some Amerindians)

Literacy: 88%

Government Type: republic

Capital: San Salvador

Independence: 15 September 1821 (from Spain)

GDP Per Capita: $8,300

Occupations: agriculture, 21%, industry 20%, services 58%

Currency: U.S. dollar (USD)

Equatorial Guinea

Long Name: Republic of Equatorial Guinea

Location: Western Africa, bordering the Bight of Biafra, between Cameroon and Gabon

Area: 10,831 sq. mi. (28,051 sq. km)

Climate: tropical; always hot, humid

Terrain: coastal plains rise to interior hills; islands are volcanic

Population: 740,743

Population Growth Rate: 2.51%

Birth Rate (per 1,000): 33.31

Death Rate (per 1,000): 8.19

Life Expectancy: 63.85 years

Ethnic Groups: Fang 85.7%, Bubi 6.5%, Mdowe 3.6%, Annobon 1.6%, Bujeba 1.1%, other 1.4% (1994 census)

Religion: nominally Christian and predominantly Roman Catholic, pagan practices

Languages: Spanish (official) 67.6%, other (includes French [official], Fang, Bubi) 32.4% (1994 census)

Literacy: 95.3%

Government Type: republic

Capital: Malabo

Independence: 12 October 1968 (from Spain)

GDP Per Capita: $33,300

Occupations: N/A

Currency: Communaute Financiere Africaine franc (XAF)

Eritrea

Long Name: State of Eritrea

Location: Eastern Africa, bordering the Red Sea, between Djibouti and Sudan

Area: 45,406 sq. mi. (117,600 sq. km)

Climate: hot, dry desert strip along Red Sea coast; cooler and wetter in the central highlands (up to 61 cm of rainfall annually, heaviest June to September); semiarid in western hills and lowlands

Terrain: dominated by extension of Ethiopian north-south trending highlands, descending on the east to a coastal desert plain, on the northwest to hilly terrain and on the southwest to flat-to-rolling plains

Population: 6,527,689

Population Growth Rate: 2.25%

Birth Rate (per 1,000): 30

Death Rate (per 1,000): 7.52

Life Expectancy: 63.81 years

Ethnic Groups: nine recognized ethnic groups: Tigrinya 55%, Tigre 30%,

Saho 4%, Kunama 2%, Rashaida 2%, Bilen 2%, other (Afar, Beni Amir, Nera) 5% (2010 est.)

Religion: Muslim, Coptic Christian, Roman Catholic, Protestant

Languages: Tigrinya (official), Arabic (official), English (official), Tigre, Kunama, Afar, other Cushitic languages

Literacy: 73.8%

Government Type: transitional government

Capital: Asmara

Independence: 24 May 1993 (from Ethiopia)

GDP Per Capita: $1,200

Occupations: agriculture 80%, industry and services 20%

Currency: nakfa (ERN)

Estonia

Long Name: Republic of Estonia

Location: Eastern Europe, bordering the Baltic Sea and Gulf of Finland, between Latvia and Russia

Area: 17,463 sq. mi. (45,228 sq. km)

Climate: maritime, wet, moderate winters, cool summers

Terrain: marshy, lowlands; flat in the north, hilly in the south

Population: 1,265,420

Population Growth Rate: -0.55%

Birth Rate (per 1,000): 10.51

Death Rate (per 1,000): 12.4

Life Expectancy: 76.47 years

Ethnic Groups: Estonian 68.7%, Russian 24.8%, Ukrainian 1.7%, Belarusian 1%, Finn 0.6%, other 1.6%, unspecified 1.6% (2011 est.)

Religion: Lutheran 9.9%, Orthodox 16.2%, other Christian (including Methodist, Seventh-Day Adventist, Roman Catholic, Pentecostal) 2.2%, other 0.9%, none 54.1%, unspecified 16.7% (2011 est.)

Languages: Estonian (official) 68.5%, Russian 29.6%, Ukrainian 0.6%, other 1.2%, unspecified 0.1% (2011 est.)

Literacy: 99.8%

Government Type: parliamentary republic

Capital: Tallinn

Independence: 20 August 1991 (from Soviet Union)

GDP Per Capita: $28,700

Occupations: agriculture 3.9%, industry 28.4%, services 67.7%

Currency: Estonian kroon (EEK)

Ethiopia

Long Name: Federal Democratic Republic of Ethiopia

Location: Eastern Africa, west of Somalia

Area: 426,373 sq. mi. (1,104,300 sq. km)

Climate: tropical monsoon with wide topographic-induced variation

Terrain: high plateau with central mountain range divided by Great Rift Valley

Population: 99,465,819

Population Growth Rate: 2.89%

Birth Rate (per 1,000): 37.27

Death Rate (per 1,000): 8.19

Life Expectancy: 61.48 years

Ethnic Groups: Oromo 34.4%, Amhara (Amara) 27%, Somali (Somalie) 6.2%, Tigray (Tigrinya) 6.1%, Sidama 4%, Gurage 2.5%, Welaita 2.3%, Hadiya 1.7%, Afar (Affar) 1.7%, Gamo 1.5%, Gedeo 1.3%, Silte 1.3%, Kefficho 1.2%, other 10.5% (2007 est.)

Religion: Ethiopian Orthodox 43.5%, Muslim 33.9%, Protestant 18.5%, traditional 2.7%, Catholic 0.7%, other 0.6% (2007 est.)

Languages: Oromo (official working language in the State of Oromiya) 33.8%, Amharic (official national language) 29.3%, Somali (official working language of the State of Sumale) 6.2%, Tigrigna (Tigrinya) (official working language of the State of Tigray) 5.9%, Sidamo 4%, Wolaytta 2.2%, Gurage 2%, Afar (official working language of the State of Afar) 1.7%, Hadiyya 1.7%, Gamo 1.5%, Gedeo 1.3%, Opuuo 1.2%, Kafa 1.1%, other 8.1%, English (major foreign language taught in schools), Arabic (2007 est.)

Literacy: 49.1%

Government Type: federal republic

Capital: Addis Ababa

Independence: oldest independent country in Africa and one of the oldest in the world—at least 2,000 years

GDP Per Capita: $1,700

Occupations: agriculture 85%, industry 5%, services 10%

Currency: birr (ETB)

Fiji

Long Name: Republic of the Fiji Islands

Location: Oceania, island group in the South Pacific Ocean, about two-thirds of the way from Hawaii to New Zealand

Area: 7,056 sq. mi. (18,274 sq. km)

Climate: tropical marine; only slight seasonal temperature variation

Terrain: mostly mountains of volcanic origin

Population: 909,389

Population Growth Rate: 0.67%

Birth Rate (per 1,000): 19.43

Death Rate (per 1,000): 6.04

Life Expectancy: 72.43 years

Ethnic Groups: iTaukei 56.8% (predominantly Melanesian with a Polynesian admixture), Indian 37.5%, Rotuman 1.2%, other 4.5% (European, part European, other Pacific Islanders, Chinese)

Religion: Protestant 45% (Methodist 34.6%, Assembly of God 5.7%, Seventh Day Adventist 3.9%, and Anglican 0.8%), Hindu 27.9%, other Christian 10.4%, Roman Catholic 9.1%, Muslim 6.3%, Sikh 0.3%, other 0.3%, none 0.8% (2007 est.)

Languages: English (official), Fijian (official), Hindustani

Literacy: 93.7%

Government Type: republic

Capital: Suva

Independence: 10 October 1970 (from United Kingdom)

GDP Per Capita: $8,800

Occupations: agriculture 70%, industry and services 30%

Currency: Fijian dollar (FJD)

Finland

Long Name: Republic of Finland

Location: Northern Europe, bordering the Baltic Sea, Gulf of Bothnia, and Gulf of Finland, between Sweden and Russia

Area: 130,558 sq. mi. (338,145 sq. km)

Climate: cold temperate; potentially subarctic but comparatively mild because of moderating influence of the North Atlantic Current, Baltic Sea, and more than 60,000 lakes

Terrain: mostly low, flat to rolling plains interspersed with lakes and low hills

Population: 5,476,922

Population Growth Rate: 0.4%

Birth Rate (per 1,000): 10.72

Death Rate (per 1,000): 9.83

Life Expectancy: 80.77 years

Ethnic Groups: Finn 93.4%, Swede 5.6%, Russian 0.5%, Estonian 0.3%, Roma 0.1%, Sami 0.1% (2006)

Religion: Lutheran 73.8%, Orthodox 1.1%, other or none 25.1% (2014 est.)

Languages: Finnish (official) 89%, Swedish (official) 5.3%, Russian 1.3%, other 4.4% (2014 est.)

Literacy: 100%

Government Type: republic

Capital: Helsinki

Independence: 6 December 1917 (from Russia)

GDP Per Capita: $41,200

Occupations: agriculture and forestry 4.4%, industry 15.5%, construction 7.1%, commerce 21.3%, finance, insurance, and business services 13.3%, transport and communications 9.9%, public services 28.5% (2011)

Currency: euro (EUR)

France

Long Name: French Republic

Location: Western Europe, bordering the Bay of Biscay and English Channel, between Belgium and Spain, southeast of the United Kingdom; bordering the Mediterranean Sea, between Italy and Spain

Area: 212,935 sq. mi. (551,500 sq. km)

Climate: generally cool winters and mild summers, but mild winters and hot summers along the Mediterranean; occasional strong, cold, dry, north-to-northwesterly wind known as mistral

Terrain: mostly flat plains or gently rolling hills in north and west; remainder is mountainous, especially Pyrenees in south, Alps in east

Population: 66,553,766

Population Growth Rate: 0.43%

Birth Rate (per 1,000): 12.38

Death Rate (per 1,000): 9.16

Life Expectancy: 81.75 years

Ethnic Groups: Celtic and Latin with Teutonic, Slavic, North African, Indochinese, Basque minorities

Religion: Christian (overwhelmingly Roman Catholic) 63-66%, Muslim 7-9%, Jewish 0.5-0.75%, Buddhist 0.5-0.75%, other 0.5-1.0%, none 23-28%

Languages: French 100%, rapidly declining regional dialects and languages (Provencal, Breton, Alsatian, Corsican, Catalan, Basque, Flemish)

Literacy: 99%

Government Type: republic

Capital: Paris

Independence: 486 (Frankish tribes unified); 843 (Western Francia established from the division of the Carolingian Empire)

GDP Per Capita: $41,400

Occupations: agriculture 3%, industry 21.3%, services 75.7%

Currency: euro (EUR)

Gabon

Long Name: Gabonese Republic

Location: Western Africa, bordering the Atlantic Ocean at the Equator, between Republic of the Congo and Equatorial Guinea

Area: 103,346 sq. mi. (267,667 sq. km)

Climate: tropical; always hot, humid

Terrain: narrow coastal plain; hilly interior; savanna in east and south

Population: 1,705,336

Population Growth Rate: 1.93%

Birth Rate (per 1,000): 34.49

Death Rate (per 1,000): 13.12

Life Expectancy: 52.04 years

Ethnic Groups: Bantu tribes, including four major tribal groupings (Fang, Bapounou, Nzebi, Obamba); other Africans and Europeans, 154,000, including 10,700 French and 11,000 persons of dual nationality

Religion: Catholic 41.9%, Protestant 13.7%, other Christian 32.4%, Muslim 6.4%, animist 0.3%, other 0.3%, none/no answer 5% (2012 est.)

Languages: French (official), Fang, Myene, Nzebi, Bapounou/Eschira, Bandjabi

Literacy: 83.2%

Government Type: republic; multiparty presidential regime

Capital: Libreville

Independence: 17 August 1960 (from France)

GDP Per Capita: $21,700

Occupations: agriculture 60%, industry 15%, services 25%

Currency: Communaute Financiere Africaine franc (XAF)

The Gambia

Long Name: Republic of The Gambia

Location: Western Africa, bordering the North Atlantic Ocean and Senegal

Area: 4,363 sq. mi. (11,300 sq. km)

Climate: tropical; hot, rainy season (June to November); cooler, dry season (November to May)

Terrain: flood plain of the Gambia River flanked by some low hills

Population: 1,967,709

Population Growth Rate: 2.16%

Birth Rate (per 1,000): 30.86

Death Rate (per 1,000): 7.15

Life Expectancy: 64.6 years

Ethnic Groups: Mandinka/Jahanka 33.8%, Fulani/Tukulur/Lorobo 22.1%, Wollof 12.2%, Jola/Karoninka 10.9%, Serahuleh 7%, Serere 3.2%, Manjago 2.1%, Bambara 1%, Creole/Aku Marabout 0.8%, other 0.9%, non-Gambian 5.2%, no answer 0.7% (2013 est.)

Religion: Muslim 95.7%, Christian 4.2%, none 0.1%, no answer 0.1% (2013 est.)

Languages: English (official), Mandinka, Wolof, Fula, other indigenous vernaculars

Literacy: 55.5%

Government Type: republic

Capital: Banjul

Independence: 18 February 1965 (from United Kingdom)

GDP Per Capita: $1,700

Occupations: agriculture 75%, industry 19%, services 6%

Currency: dalasi (GMD)

Georgia

Long Name: Georgia

Location: Southwestern Asia, bordering the Black Sea, between Turkey and Russia

Area: 26,911 sq. mi. (69,700 sq. km)

Climate: warm and pleasant; Mediterranean-like on Black Sea coast

Terrain: largely mountainous with Great Caucasus Mountains in the north and Lesser Caucasus Mountains in the south; Kolkhet'is Dablobi (Kolkhida Lowland) opens to the Black Sea in the west; Mtkvari River Basin in the east; good soils in river valley flood plains, foothills of Kolkhida Lowland

Population: 4,931,226

Population Growth Rate: -0.08%

Birth Rate (per 1,000): 12.74

Death Rate (per 1,000): 10.82

Life Expectancy: 75.95 years

Ethnic Groups: Georgian 83.8%, Azeri 6.5%, Armenian 5.7%, Russian 1.5%, other 2.5% (2002 est.)

Religion: Orthodox Christian (official) 83.9%, Muslim 9.9%, Armenian-Gregorian 3.9%, Catholic 0.8%, other 0.8%, none 0.7% (2002 census)

Languages: Georgian 71% (official), Russian 9%, Armenian 7%, Azeri 6%, other 7%

Literacy: 99.8%

Government Type: republic

Capital: T'bilisi

Independence: 9 April 1991 (from Soviet Union)

GDP Per Capita: $9,500

Occupations: agriculture 55.6%, industry 8.9%, services 35.5%

Currency: lari (GEL)

Germany

Long Name: Federal Republic of Germany

Location: Central Europe, bordering the Baltic Sea and the North Sea, between the Netherlands and Poland, south of Denmark

Area: 137,847 sq. mi. (357,022 sq. km)

Climate: temperate and marine; cool, cloudy, wet winters and summers; occasional warm mountain (foehn) wind

Terrain: lowlands in north, uplands in center, Bavarian Alps in south

Population: 80,854,408

Population Growth Rate: -0.17%

Birth Rate (per 1,000): 8.47

Death Rate (per 1,000): 11.42

Life Expectancy: 80.57 years

Ethnic Groups: German 91.5%, Turkish 2.4%, other 6.1% (made up largely of Greek, Italian, Polish, Russian, Serbo-Croatian, Spanish)

Religion: Protestant 34%, Roman Catholic 34%, Muslim 3.7%, unaffiliated or other 28.3%

Languages: German

Literacy: 99%

Government Type: federal republic

Capital: Berlin

Independence: 18 January 1871 (German Empire unification); divided into four zones of occupation (United Kingdom, US, USSR, and later, France) in 1945 following World War II; Federal Republic of Germany (FRG or West Germany) proclaimed 23 May 1949 and included the former United Kingdom, US, and French zones; German Democratic Republic (GDR or East Germany) proclaimed 7 October 1949 and included the former USSR zone; unification of West Germany and East Germany took place 3 October 1990; all four powers formally relinquished rights 15 March 1991

GDP Per Capita: $47,400

Occupations: agriculture 1.6%, industry 24.6%, services 73.8%

Currency: euro (EUR)

Ghana

Long Name: Republic of Ghana

Location: Western Africa, bordering the Gulf of Guinea, between Cote d'Ivoire and Togo

Area: 92,098 sq. mi. (238,533 sq. km)

Climate: tropical; warm and comparatively dry along southeast coast; hot and humid in southwest; hot and dry in north

Terrain: mostly low plains with dissected plateau in south-central area

Population: 26,327,649

Population Growth Rate: 2.18%

Birth Rate (per 1,000): 31.09

Death Rate (per 1,000): 7.22

Life Expectancy: 66.18 years

Ethnic Groups: Akan 47.5%, Mole-Dagbon 16.6%, Ewe 13.9%, Ga-Dangme 7.4%, Gurma 5.7%, Guan 3.7%, Grusi 2.5%, Mande 1.1%, other 1.4% (2010 est.)

Religion: Christian 71.2% (Pentecostal/Charismatic 28.3%, Protestant 18.4%, Catholic 13.1%, other 11.4%), Muslim 17.6%, traditional 5.2%, other 0.8%, none 5.2% (2010 est.)

Languages: Asante 16%, Ewe 14%, Fante 11.6%, Boron (Brong) 4.9%, Dagomba 4.4%, Dangme 4.2%, Dagarte (Dagaba) 3.9%, Kokomba 3.5%, Akyem 3.2%, Ga 3.1%, other 31.2%

Literacy: 76.6%

Government Type: constitutional democracy

Capital: Accra

Independence: 6 March 1957 (from United Kingdom)

GDP Per Capita: $4,300

Occupations: agriculture 44.7%, industry 14.4%, services 40.9%

Currency: Ghana cedi (GHC)

Greece

Long Name: Hellenic Republic

Location: Southern Europe, bordering the Aegean Sea, Ionian Sea, and the Mediterranean Sea, between Albania and Turkey

Area: 50,949 sq. mi. (131,957 sq. km)

Climate: temperate; mild, wet winters; hot, dry summers

Terrain: mostly mountains with ranges extending into the sea as peninsulas or chains of islands

Population: 10,775,643

Population Growth Rate: –0.01%

Birth Rate (per 1,000): 8.66

Death Rate (per 1,000): 11.09

Life Expectancy: 80.43 years

Ethnic Groups: population: Greek 93%, other (foreign citizens) 7% (2001 census)

Religion: Greek Orthodox 98%, Muslim 1.3%, other 0.7%

Languages: Greek 99% (official), other 1% (includes English and French)

Literacy: 97.7%

Government Type: parliamentary republic

Capital: Athens

Independence: 1829 (from the Ottoman Empire)

GDP Per Capita: $25,600

Occupations: agriculture 12.5%, industry 13.9%, services 73.6%

Currency: euro (EUR)

Grenada

Long Name: Grenada

Location: Caribbean, island between the Caribbean Sea and Atlantic Ocean, north of Trinidad and Tobago

Area: 133 sq. mi. (344 sq. km)

Climate: tropical; tempered by northeast trade winds

Terrain: volcanic in origin with central mountains

Population: 110,694

Population Growth Rate: 0.48%

Birth Rate (per 1,000): 16.03

Death Rate (per 1,000): 8.08

Life Expectancy: 74.05 years

Ethnic Groups: African descent 89.4%, mixed 8.2%, East Indian 1.6%, other 0.9% (includes indigenous) (2001 est.)

Religion: Roman Catholic 44.6%, Protestant 43.5% (includes Anglican 11.5%, Pentecostal 11.3%, Seventh Day Adventist 10.5%, Baptist 2.9%, Church of God 2.6%, Methodist 1.8%, Evangelical 1.6%, other 1.3%), Jehovah's Witness 1.1%, Rastafarian 1.1%, other 6.2%, none 3.6%

Languages: English (official), French patois

Literacy: 96%

Government Type: parliamentary democracy

Capital: Saint George's

Independence: 7 February 1974 (from United Kingdom)

GDP Per Capita: $13,000

Occupations: agriculture 11%, industry 20%, services 69%

Currency: East Caribbean dollar (XCD)

Guatemala

Long Name: Republic of Guatemala

Location: Central America, bordering the North Pacific Ocean, between El Salvador and Mexico, and bordering the Gulf of Honduras (Caribbean Sea) between Honduras and Belize

Area: 42,041 sq. mi. (108,889 sq. km)

Climate: tropical; hot, humid in lowlands; cooler in highlands

Terrain: mostly mountains with narrow coastal plains and rolling limestone plateau

Population: 14,918,999

Population Growth Rate: 1.82%

Birth Rate (per 1,000): 24.89

Death Rate (per 1,000): 4.77

Life Expectancy: 72.02 years

Ethnic Groups: mestizo (mixed Amerindian-Spanish-in local Spanish called Ladino) and European 59.4%, K'iche 9.1%, Kaqchikel 8.4%, Mam 7.9%, Q'eqchi 6.3%, other Mayan 8.6%, indigenous non-Mayan 0.2%, other 0.1% (2001 census)

Religion: Roman Catholic, Protestant, indigenous Mayan beliefs

Languages: Spanish 60%, Amerindian languages 40% (23 officially recognized Amerindian languages, including Quiche, Cakchiquel, Kekchi, Mam, Garifuna, and Xinca)

Literacy: 81.5%

Government Type: constitutional democratic republic

Capital: Guatemala

Independence: 15 September 1821 (from Spain)

GDP Per Capita: $7,900

Occupations: agriculture 38%, industry 14%, services 48%

Currency: quetzal (GTQ), U.S. dollar (USD), others allowed

Guinea

Long Name: Republic of Guinea

Location: Western Africa, bordering the North Atlantic Ocean, between Guinea-Bissau and Sierra Leone

Area: 94,925 sq. mi. (245,857 sq. km)

Climate: generally hot and humid; monsoonal-type rainy season (June to November) with southwesterly winds; dry season (December to May) with northeasterly harmattan winds

Terrain: generally flat coastal plain, hilly to mountainous interior

Population: 11,780,162

Population Growth Rate: 2.63%

Birth Rate (per 1,000): 35.74

Death Rate (per 1,000): 9.46

Life Expectancy: 60.08 years

Ethnic Groups: Fulani (Peul) 33.9%, Malinke 31.1%, Soussou 19.1%, Guerze 6%, Kissi 4.7%, Toma 2.6%, other/no answer 2.7% (2012 est.)

Religion: Muslim 86.7%, Christian 8.9%, animist/other/none 7.8% (2012 est.)

Languages: French (official)

Literacy: 30.4%

Government Type: republic

Capital: Conakry

Independence: 2 October 1958 (from France)

GDP Per Capita: $1,300

Occupations: agriculture 76%, industry and services 24%

Currency: Guinean franc (GNF)

Guinea-Bissau

Long Name: Republic of Guinea-Bissau

Location: Western Africa, bordering the North Atlantic Ocean, between Guinea and Senegal

Area: 13,948 sq. mi. (36,125 sq. km)

Climate: tropical; generally hot and humid; monsoonal-type rainy season (June to November) with southwesterly winds; dry season (December to May) with northeasterly harmattan winds

Terrain: mostly low coastal plain rising to savanna in east

Population: 1,726,170

Population Growth Rate: 1.91%

Birth Rate (per 1,000): 33.38

Death Rate (per 1,000): 14.33

Life Expectancy: 50.23 years

Ethnic Groups: Fulani 28.5%, Balanta 22.5%, Mandinga 14.7%, Papel 9.1%, Manjaco 8.3%, Beafada 3.5%, Mancanha 3.1%, Bijago 2.1%, Felupe 1.7%, Mansoanca 1.4%, Balanta Mane 1%, other 1.8%, none 2.2% (2008 est.)

Religion: Muslim 45.1%, Christian 22.1%, animist 14.9%, none 2%, unspecified 15.9% (2008 est.)

Languages: Crioulo 90.4%, Portuguese 27.1% (official), French 5.1%, English 2.9%, other 2.4%

Literacy: 59.9%

Government Type: republic

Capital: Bissau

Independence: 24 September 1973 (declared); 10 September 1974 (from Portugal)

GDP Per Capita: $1,500

Occupations: agriculture 82%, industry and services 18%

Currency: Communaute Financiere Africaine franc (XOF)

Guyana

Long Name: Cooperative Republic of Guyana

Location: Northern South America, bordering the North Atlantic Ocean, between Suriname and Venezuela

Area: 83,000 sq. mi. (214,969 sq. km)

Climate: tropical; hot, humid, moderated by northeast trade winds; two rainy seasons (May to August, November to January)

Terrain: mostly rolling highlands; low coastal plain; savanna in south

Population: 735,222

Population Growth Rate: 0.02%

Birth Rate (per 1,000): 15.59

Death Rate (per 1,000): 7.32

Life Expectancy: 68.09 years

Ethnic Groups: East Indian 43.5%, black (African) 30.2%, mixed 16.7%, Amerindian 9.1%, other 0.5% (includes Portuguese, Chinese, white; 2002 est.)

Religion: Protestant 30.5% (Pentecostal 16.9%, Anglican 6.9%, Seventh Day Adventist 5%, Methodist 1.7%), Hindu 28.4%, Roman Catholic 8.1%, Muslim 7.2%, Jehovah's Witness 1.1%, other Christian 17.7%, other 1.9%, none 4.3%, unspecified 0.9% (2002 est.)

Languages: English, Amerindian dialects, Creole, Caribbean Hindustani (a dialect of Hindi), Urd

Literacy: 88.5%

Government Type: republic

Capital: Georgetown

Independence: 26 May 1966 (from United Kingdom)

GDP Per Capita: $7,200

Occupations: N/A

Currency: Guyanese dollar (GYD)

Haiti

Long Name: Republic of Haiti

Location: Caribbean, western one-third of the island of Hispaniola, between the Caribbean Sea and the North Atlantic Ocean, west of the Dominican Republic

Area: 10,714 sq. mi. (27,750 sq. km)

Climate: tropical; semiarid where mountains in east cut off trade winds

Terrain: mostly rough and mountainous

Population: 10,110,019

Population Growth Rate: 1.17%

Birth Rate (per 1,000): 22.31

Death Rate (per 1,000): 7.83

Life Expectancy: 63.51 years

Ethnic Groups: black 95%, mulatto and white 5%

Religion: Roman Catholic 80%, Protestant 16% (Baptist 10%, Pentecostal 4%, Adventist 1%, other 1%), none 1%, other 3% (voodoo practiced by about half the population)

Languages: French (official), Creole (official)

Literacy: 60.7%

Government Type: republic

Capital: Port-au-Prince

Independence: 1 January 1804 (from France)

GDP Per Capita: $1,800

Occupations: agriculture 38.1%, industry 11.5%, services 50.4%

Currency: gourde (HTG)

Honduras

Long Name: Republic of Honduras

Location: Central America, bordering the Caribbean Sea, between Guatemala and Nicaragua and bordering the Gulf of Fonseca (North Pacific Ocean), between El Salvador and Nicaragua

Area: 43,278 sq. mi. (112,090 sq. km)

Climate: subtropical in lowlands, temperate in mountains

Terrain: mostly mountains in interior, narrow coastal plains

Population: 8,746,673

Population Growth Rate: 1.68%

Birth Rate (per 1,000): 23.14

Death Rate (per 1,000): 5.17

Life Expectancy: 71 years

Ethnic Groups: mestizo (mixed Amerindian and European) 90%, Amerindian 7%, black 2%, white 1%

Religion: Roman Catholic 97%, Protestant 3%

Languages: Spanish, Amerindian dialects

Literacy: 88.5%

Government Type: democratic constitutional republic

Capital: Tegucigalpa

Independence: 15 September 1821 (from Spain)

GDP Per Capita: $5,000

Occupations: agriculture 39.2%, industry 20.9%, services 39.8%

Currency: lempira (HNL)

Hungary

Long Name: Republic of Hungary

Location: Central Europe, northwest of Romania

Area: 35,918 sq. mi. (93,028 sq. km)

Climate: temperate; cold, cloudy, humid winters; warm summers

Terrain: mostly flat to rolling plains; hills and low mountains on the Slovakian border

Population: 9,897,541

Population Growth Rate: -0.22%

Birth Rate (per 1,000): 9.16

Death Rate (per 1,000): 12.73

Life Expectancy: 75.69 years

Ethnic Groups: Hungarian 85.6%, Roma 3.2%, German 1.9%, other 2.6%, unspecified 14.1%

Religion: Roman Catholic 37.2%, Calvinist 11.6%, Lutheran 2.2%, Greek Catholic 1.8%, other 1.9%, none 18.2%, unspecified 27.2% (2011 est.)

Languages: Hungarian (official) 99.6%, English 16%, German 11.2%, Russian 1.6%, Romanian 1.3%, French 1.2%, other 4.2%

Literacy: 99.1%

Government Type: parliamentary democracy

Capital: Budapest

Independence: 25 December 1000 (crowning of King Stephen I, traditional founding date)

GDP Per Capita: $26,000

Occupations: agriculture 7.1%, industry 29.7%, services 63.2%

Currency: forint (HUF)

Iceland

Long Name: Republic of Iceland

Location: Northern Europe, island between the Greenland Sea and the North Atlantic Ocean, northwest of the United Kingdom

Area: 39,768 sq. mi. (103,000 sq. km)

Climate: temperate; moderated by North Atlantic Current; mild, windy winters; damp, cool summers

Terrain: mostly plateau interspersed with mountain peaks, icefields; coast deeply indented by bays and fiords

Population: 331,918

Population Growth Rate: 1.21%

Birth Rate (per 1,000): 13.91

Death Rate (per 1,000): 6.28

Life Expectancy: 82.97 years

Ethnic Groups: homogeneous mixture of descendants of Norse and Celts 94%, population of foreign origin 6%

Religion: Evangelical Lutheran Church of Iceland (official) 73.8%, Roman Catholic 3.6%, Reykjavik Free Church 2.9%, Hafnarfjorour Free Church 2%, The Independent Congregation 1%, other religions 3.9% (includes Pentecostal and Asatru Association), none 5.6%, other or unspecified 7.2% (2015 est.)

Languages: Icelandic, English, Nordic languages, German widely spoken

Literacy: 99%

Government Type: constitutional republic

Capital: Reykjavik

Independence: 1 December 1918 (became a sovereign state under the Danish Crown); 17 June 1944 (from Denmark)

GDP Per Capita: $46,600

Occupations: agriculture 4.8%, industry 22.2%, services 73%

Currency: Icelandic krona (ISK)

India

Long Name: Republic of India

Location: Southern Asia, bordering the Arabian Sea and the Bay of Bengal, between Burma and Pakistan

Area: 1,269,291 sq. mi. (3,287,263 sq. km)

Climate: varies from tropical monsoon in south to temperate in north

Terrain: upland plain (Deccan Plateau) in south, flat to rolling plain along the Ganges, deserts in west, Himalayas in north

Population: 1,251,695,584

Population Growth Rate: 1.2%

Birth Rate (per 1,000): 19.55

Death Rate (per 1,000): 7.32

Life Expectancy: 68.13 years

Ethnic Groups: Indo-Aryan 72%, Dravidian 25%, Mongoloid and other 3%

Religion: Hindu 79.8%, Muslim 14.2%, Christian 2.3%, Sikh 1.7%, other and unspecified 2% (2011 est.)

Languages: Hindi 41%, Bengali 8.1%, Telugu 7.2%, Marathi 7%, Tamil 5.9%, Urdu 5%, Gujarati 4.5%, Kannada 3.7%, Malayalam 3.2%, Oriya 3.2%, Punjabi 2.8%, Assamese 1.3%, Maithili 1.2%, other 5.9%

Literacy: 71.2%

Government Type: federal republic

Capital: New Delhi

Independence: 15 August 1947 (from United Kingdom)

GDP Per Capita: $6,300

Occupations: agriculture 49%, industry 20%, services 31%

Currency: Indian rupee (INR)

Indonesia

Long Name: Republic of Indonesia

Location: Southeastern Asia, archipelago between the Indian Ocean and the Pacific Ocean

Area: 735,358 sq. mi. (1,904,569 sq. km)

Climate: tropical; hot, humid; more moderate in highlands

Terrain: mostly coastal lowlands; larger islands have interior mountains

Population: 255,993,674

Population Growth Rate: 0.92%

Birth Rate (per 1,000): 16.72

Death Rate (per 1,000): 6.37

Life Expectancy: 72.45 years

Ethnic Groups: Javanese 40.1%, Sundanese 15.5%, Malay 3.7%, Batak 3.6%, Madurese 3%, Betawi 2.9%, Minangkabau 2.7%, Buginese 2.7%, Bantenese 2%, Banjarese 1.7%, Balinese 1.7%, Acehnese 1.4%, Dayak 1.4%, Sasak 1.3%, Chinese 1.2%, other 15% (2010 est.)

Religion: Muslim 87.2%, Christian 7%, Roman Catholic 2.9%, Hindu 1.7%, other 0.9% (includes Buddhist and Confucian), unspecified 0.4% (2010 est.)

Languages: Bahasa Indonesia (official, modified form of Malay), English, Dutch, local dialects (the most widely spoken of which is Javanese)

Literacy: 93.9%

Government Type: republic

Capital: Jakarta

Independence: 17 August 1945

GDP Per Capita: $11,300

Occupations: agriculture 38.9%, industry 13.2%, services 47.9%

Currency: Indonesian rupiah (IDR)

Iran

Long Name: Islamic Republic of Iran

Location: Middle East, bordering the Gulf of Oman, the Persian Gulf, and the Caspian Sea, between Iraq and Pakistan

Area: 636,372 sq. mi. (1,648,195 sq. km)

Climate: mostly arid or semiarid, subtropical along Caspian coast

Terrain: rugged, mountainous rim; high, central basin with deserts, mountains; small, discontinuous plains along both coasts

Population: 81,824,270

Population Growth Rate: 1.2%

Birth Rate (per 1,000): 17.99

Death Rate (per 1,000): 5.94

Life Expectancy: 71.15 years

Ethnic Groups: Persian, Azeri, Kurd, Lur, Baloch, Arab, Turkmen and Turkic tribes

Religion: Muslim (official) 99.4% (Shia 90-95%, Sunni 5-10%), other (includes Zoroastrian, Jewish, and Christian) 0.3%, unspecified 0.4% (2011 est.)

Literacy: 86.8%

Government Type: theocratic republic

Capital: Tehran

Independence: 1 April 1979 (Islamic Republic of Iran proclaimed)

GDP Per Capita: $17,800

Occupations: agriculture 16.3%, industry 35.1%, services 48.6%

Currency: Iranian rial (IRR)

Iraq

Long Name: Republic of Iraq

Location: Middle East, bordering the Persian Gulf, between Iran and Kuwait

Area: 169,235 sq. mi. (438,317 sq. km)

Climate: mostly desert; mild to cool winters with dry, hot, cloudless summers; northern mountainous regions along Iranian and Turkish borders experience cold winters with occasionally heavy snows that melt in early spring, sometimes causing extensive flooding in central and southern Iraq

Terrain: mostly broad plains; reedy marshes along Iranian border in south with large flooded areas; mountains along borders with Iran and Turkey

Population: 37,056,169

Population Growth Rate: 2.93%

Birth Rate (per 1,000): 31.45

Death Rate (per 1,000): 3.77

Life Expectancy: 74.85 years

Ethnic Groups: Arab 75%-80%, Kurdish 15%-20%, Turkoman, Assyrian, or other 5%

Religion: Muslim (official) 99% (Shia 60%-65%, Sunni 32%-37%), Christian 0.8%, Hindu <0.1, Buddhist <0.1, Jewish <0.1, folk religion <0.1, unafilliated 0.1, other <0.1

Languages: Arabic (official), Kurdish (official), Turkmen (a Turkish dialect) and Assyrian (Neo-Aramaic) are official in areas where they constitute a majority of the population), Armenian

Literacy: 79.7%

Government Type: parliamentary democracy

Capital: Baghdad

Independence: 3 October 1932 (from League of Nations mandate under British administration)

GDP Per Capita: $15,500

Occupations: N/A

Currency: New Iraqi dinar (NID)

Ireland

Long Name: Ireland

Location: Western Europe, occupying five-sixths of the island of Ireland in the North Atlantic Ocean, west of Great Britain

Area: 27,133 sq. mi. (70,273 sq. km)

Climate: temperate maritime; modified by North Atlantic Current; mild winters, cool summers; consistently humid; overcast about half the time

Terrain: mostly level to rolling interior plain surrounded by rugged hills and low mountains; sea cliffs on west coast

Population: 4,892,305

Population Growth Rate: 1.25%

Birth Rate (per 1,000): 14.84

Death Rate (per 1,000): 6.48

Life Expectancy: 80.68 years

Ethnic Groups: Irish 84.5%, other white 9.8%, Asian 1.9%, black 1.4%, mixed and other 0.9%, unspecified 1.6% (2011 est.)

Religion: Roman Catholic 84.7%, Church of Ireland 2.7%, other Christian 2.7%, Muslim 1.1%, other 1.7%, unspecified 1.5%, none 5.7% (2011 est.)

Languages: English (official) is the language generally used, Irish (Gaelic or Gaeilge) (official) spoken mainly in areas located along the western seaboard

Literacy: 99%

Government Type: republic, parliamentary democracy

Capital: Dublin

Independence: 6 December 1921 (from United Kingdom by treaty)

GDP Per Capita: $54,300

Occupations: agriculture 5%, industry 19%, services 76%

Currency: euro (EUR)

Israel

Long Name: State of Israel

Location: Middle East, bordering the Mediterranean Sea, between Egypt and Lebanon

Area: 8,019 sq. mi. (20,770 sq. km)

Climate: temperate; hot and dry in southern and eastern desert areas

Terrain: Negev desert in the south; low coastal plain; central mountains; Jordan Rift Valley

Population: 8,049,314

Population Growth Rate: 1.56%

Birth Rate (per 1,000): 18.48

Death Rate (per 1,000): 5.15

Life Expectancy: 82.27 years

Ethnic Groups: Jewish 75% (of which Israel-born 74.4%, Europe/America/Oceania-born 17.4%, Africa-born 5.1%, Asia-born 3.1%), non-Jewish 25% (mostly Arab) (2013 est.)

Religion: Jewish 75%, Muslim 17.5%, Christian 2%, Druze 1.6%, other 3.9% (2013 est.)

Languages: Hebrew (official), Arabic used officially for Arab minority, English most commonly used foreign language

Literacy: 97.8%

Government Type: parliamentary democracy

Capital: Jerusalem

Independence: 14 May 1948 (from League of Nations mandate under British administration)

GDP Per Capita: $34,300

Occupations: agriculture 1.1%, industry 17.3%, services 81.6%, other 7.8%

Currency: new Israeli shekel (ILS)

Italy

Long Name: Italian Republic

Location: Southern Europe, a peninsula extending into the central Mediterranean Sea, northeast of Tunisia

Area: 116,348 sq. mi. (301,340 sq. km)

Climate: predominantly Mediterranean; Alpine in far north; hot, dry in south

Terrain: mostly rugged and mountainous; some plains, coastal lowlands

Population: 61,855,120

Population Growth Rate: 0.27%

Birth Rate (per 1,000): 8.74

Death Rate (per 1,000): 10.19

Life Expectancy: 82.12 years

Ethnic Groups: Italian (includes small clusters of German-, French-, and Slovene-Italians in the north and Albanian-Italians and Greek-Italians in the south)

Religion: Christian 80% (overwhelmingly Roman Catholic with very small groups of Jehovah's Witnesses and Protestants), Muslim (about 800,000 to 1 million), Atheist and Agnostic 20%

Languages: Italian (official), German (parts of Trentino-Alto Adige region are predominantly German speaking), French (small French-speaking minority in Valle d'Aosta region), Slovene (Slovene-speaking minority in the Trieste-Gorizia area)

Literacy: 99.2%

Government Type: republic

Capital: Rome

Independence: 17 March 1861 (Kingdom of Italy proclaimed; Italy was not finally unified until 1870)

GDP Per Capita: $35,800

Occupations: agriculture 3.9%, industry 28.3%, services 67.8%

Currency: euro (EUR)

Jamaica

Long Name: Jamaica

Location: Caribbean, island in the Caribbean Sea, south of Cuba

Area: 4,244 sq. mi. (10,991 sq. km)

Climate: tropical; hot, humid; temperate interior

Terrain: mostly mountains, with narrow, discontinuous coastal plain

Population: 2,950,210

Population Growth Rate: 0.68%

Birth Rate (per 1,000): 18.16

Death Rate (per 1,000): 6.17

Life Expectancy: 73.55 years

Ethnic Groups: black 92.1%, mixed 6.1%, East Indian 0.8%, other 0.4%, unspecified 0.7% (2011 est.)

Religion: Protestant 64.8% (includes Seventh Day Adventist 12.0%, Pentecostal 11.0%, Other Church of God 9.2%, New Testament Church of God 7.2%, Baptist 6.7%, Church of God in Jamaica 4.8%, Church of God of Prophecy 4.5%, Anglican 2.8%, United Church 2.1%, Methodist 1.6%, Revived 1.4%, Brethren 0.9%, and Moravian 0.7%), Roman Catholic 2.2%, Jehovah's Witness 1.9%, Rastafarian 1.1%, other 6.5%, none 21.3%, unspecified 2.3% (2011 est.)

Languages: English, English patois

Literacy: 88.7%

Government Type: constitutional parliamentary democracy

Capital: Kingston

Independence: 6 August 1962 (from United Kingdom)

GDP Per Capita: $8,800

Occupations: agriculture 17%, industry 19%, services 64%

Currency: Jamaican dollar (JMD)

Japan

Long Name: Japan

Location: Eastern Asia, island chain between the North Pacific Ocean and the Sea of Japan, east of the Korean Peninsula

Area: 145,914 sq. mi. (377,915 sq. km)

Climate: varies from tropical in south to cool temperate in north

Terrain: mostly rugged and mountainous

Population: 126,919,659

Population Growth Rate: -0.16%

Birth Rate (per 1,000): 7.93

Death Rate (per 1,000): 9.51

Life Expectancy: 84.74 years

Ethnic Groups: Japanese 98.5%, Koreans 0.5%, Chinese 0.4%, other 0.6%

Religion: Shintoism 79.2%, Buddhism 66.8%, Christianity 1.5%, other 7.1%

Languages: Japanese

Literacy: 99%

Government Type: constitutional monarchy with a parliamentary government

Capital: Tokyo

Independence: 660 B.C.E. (traditional founding by Emperor Jimmu)

GDP Per Capita: $38,200

Occupations: agriculture 2.9%, industry 26.2%, services 70.9%

Currency: yen (JPY)

Jordan

Long Name: Hashemite Kingdom of Jordan

Location: Middle East, northwest of Saudi Arabia

Area: 34,495 sq. mi. (89,342 sq. km)

Climate: mostly arid desert; rainy season in west (November to April)

Terrain: mostly desert plateau in east, highland area in west; Great Rift Valley separates East and West Banks of the Jordan River

Population: 8,117,564

Population Growth Rate: 0.83%

Birth Rate (per 1,000): 25.37

Death Rate (per 1,000): 3.79

Life Expectancy: 74.35 years

Ethnic Groups: Arab 98%, Circassian 1%, Armenian 1%

Religion: Muslim 97.2% (official; predominantly Sunni), Christian 2.2% (majority Greek Orthodox, but some Greek and Roman Catholics, Syrian Orthodox, Coptic Orthodox, Armenian Orthodox, and Protestant denominations), Buddhist 0.4%, Hindu 0.1%, Jewish <0.1, folk religion <0.1, unaffiliated <0.1, other <0.1 (2010 est.)

Languages: Arabic (official), English widely understood among upper and middle classes

Literacy: 95.4%

Government Type: constitutional monarchy

Capital: Amman

Independence: 25 May 1946 (from League of Nations mandate under British administration)

GDP Per Capita: $12,400

Occupations: agriculture 2%, industry 20%, services 78%

Currency: Jordanian dinar (JOD)

Kazakhstan

Long Name: Republic of Kazakhstan

Location: Central Asia, northwest of China; a small portion west of the Ural River in eastern-most Europe

Area: 1,052,090 sq. mi. (2,724,900 sq. km)

Climate: continental, cold winters and hot summers, arid and semiarid

Terrain: extends from the Volga to the Altai Mountains and from the plains in western Siberia to oases and desert in Central Asia

Population: 18,157,122

Population Growth Rate: 1.14%

Birth Rate (per 1,000): 19.15

Death Rate (per 1,000): 8.21

Life Expectancy: 70.55 years

Ethnic Groups: Kazakh (Qazaq) 63.1%, Russian 23.7%, Uzbek 2.9%, Ukrainian 2.1%, Uighur 1.4%, Tatar 1.3%, German 1.1%, other 4.4% (2009 est.)

Religion: Muslim 70.2%, Christian 26.2% (mainly Russian Orthodox), other 0.2%, atheist 2.8%, unspecified 0.5% (2009 est.)

Languages: Kazakh (official, Qazaq) 74% (understand spoken language), Russian (official, used in everyday business, designated the "language of interethnic communication") 94.4% (understand spoken language) (2009 est.)

Literacy: 99.8%

Government Type: republic; authoritarian presidential rule, with little power outside the executive branch

Capital: Astana

Independence: 16 December 1991 (from Soviet Union)

GDP Per Capita: $24,700

Occupations: agriculture 25.8%, industry 11.9%, services 62.3%

Currency: tenge (KZT)

Kenya

Long Name: Republic of Kenya

Location: Eastern Africa, bordering the Indian Ocean, between Somalia and Tanzania

Area: 224,081 sq. mi. (580,367 sq. km)

Climate: varies from tropical along coast to arid in interior

Terrain: low plains rise to central highlands bisected by Great Rift Valley; fertile plateau in west

Population: 45,925,301

Population Growth Rate: 1.93%

Birth Rate (per 1,000): 26.4

Death Rate (per 1,000): 6.89

Life Expectancy: 63.77 years

Ethnic Groups: Kikuyu 22%, Luhya 14%, Luo 13%, Kalenjin 12%, Kamba 11%, Kisii 6%, Meru 6%, other African 15%, non-African (Asian, European, and Arab) 1%

Religion: Christian 82.5% (Protestant 47.4%, Catholic 23.3%, other 11.8%), Muslim 11.1%, Traditionalists 1.6%, other 1.7%, none 2.4%, unspecified 0.7% (2009 census)

Languages: English (official), Kiswahili (official), numerous indigenous languages

Literacy: 78%

Government Type: republic

Capital: Nairobi

Independence: 12 December 1963 (from United Kingdom)

GDP Per Capita: $3,300

Occupations: agriculture 75%, industry and services 25%

Currency: Kenyan shilling (KES)

Kiribati

Long Name: Republic of Kiribati

Location: Oceania, group of 33 coral atolls in the Pacific Ocean, straddling the Equator; the capital Tarawa is about half way between Hawaii and Australia; note-on 1 January 1995, Kiribati proclaimed that all of its territory lies in the same time zone as its Gilbert Islands group (UTC +12) even though the Phoenix Islands and the Line Islands under its jurisdiction lie on the other side of the International Date Line

Area: 313 sq. mi. (811 sq. km)

Climate: tropical; marine, hot and humid, moderated by trade winds

Terrain: mostly low-lying coral atolls surrounded by extensive reefs

Population: 105,711

Population Growth Rate: 1.15%

Birth Rate (per 1,000): 21.46

Death Rate (per 1,000): 7.12

Life Expectancy: 65.81 years

Ethnic Groups: I-Kiribati 89.5%, I-Kiribati/mixed 9.7%, Tuvaluan 0.1%, other 0.8% (2010 est.)

Religion: Roman Catholic 55.8%, Kempsville Presbyterian Church 33.5%, Mormon 4.7%, Baha'i 2.3%, Seventh Day Adventist 2%, other 1.5%, none 0.2%, unspecified 0.05% (2010 est.)

Languages: I-Kiribati, English (official)

Literacy: N/A

Government Type: republic

Capital: Tarawa

Independence: 12 July 1979 (from United Kingdom)

GDP Per Capita: $2,200

Occupations: agriculture 15%, industry 10%, services 75%

Currency: Australian dollar (AUD)

Kosovo

Long Name: Republic of Kosovo

Location: Southeast Europe, between Serbia and Macedonia

Area: 4,203 sq. mi. (10,887 sq. km)

Climate: influenced by continental air masses resulting in relatively cold winters with heavy snowfall and hot, dry summers and autumns; Mediterranean and alpine influences create regional variation; maximum rainfall between October and December **Terrain:** flat fluvial basin with an elevation of 400-700 m above sea level surrounded by several high mountain ranges with elevations of 2,000 to 2,500 m

Population: 1,870,981

Population Growth Rate: N/A

Birth Rate (per 1,000): N/A

Death Rate (per 1,000): N/A

Life Expectancy: N/A

Ethnic Groups: Albanians 92.9%, Bosniaks 1.6%, Serbs 1.5%, Turk 1.1%, Ashkali 0.9%, Egyptian 0.7%, Gorani 0.6%, Roma 0.5%, other/unspecified 0.2%

Religion: Muslim 95.6%, Orthodox 1.5%, Roman Catholic 2.2%, other 0.07%, none 0.07%, unspecified 0.6% (2011 est.)

Languages: Albanian (official) 94.5%, Bosnian 1.7%, Serbian (official) 1.6%, Turkish 1.1%, other 0.9% (includes Romani), unspecified 0.1%

Literacy: 91.9%

Government Type: republic

Capital: Pristina

Independence: 17 February 2008

GDP Per Capita: $0

Occupations: agriculture 5.9%, industry 16.8%, services 64.5%

Currency: euro (EUR); Serbian Dinar (RSD) is also in circulation

Kuwait

Long Name: State of Kuwait

Location: Middle East, bordering the Persian Gulf, between Iraq and Saudi Arabia

Area: 6,880 sq. mi. (17,818 sq. km)

Climate: dry desert; intensely hot summers; short, cool winters

Terrain: flat to slightly undulating desert plain

Population: 2,788,534

Population Growth Rate: 1.62% (reflects a return to pre-Gulf crisis immigration of expatriates)

Birth Rate (per 1,000): 19.91

Death Rate (per 1,000): 2.18

Life Expectancy: 77.82 years

Ethnic Groups: Kuwaiti 31.3%, other Arab 27.9%, Asian 37.8%, African 1.9%, other 1.1% (includes European, North American, South American, and Australian) (2013 est.)

Religion: Muslim (official) 76.7%, Christian 17.3%, other and unspecified 5.9%

Languages: Arabic (official), English widely spoken

Literacy: 96.3%

Government Type: constitutional emirate

Capital: Kuwait

Independence: 19 June 1961 (from United Kingdom)

GDP Per Capita: $72,200

Occupations: N/A

Currency: Kuwaiti dinar (KD)

Kyrgyzstan

Long Name: Kyrgyz Republic

Location: Central Asia, west of China

Area: 77,202 sq. mi. (199,951 sq. km)

Climate: dry continental to polar in high Tien Shan; subtropical in southwest (Fergana Valley); temperate in northern foothill zone

Terrain: peaks of Tien Shan and associated valleys and basins encompass entire nation

Population: 5,664,939

Population Growth Rate: 1.11%

Birth Rate (per 1,000): 22.98

Death Rate (per 1,000): 6.65

Life Expectancy: 70.36 years

Ethnic Groups: Kyrgyz 70.9%, Uzbek 14.3%, Russian 7.7%, Dungan 1.1%, other 5.9% (includes Uyghur, Tajik, Turk, Kazakh, Tatar, Ukrainian, Korean, German) (2009 est.)

Religion: Muslim 75%, Russian Orthodox 20%, other 5%

Languages: Kyrgyz 64.7% (official), Uzbek 13.6%, Russian 12.5% (official), Dungun 1%, other 8.2%

Literacy: 99.5%

Government Type: republic

Capital: Bishkek

Independence: 31 August 1991 (from Soviet Union)

GDP Per Capita: $3,400

Occupations: agriculture 48%, industry 12.5%, services 39.5%

Currency: som (KGS)

Laos

Long Name: Lao People's Democratic Republic

Location: Southeastern Asia, northeast of Thailand, west of Vietnam

Area: 91,429 sq. mi. (236,800 sq. km)

Climate: tropical monsoon; rainy season (May to November); dry season (December to April)

Terrain: mostly rugged mountains; some plains and plateaus

Population: 6,911,544

Population Growth Rate: 1.55%

Birth Rate (per 1,000): 24.25

Death Rate (per 1,000): 7.63

Life Expectancy: 63.88 years

Ethnic Groups: Lao 54.6%, Khmou 10.9%, Hmong 8%, Tai 3.8%, Phuthai 3.3%, Lue 2.2%, Katang 2.1%, Makong 2.1%, Akha 1.6%, other 10.4%, unspecified 1% (2005 est.)

Religion: Buddhist 66.8%, Christian 1.5%, other 31%, unspecified 0.7% (2005 est.)

Languages: Lao (official), French, English, and various ethnic languages

Literacy: 79.9%

Government Type: Communist state

Capital: Vientiane

Independence: 19 July 1949 (from France)

GDP Per Capita: $5,400

Occupations: agriculture 73.1%, industry 6.1%, services 20.6%

Currency: kip (LAK)

Latvia

Long Name: Republic of Latvia

Location: Eastern Europe, bordering the Baltic Sea, between Estonia and Lithuania

Area: 24,938 sq. mi. (64,589 sq. km)

Climate: maritime; wet, moderate winters

Terrain: low plain

Population: 1,986,705

Population Growth Rate: -1.06%

Birth Rate (per 1,000): 10

Death Rate (per 1,000): 14.31

Life Expectancy: 74.23 years

Ethnic Groups: Latvian 61.1%, Russian 26.2%, Belarusian 3.5%, Ukrainian 2.3%, Polish 2.2%, Lithuanian 1.3%, other 3.4% (2013 est.)

Religion: Lutheran 19.6%, Orthodox 15.3%, other Christian 1%, other 0.4%, unspecified 63.7% (2006)

Languages: Latvian (official) 56.3%, Russian 33.8%, other 0.6% (includes Polish, Ukrainian, and Belarusian), unspecified 9.4%

Literacy: 99.9%

Government Type: parliamentary democracy

Capital: Riga

Independence: 18 November 1918 (from Soviet Russia); 4 May 1990 is when it declared the renewal of independence; 21 August 1991 was the date of *de facto* independence from the Soviet Union

GDP Per Capita: $24,500

Occupations: agriculture 8.8%, industry 24%, services 67.2%

Currency: lat (LVL)

Lebanon

Long Name: Lebanese Republic

Location: Middle East, bordering the Mediterranean Sea, between Israel and Syria

Area: 4,015 sq. mi. (10,400 sq. km)

Climate: Mediterranean; mild to cool, wet winters with hot, dry summers; Lebanon mountains experience heavy winter snows

Terrain: narrow coastal plain; El Beqaa (Bekaa Valley) separates Lebanon and Anti-Lebanon Mountains

Population: 6,184,701

Population Growth Rate: 0.86%

Birth Rate (per 1,000): 14.59

Death Rate (per 1,000): 4.88

Life Expectancy: 77.4 years

Ethnic Groups: Arab 95%, Armenian 4%, other 1%

Religion: Muslim 54% (27% Sunni, 27% Shia), Christian 40.5% (includes 21% Maronite Catholic, 8% Greek Orthodox, 5% Greek Catholic, 6.5% other Christian), Druze 5.6%, very small numbers of Jews, Baha'is, Buddhists, Hindus, and Mormons

Languages: Arabic (official), French, English, Armenian

Literacy: 93.9%

Government Type: republic

Capital: Beirut

Independence: 22 November 1943 (from League of Nations mandate under French administration)

GDP Per Capita: $18,600

Occupations: N/A

Currency: Lebanese pound (LBP)

Lesotho

Long Name: Kingdom of Lesotho

Location: Southern Africa, an enclave of South Africa

Area: 11,720 sq. mi. (30,355 sq. km)

Climate: temperate; cool to cold, dry winters; hot, wet summers

Terrain: mostly highland with plateaus, hills, and mountains

Population: 1,947,701

Population Growth Rate: 0.32%

Birth Rate (per 1,000): 25.47

Death Rate (per 1,000): 14.89

Life Expectancy: 52.86 years

Ethnic Groups: Sotho 99.7%, Europeans, Asians, and other 0.3%

Religion: Christian 80%, indigenous beliefs 20%

Languages: Sesotho (southern Sotho), English (official), Zulu, Xhosa

Literacy: 79.4%

Government Type: parliamentary constitutional monarchy

Capital: Maseru

Independence: 4 October 1966 (from United Kingdom)

GDP Per Capita: $3,000

Occupations: agriculture 86% of resident population engaged in subsistence agriculture, roughly 35% of the active male wage earners work in South Africa, industry and services 14%

Currency: loti (LSL); South African rand (ZAR)

Liberia

Long Name: Republic of Liberia

Location: Western Africa, bordering the North Atlantic Ocean, between Cote d'Ivoire and Sierra Leone

Area: 43,000 sq. mi. (111,369 sq. km)

Climate: tropical; hot, humid; dry winters with hot days and cool to cold nights; wet, cloudy summers with frequent heavy showers

Terrain: mostly flat to rolling coastal plains rising to rolling plateau and low mountains in northeast

Population: 4,195,666

Population Growth Rate: 2.47%

Birth Rate (per 1,000): 34.41

Death Rate (per 1,000): 9.69

Life Expectancy: 58.6 years

Ethnic Groups: Kpelle 20.3%, Bassa 13.4%, Grebo 10%, Gio 8%, Mano 7.9%, Kru 6%, Lorma 5.1%, Kissi 4.8%, Gola 4.4%, other 20.1% (2008 Census)

Religion: Christian 85.6%, Muslim 12.2%, Traditional 0.6%, other 0.2%, none 1.4% (2008 Census)

Languages: English 20% (official), some 20 ethnic group languages, of which a few can be written and are used in correspondence

Literacy: 47.6%

Government Type: republic

Capital: Monrovia

Independence: 26 July 1847

GDP Per Capita: $900

Occupations: agriculture 70%, industry 8%, services 22%

Currency: Liberian dollar (LRD)

Libya

Long Name: Great Socialist People's Libyan Arab Jamahiriya

Location: Northern Africa, bordering the Mediterranean Sea, between Egypt and Tunisia

Area: 679,359 sq. mi. (1,759,540 sq. km)

Climate: Mediterranean along coast; dry, extreme desert interior

Terrain: mostly barren, flat to undulating plains, plateaus, depressions

Population: 6,411,776

Population Growth Rate: 2.23%

Birth Rate (per 1,000): 18.03

Death Rate (per 1,000): 3.58

Life Expectancy: 76.26 years

Ethnic Groups: Berber and Arab 97%, other 3% (includes Greeks, Maltese, Italians, Egyptians, Pakistanis, Turks, Indians, and Tunisians)

Religion: Muslim (official; virtually all Sunni) 96.6%, Christian 2.7%, Buddhist 0.3%, Hindu <0.1, Jewish <0.1, folk religion <0.1, unafilliated 0.2%, other <0.1

Languages: Arabic (official), Italian, English (all widely understood in the major cities); Berber (Nafusi, Ghadamis, Suknah, Awjilah, Tamasheq)

Literacy: 91%

Government Type: Jamahiriya (a state of the masses) in theory, governed by the populace through local councils; in practice, an authoritarian state

Capital: Tripoli

Independence: 24 December 1951 (from UN trusteeship)

GDP Per Capita: $15,100

Occupations: agriculture 17%, industry 23%, services 59%

Currency: Libyan dinar (LYD)

Liechtenstein

Long Name: Principality of Liechtenstein

Location: Central Europe, between Austria and Switzerland

Area: 62 sq. mi. (160 sq. km)

Climate: continental; cold, cloudy winters with frequent snow or rain; cool to moderately warm, cloudy, humid summers

Terrain: mostly mountainous (Alps) with Rhine Valley in western third

Population: 37,624

Population Growth Rate: 0.84%

Birth Rate (per 1,000): 10.45

Death Rate (per 1,000): 7.12

Life Expectancy: 81.77 years

Ethnic Groups: Liechtensteiner 66.3%, other 33.7% (2013 est.)

Religion: Roman Catholic (official) 75.9%, Protestant Reformed 6.5%, Muslim 5.4%, Lutheran 1.3%, other 2.9%, none 5.4%, unspecified 2.6% (2010 est.)

Languages: German 94.5% (official) (Alemannic is the main dialect), Italian 1.1%, other 4.3% (2010 est.)

Literacy: 100%

Government Type: constitutional monarchy

Capital: Vaduz

Independence: 23 January 1719 (Principality of Liechtenstein established); 12 July 1806 (independence from the Holy Roman Empire)

GDP Per Capita: $89,400

Occupations: agriculture 0.8%, industry 39.4%, services 59.9%

Currency: Swiss franc (CHF)

Lithuania

Long Name: Republic of Lithuania

Location: Eastern Europe, bordering the Baltic Sea, between Latvia and Russia

Area: 25,212 sq. mi. (65,300 sq. km)

Climate: transitional, between maritime and continental; wet, moderate winters and summers

Terrain: lowland, many scattered small lakes, fertile soil

Population: 2,884,433

Population Growth Rate: -1.04%

Birth Rate (per 1,000): 10.1

Death Rate (per 1,000): 14.27

Life Expectancy: 74.69 years

Ethnic Groups: Lithuanian 84.1%, Polish 6.6%, Russian 5.8%, Belarusian 1.2%, other 1.1%, unspecified 1.2% (2011 est.)

Religion: Roman Catholic 77.2%, Russian Orthodox 4.1%, Old Believer 0.8%, Evangelical Lutheran 0.6%, Evangelical Reformist 0.2%, other (including Sunni Muslim, Jewish, Greek Catholic, and Karaite) 0.8%, none 6.1%, unspecified 10.1% (2011 est.)

Languages: Lithuanian (official) 82%, Russian 8%, Polish 5.6%, other 0.9%, unspecified 3.5% (2011 est.)

Literacy: 99.8%

Government Type: parliamentary democracy

Capital: Vilnius

Independence: 11 March 1990 (declared); 6 September 1991 (recognized by Soviet Union)

GDP Per Capita: $28,000

Occupations: agriculture 7.9%, industry 19.6%, services 72.5%

Currency: litas (LTL)

Luxembourg

Long Name: Grand Duchy of Luxembourg

Location: Western Europe, between France and Germany

Area: 998 sq. mi. (2,586 sq. km)

Climate: modified continental with mild winters, cool summers

Terrain: mostly gently rolling uplands with broad, shallow valleys; uplands to slightly mountainous in the north; steep slope down to Moselle flood plain in the southeast

Population: 570,252

Population Growth Rate: 2.13%

Birth Rate (per 1,000): 11.37

Death Rate (per 1,000): 7.24

Life Expectancy: 82.17 years

Ethnic Groups: Luxembourger 54.1%, Portuguese 16.4%, French 7%, Italian 3.5%, Belgian 3.3%, German 2.3%, British 1.1%, other 12.3%

Languages: Luxembourgish (official administrative and judicial language and national language [spoken vernacular]) 88.8%, French (official administrative, judicial, and legislative language) 4.2%, Portuguese 2.3%, German (official administrative and judicial language) 1.1%, other 3.5% (2011 est.)

Literacy: 100%

Government Type: constitutional monarchy

Capital: Luxembourg

Independence: 1839 (from the Netherlands)

GDP Per Capita: $102,900

Occupations: agriculture 1.1%, industry 20%, services 78.9%

Currency: euro (EUR)

Macedonia

Long Name: Republic of Macedonia

Location: Southeastern Europe, north of Greece

Area: 9,928 sq. mi. (25,713 sq. km)

Climate: warm, dry summers and autumns; relatively cold winters with heavy snowfall

Terrain: mountainous territory covered with deep basins and valleys; three large lakes, each divided by a frontier line; country bisected by the Vardar River

Population: 2,096,015

Population Growth Rate: 0.2%

Birth Rate (per 1,000): 11.5

Death Rate (per 1,000): 9.08

Life Expectancy: 76.02 years

Ethnic Groups: Macedonian 64.2%, Albanian 25.2%, Turkish 3.9%, Roma (Gypsy) 2.7%, Serb 1.8%, other 2.2% (2002 est.)

Religion: Macedonian Orthodox 64.8%, Muslim 33.3%, other Christian 0.4%, other and unspecified 1.5% (2002 est.)

Languages: Macedonian (official) 66.5%, Albanian (official) 25.1%, Turkish 3.5%, Roma 1.9%, Serbian 1.2%, other 1.8% (2002 est.)

Literacy: 97.8%

Government Type: parliamentary democracy

Capital: Skopje

Independence: 8 September 1991 (referendum by registered voters endorsed independence from Yugoslavia)

GDP Per Capita: $14,000

Occupations: agriculture 18.3%, industry 29.1%, services 52.6%

Currency: Macedonian denar (MKD)

Madagascar

Long Name: Republic of Madagascar

Location: Southern Africa, island in the Indian Ocean, east of Mozambique

Area: 226,658 sq. mi. (587,041 sq. km)

Climate: tropical along coast, temperate inland, arid in south

Terrain: narrow coastal plain, high plateau and mountains in center

Population: 23,812,681

Population Growth Rate: 2.58%

Birth Rate (per 1,000): 32.61

Death Rate (per 1,000): 6.81

Life Expectancy: 65.55 years

Ethnic Groups: Malayo-Indonesian (Merina and related Betsileo), Cotiers (mixed African, Malayo-Indonesian, and Arab ancestry-Betsimisaraka, Tsimihety, Antaisaka, Sakalava), French, Indian, Creole, Comoran

Religion: indigenous beliefs 52%, Christian 41%, Muslim 7%

Languages: English (official), French (official), Malagasy (official)

Literacy: 64.7%

Government Type: republic

Capital: Antananarivo

Independence: 26 June 1960 (from France)

GDP Per Capita: $1,500

Occupations: N/A

Currency: ariary (MGA)

Malawi

Long Name: Republic of Malawi

Location: Southern Africa, east of Zambia

Area: 45,747 sq. mi. (118,484 sq. km)

Climate: sub-tropical; rainy season (November to May); dry season (May to November)

Terrain: narrow elongated plateau with rolling plains, rounded hills, some mountains

Population: 17,964,697

Population Growth Rate: 3.32%

Birth Rate (per 1,000): 41.56

Death Rate (per 1,000): 8.41

Life Expectancy: 60.66 years

Ethnic Groups: Chewa 32.6%, Lomwe 17.6%, Yao 13.5%, Ngoni 11.5%, Tumbuka 8.8%, Nyanja 5.8%, Sena 3.6%, Tonga 2.1%, Ngonde 1%, other 3.5%

Religion: Christian 82.6%, Muslim 13%, other 1.9%, none 2.5% (2008 est.)

Languages: Chichewa 57.2% (official), Chinyanja 12.8%, Chiyao 10.1%, Chitumbuka 9.5%, Chisena 2.7%, Chilomwe 2.4%, Chitonga 1.7%, other 3.6%

Literacy: 65.8%

Government Type: multiparty democracy

Capital: Lilongwe

Independence: 6 July 1964 (from United Kingdom)

GDP Per Capita: $1,200

Occupations: agriculture 90%, industry and services 10%

Currency: Malawian kwacha (MWK)

Malaysia

Long Name: Malaysia

Location: Southeastern Asia, peninsula bordering Thailand and northern one-third of the island of Borneo, bordering Indonesia, Brunei, and the South China Sea, south of Vietnam

Area: 127,355 sq. mi. (329,847 sq. km)

Climate: tropical; annual southwest (April to October) and northeast (October to February) monsoons

Terrain: coastal plains rising to hills and mountains

Population: 30,513,848

Population Growth Rate: 1.44%

Birth Rate (per 1,000): 19.71

Death Rate (per 1,000): 5.03

Life Expectancy: 74.75 years

Ethnic Groups: Malay 50.1%, Chinese 22.6%, indigenous 11.8%, Indian 6.7%, other 0.7%, non-citizens 8.2% (2010 est.)

Religion: Muslim (official) 61.3%, Buddhist 19.8%, Christian 9.2%, Hindu 6.3%, Confucianism, Taoism, other traditional Chinese religions 1.3%, other 0.4%, none 0.8%, unspecified 1% (2010 est.)

Languages: Bahasa Malaysia (official), English, Chinese (Cantonese, Mandarin, Hokkien, Hakka, Hainan, Foochow), Tamil, Telugu, Malayalam, Panjabi, Thai, in East Malaysia there are several indigenous languages; most widely spoken are Iban and Kadazan

Literacy: 94.6%

Government Type: constitutional monarchy

Capital: Kuala Lumpur

Independence: 31 August 1957 (from United Kingdom)

GDP Per Capita: $26,600

Occupations: agriculture 11%, industry 36%, services 53%

Currency: ringgit (MYR)

Maldives

Long Name: Republic of Maldives

Location: Southern Asia, group of atolls in the Indian Ocean, south-southwest of India

Area: 115 sq. mi. (298 sq. km)

Climate: tropical; hot, humid; dry, northeast monsoon (November to March); rainy, southwest monsoon (June to August)

Terrain: flat, with white sandy beaches

Population: 393,253

Population Growth Rate: –0.08%

Birth Rate (per 1,000): 15.75

Death Rate (per 1,000): 3.89

Life Expectancy: 75.37 years

Ethnic Groups: South Indians, Sinhalese, Arabs

Religion: Sunni Muslim

Languages: Maldivian Dhivehi (dialect of Sinhala, script derived from Arabic), English spoken by most government officials

Literacy: 99.3%

Government Type: republic

Capital: Male

Independence: 26 July 1965 (from United Kingdom)

GDP Per Capita: $13,600

Occupations: agriculture 15%, industry 15%, services 70%

Currency: rufiyaa (MVR)

Mali

Long Name: Republic of Mali

Location: Western Africa, southwest of Algeria

Area: 478,841 sq. mi. (1,240,192 sq. km)

Climate: subtropical to arid; hot and dry (February to June); rainy, humid, and mild (June to November); cool and dry (November to February)

Terrain: mostly flat to rolling northern plains covered by sand; savanna in south, rugged hills in northeast

Population: 16,955,536

Population Growth Rate: 2.98%

Birth Rate (per 1,000): 44.99

Death Rate (per 1,000): 12.89

Life Expectancy: 55.34 years

Ethnic Groups: Bambara 34.1%, Fulani (Peul) 14.7%, Sarakole 10.8%, Senufo 10.5%, Dogon 8.9%, Malinke 8.7%, Bobo 2.9%, Songhai 1.6%, Tuareg 0.9%, other Malian 6.1%, from member of Economic Community of West African States 0.3%, other 0.4% (2012-13 est.)

Religion: Muslim 94.8%, Christian 2.4%, Animist 2%, none 0.5%, unspecified 0.3% (2009 est.)

Languages: French (official), Bambara 46.3%, Peul/Foulfoulbe 9.4%, Dogon 7.2%, Maraka/Soninke 6.4%, Malinke 5.6%, Sonrhai/Djerma 5.6%, Minianka 4.3%, Tamacheq 3.5%, Senoufo 2.6%, Bobo 2.1%, unspecified 0.7%, other 6.3%

Literacy: 38.7%

Government Type: republic

Capital: Bamako

Independence: 22 September 1960 (from France)

GDP Per Capita: $1,800

Occupations: agriculture 80%, industry and services 20%

Currency: Communaute Financiere Africaine franc (XOF)

Malta

Long Name: Republic of Malta

Location: Southern Europe, islands in the Mediterranean Sea, south of Sicily (Italy)

Area: 122 sq. mi. (316 sq. km)

Climate: Mediterranean; mild, rainy winters; hot, dry summers

Terrain: mostly low, rocky, flat to dissected plains; many coastal cliffs

Population: 413,965

Population Growth Rate: 0.31%

Birth Rate (per 1,000): 10.18

Death Rate (per 1,000): 9.09

Life Expectancy: 80.25 years

Ethnic Groups: Maltese (descendants of ancient Carthaginians and Phoenicians with strong elements of Italian and other Mediterranean stock)

Religion: Roman Catholic (official) more than 90% (2011 est.)

Languages: Maltese (official) 90.1%, English (official) 6%, multilingual 3%, other 0.9% (2005 est.)

Literacy: 94.4%

Government Type: republic

Capital: Valletta

Independence: 21 September 1964 (from United Kingdom)

GDP Per Capita: $34,700

Occupations: agriculture 1.5%, industry 25.7%, services 72.8%

Currency: euro (EUR)

Marshall Islands

Long Name: Republic of the Marshall Islands

Location: Oceania, two archipelagic island chains of 29 atolls, each made up of many small islets, and five single islands in the North Pacific Ocean, about half way between Hawaii and Australia

Area: 70 sq. mi. (181 sq. km)

Climate: tropical; hot and humid; wet season May to November; islands border typhoon belt

Terrain: low coral limestone and sand islands

Population: 72,191

Population Growth Rate: 1.66%

Birth Rate (per 1,000): 25.6

Death Rate (per 1,000): 4.21

Life Expectancy: 72.84 years

Ethnic Groups: Marshallese 92.1%, mixed Marshallese 5.9%, other 2% (2006)

Religion: Protestant 54.8%, Assembly of God 25.8%, Roman Catholic 8.4%, Bukot nan Jesus 2.8%, Mormon 2.1%, other Christian 3.6%, other 1%, none 1.5% (1999 census)

Languages: Marshallese (official) 98.2%, other languages 1.8%, English widely spoken as a second language

Literacy: 93.7%

Government Type: constitutional government in free association with the US; the Compact of Free Association entered into force 21 October 1986 and the Amended Compact entered into force in May 2004

Capital: Majuro

Independence: 21 October 1986 (from the US-administered UN trusteeship)

GDP Per Capita: $3,400

Occupations: agriculture 11%, industry 16.3%, services 72.7%

Currency: U.S. dollar (USD)

Mauritania

Long Name: Islamic Republic of Mauritania

Location: Northern Africa, bordering the North Atlantic Ocean, between Senegal and Western Sahara

Area: 397,954 sq. mi. (1,030,700 sq. km)

Climate: desert; constantly hot, dry, dusty

Terrain: mostly barren, flat plains of the Sahara; some central hills

Population: 3,596,702

Population Growth Rate: 2.23%

Birth Rate (per 1,000): 31.34

Death Rate (per 1,000): 8.2

Life Expectancy: 62.65 years

Ethnic Groups: black Moors (Haratines-Arab-speaking slaves, former slaves, and their descendants of African origin, enslaved by white Moors) 40%, white Moors (of Arab-Berber descent, known as Bidhan) 30%, black Africans (non-Arabic speaking, Halpulaar, Soninke, Wolof, and Bamara ethnic groups) 30%

Religion: Muslim 100%

Languages: Arabic (official and national), Pulaar, Soninke, Wolof (all national languages), French

Literacy: 52.1%

Government Type: Democratic Republic

Capital: Nouakchott

Independence: 28 November 1960 (from France)

GDP Per Capita: $4,500

Occupations: agriculture 50%, industry 2%, services 48%

Currency: ouguiya (MRO)

Mauritius

Long Name: Republic of Mauritius

Location: Southern Africa, island in the Indian Ocean, east of Madagascar

Area: 788 sq. mi. (2,040 sq. km)

Climate: tropical, modified by southeast trade winds; warm, dry winter (May to November); hot, wet, humid summer (November to May)

Terrain: small coastal plain rising to discontinuous mountains encircling central plateau

Population: 1,339,827

Population Growth Rate: 0.64%

Birth Rate (per 1,000): 13.29

Death Rate (per 1,000): 6.91

Life Expectancy: 75.4 years

Ethnic Groups: Indo-Mauritian 68%, Creole 27%, Sino-Mauritian 3%, Franco-Mauritian 2%

Religion: Hindu 48.5%, Roman Catholic 26.3%, Muslim 17.3%, other Christian 6.4%, other 0.6%, none 0.7%, unspecified 0.1% (2011 est.)

Languages: Creole 86.5%, Bhojpuri 5.3%, French 4.1%, two languages 1.4%, other 2.6% (includes English, the official language, which is spoken by less than 1% of the population), unspecified 0.1% (2011 est.)

Literacy: 90.6%

Government Type: parliamentary democracy

Capital: Port Louis

Independence: 12 March 1968 (from United Kingdom)

GDP Per Capita: $19,500

Occupations: agriculture and fishing 9%, construction and industry 30%, transportation and communication 7%, trade, restaurants, hotels 22%, finance 6%, other services 25%

Currency: Mauritian rupee (MUR)

Mexico

Long Name: United Mexican States

Location: Middle America, bordering the Caribbean Sea and the Gulf of Mexico, between Belize and the US and bordering the North Pacific Ocean, between Guatemala and the US

Area: 758,449 sq. mi. (1,964,375 sq. km)

Climate: varies from tropical to desert

Terrain: high, rugged mountains; low coastal plains; high plateaus; desert

Population: 121,736,809

Population Growth Rate: 1.18%

Birth Rate (per 1,000): 18.78

Death Rate (per 1,000): 5.26

Life Expectancy: 75.65 years

Ethnic Groups: mestizo (Amerindian-Spanish) 62%, predominantly Amerindian 21%, Amerindian 7%, other 10% (mostly European)

Religion: Roman Catholic 82.7%, Pentecostal 1.6%, Jehovah's Witnesses 1.4%, other Evangelical Churches 5%, other 1.9%, none 4.7%, unspecified 2.7% (2010 est.)

Languages: Spanish only 92.7%, Spanish and indigenous languages 5.7%, indigenous only 0.8%, unspecified 0.8%; note-indigenous languages include various Mayan, Nahuatl, and other regional languages

Literacy: 95.1%

Government Type: federal republic

Capital: Mexico (Distrito Federal)

Independence: 16 September 1810 (declared); 27 September 1821 (recognized by Spain)

GDP Per Capita: $18,500

Occupations: agriculture 13.4%, industry 24.1%, services 61.9%

Currency: Mexican peso (MXN)

Micronesia

Long Name: Federated States of Micronesia

Location: Oceania, island group in the North Pacific Ocean, about three-quarters of the way from Hawaii to Indonesia

Area: 271 sq. mi. (702 sq. km)

Climate: tropical; heavy year-round rainfall, especially in the eastern islands; located on southern edge of the typhoon belt with occasionally severe damage

Terrain: islands vary geologically from high mountainous islands to low, coral atolls; volcanic outcroppings on Pohnpei, Kosrae, and Chuuk

Population: 105,216

Population Growth Rate: -0.46%

Birth Rate (per 1,000): 20.54

Death Rate (per 1,000): 4.23

Life Expectancy: 72.62 years

Ethnic Groups: Chuukese/Mortlockese 49.3%, Pohnpeian 29.8%, Kosraean 6.3%, Yapese 5.7%, Yap outer islanders 5.1%, Polynesian 1.6%, Asian 1.4%, other 0.8% (2010 est.)

Religion: Roman Catholic 54.7%, Protestant 41.1% (includes Congregational 38.5%, Baptist 1.1%, Seventh Day Adventist 0.8%, Assembly of God 0.7%), Mormon 1.5%, other 1.9%, none 0.7%, unspecified 0.1% (2010 est.)

Languages: English (official and common language), Chuukese, Kosrean, Pohnpeian, Yapese, Ulithian, Woleaian, Nukuoro, Kapingamarangi

Literacy: 89%

Government Type: constitutional government in free association with the US; the Compact of Free Association entered into force 3 November 1986 and the Amended Compact entered into force May 2004

Capital: Palikir

Independence: 3 November 1986 (from the US-administered UN trusteeship)

GDP Per Capita: $3,000

Occupations: agriculture 0.9%, industry 5.2%, services 93.9%

Currency: U.S. dollar (USD)

Moldova

Long Name: Republic of Moldova

Location: Eastern Europe, northeast of Romania

Area: 13,070 sq. mi. (33,851 sq. km)

Climate: moderate winters, warm summers

Terrain: rolling steppe, gradual slope south to Black Sea

Population: 3,546,847

Population Growth Rate: -1.03%

Birth Rate (per 1,000): 12

Death Rate (per 1,000): 12.59

Life Expectancy: 70.42 years

Ethnic Groups: Moldovan 75.8%, Ukrainian 8.4%, Russian 5.9%, Gagauz 4.4%, Romanian 2.2%, Bulgarian 1.9%, other 1%, unspecified 0.4%

Religion: Orthodox 93.3%, Baptist 1%, other Christian 1.2%, other 0.9%, atheist 0.4%, none 1%, unspecified 2.2% (2004 est.)

Languages: Moldovan 58.8% (official; virtually the same as the Romanian language), Romanian 16.4%, Russian 16%, Ukrainian 3.8%, Gagauz 3.1% (a Turkish language), Bulgarian 1.1%, other 0.3%, unspecified 0.4%

Literacy: 99.4%

Government Type: republic

Capital: Chisinau

Independence: 27 August 1991 (from Soviet Union)

GDP Per Capita: $5,000

Occupations: agriculture 26.4%, industry 13.2%, services 60.4%

Currency: Moldovan leu (MDL)

Monaco

Long Name: Principality of Monaco

Location: Western Europe, bordering the Mediterranean Sea on the southern coast of France, near the border with Italy

Area: 0.77 sq. mi. (2 sq. km)

Climate: Mediterranean with mild, wet winters and hot, dry summers

Terrain: hilly, rugged, rocky

Population: 37,731

Population Growth Rate: 0.12%

Birth Rate (per 1,000): 6.65

Death Rate (per 1,000): 9.24

Life Expectancy: 89.52 years

Ethnic Groups: French 47%, Monegasque 16%, Italian 16%, other 21%

Religion: Roman Catholic 90%, other 10%

Languages: French (official), English, Italian, Monegasque

Literacy: 99%

Government Type: constitutional monarchy

Capital: Monaco

Independence: 1419 (beginning of rule by the House of Grimaldi)

GDP Per Capita: $78,700

Occupations: agriculture 0%, industry 16.1%, services 83.9%

Currency: euro (EUR)

Mongolia

Long Name: Mongolia

Location: Northern Asia, between China and Russia

Area: 603,906 sq. mi. (1,564,116 sq. km)

Climate: desert; continental (large daily and seasonal temperature ranges)

Terrain: vast semidesert and desert plains, grassy steppe, mountains in west and southwest; Gobi Desert in south-central

Population: 2,992,908

Population Growth Rate: 1.31%

Birth Rate (per 1,000): 20.25

Death Rate (per 1,000): 6.35

Life Expectancy: 69.29 years

Ethnic Groups: Khalkh 81.9%, Kazak 3.8%, Dorvod 2.7%, Bayad 2.1%, Buryat-Bouriates 1.7%, Zakhchin 1.2%, Dariganga 1%, Uriankhai 1%, other 4.6% (2010 est.)

Religion: Buddhist 53%, Muslim 3%, Christian 2.2%, Shamanist 2.9%, other 0.4%, none 38.6% (2010 est.)

Languages: Khalkha Mongol 90%, Turkic, Russian

Literacy: 98.4%

Government Type: mixed parliamentary/presidential

Capital: Ulaanbaatar

Independence: 11 July 1921 (from China)

GDP Per Capita: $12,500

Occupations: agriculture 28.6%, industry 21%, services 50.4%

Currency: togrog/tugrik (MNT)

Montenegro

Long Name: Montenegro

Location: Southeastern Europe, between the Adriatic Sea and Serbia

Area: 5,333 sq. mi. (13,812 sq. km)

Climate: Mediterranean climate, hot dry summers and autumns and relatively cold winters with heavy snowfalls inland

Terrain: highly indented coastline with narrow coastal plain backed by rugged high limestone mountains and plateaus

Population: 647,073

Population Growth Rate: -0.42%

Birth Rate (per 1,000): 10.42

Death Rate (per 1,000): 9.43

Life Expectancy: N/A

Ethnic Groups: Montenegrin 45%, Serbian 28.7%, Bosniak 8.7%, Albanian 4.9%, Muslim 3.3%, Roma 1%, Croat 1%, other 2.6%, unspecified 4.9% (2011 est.)

Religion: Orthodox 72.1%, Muslim 19.1%, Catholic 3.4%, atheist 1.2%, other 1.5%, unspecified 2.6% (2011 est.)

Languages: Serbian 42.9%, Montenegrin (official) 37%, Bosnian 5.3%, Albanian 5.3%, Serbo-Croat 2%, other 3.5%, unspecified 4% (2011 est.)

Literacy: 98.7%

Government Type: republic

Capital: Podgorica

Independence: 3 June 2006 (from Serbia and Montenegro)

GDP Per Capita: $15,700

Occupations: agriculture 5.3%, industry 17.9%, services 76.8%

Currency: euro (EUR)

Morocco

Long Name: Kingdom of Morocco

Location: Northern Africa, bordering the North Atlantic Ocean and the Mediterranean Sea, between Algeria and Western Sahara

Area: 172,413 sq. mi. (446,550 sq. km)

Climate: Mediterranean, becoming more extreme in the interior

Terrain: northern coast and interior are mountainous with large areas of bordering plateaus, intermontane valleys, and rich coastal plains

Population: 33,322,699

Population Growth Rate: 1%

Birth Rate (per 1,000): 18.2

Death Rate (per 1,000): 4.81

Life Expectancy: 76.71 years

Ethnic Groups: Arab-Berber 99%, other 1%

Religion: Muslim 99% (official; virtually all Sunni, <0.1% Shia), other 1% (includes Christian, Jewish, and Baha'i), Jewish about 6,000 (2010 est.)

Languages: Arabic (official), Berber languages (Tamazight (official), Tachelhit, Tarifit), French (often the language of business, government, and diplomacy)

Literacy: 68.5%

Government Type: constitutional monarchy

Capital: Rabat

Independence: 2 March 1956 (from France)

GDP Per Capita: $8,300

Occupations: agriculture 39.1%, industry 20.3%, services 40.5%

Currency: Moroccan dirham (MAD)

Mozambique

Long Name: Republic of Mozambique

Location: Southeastern Africa, bordering the Mozambique Channel, between South Africa and Tanzania

Area: 308,642 sq. mi. (799,380 sq. km)

Climate: tropical to subtropical

Terrain: mostly coastal lowlands, uplands in center, high plateaus in northwest, mountains in west

Population: 25,303,113

Population Growth Rate: 2.45%

Birth Rate (per 1,000): 38.58

Death Rate (per 1,000): 12.1

Life Expectancy: 52.94 years

Ethnic Groups: African 99.66% (Makhuwa, Tsonga, Lomwe, Sena, and others), Europeans 0.06%, Euro-Africans 0.2%, Indians 0.08%

Religion: Roman Catholic 28.4%, Muslim 17.9%, Zionist Christian 15.5%, Protestant 12.2% (includes Pentecostal 10.9% and Anglican 1.3%), other 6.7%, none 18.7%, unspecified 0.7% (2007 est.)

Languages: Emakhuwa 25.3%, Portuguese (official) 10.7%, Xichangana 10.3%, Cisena 7.5%, Elomwe 7%, Echuwabo 5.1%, other Mozambican languages 30.1%, other 4% (1997 census)

Literacy: 58.8%

Government Type: republic

Capital: Maputo

Independence: 25 June 1975 (from Portugal)

GDP Per Capita: $1,300

Occupations: agriculture 81% , industry 6%, services 13%

Currency: metical (MZM)

Myanmar

Long Name: Union of Myanmar

Location: Southeastern Asia, bordering the Andaman Sea and the Bay of Bengal, between Bangladesh and Thailand

Area: 261,969 sq. mi. (678,500 sq. km)

Climate: tropical monsoon; cloudy, rainy, hot, humid summers (southwest monsoon, June to September); less cloudy, scant rainfall, mild temperatures, lower humidity during winter (northeast monsoon, December to April)

Terrain: central lowlands ringed by steep, rugged highlands

Population: 47,758,180

Population Growth Rate: 0.8%

Birth Rate (per 1,000): 17.23

Death Rate (per 1,000): 9.23

Life Expectancy: 62.94 years

Ethnic Groups: Burman 68%, Shan 9%, Karen 7%, Rakhine 4%, Chinese 3%, Indian 2%, Mon 2%, other 5%

Religion: Buddhist 89%, Christian 4% (Baptist 3%, Roman Catholic 1%), Muslim 4%, animist 1%, other 2%

Languages: Burmese, minority ethnic groups have their own languages

Literacy: 89.9%

Government Type: military junta

Capital: Rangoon

Independence: 4 January 1948 (from United Kingdom)

GDP Per Capita: $1,900

Occupations: agriculture 70%, industry 7%, services 23%

Currency: kyat (MMK)

Namibia

Long Name: Republic of Namibia

Location: Southern Africa, bordering the South Atlantic Ocean, between Angola and South Africa

Area: 318,261 sq. mi. (824,292 sq. km)

Climate: desert; hot, dry; rainfall sparse and erratic

Terrain: mostly high plateau; Namib Desert along coast; Kalahari Desert in east

Population: 2,212,307

Population Growth Rate: 0.59%

Birth Rate (per 1,000): 19.8

Death Rate (per 1,000): 13.91

Life Expectancy: 51.62 years

Ethnic Groups: black 87.5%, white 6%, mixed 6.5% (about 50% of the population belong to the Ovambo tribe and 9% to the Kavangos tribe; other ethnic groups include Herero 7%, Damara 7%, Nama 5%, Caprivian 4%, Bushmen 3%, Baster 2%, Tswana 0.5%)

Religion: Christian 80% to 90% (Lutheran 50% at least), indigenous beliefs 10% to 20%

Languages: Oshiwambo languages 48.9%, Nama/Damara 11.3%, Afrikaans 10.4% (common language of most of the population and about 60% of the white population), Otjiherero languages 8.6%, Kavango languages 8.5%, Caprivi languages 4.8%, English (official) 3.4%, other African languages 2.3%, other 1.7%

Literacy: 81.9%

Government Type: republic

Capital: Windhoek

Independence: 21 March 1990 (from South African mandate)

GDP Per Capita: $11,300

Occupations: agriculture 16.3%, industry 22.4%, services 61.3%

Currency: Namibian dollar (NAD); South African rand (ZAR)

Nauru

Long Name: Republic of Nauru

Location: Oceania, island in the South Pacific Ocean, south of the Marshall Islands

Area: 8 sq. mi. (21 sq. km)

Climate: tropical with a monsoonal pattern; rainy season (November to February)

Terrain: sandy beach rises to fertile ring around raised coral reefs with phosphate plateau in center

Population: 9,540

Population Growth Rate: 0.55%

Birth Rate (per 1,000): 24.95

Death Rate (per 1,000): 5.87

Life Expectancy: 66.75 years

Ethnic Groups: Nauruan 58%, other Pacific Islander 26%, Chinese 8%, European 8%

Religion: Protestant 60.4% (includes Nauru Congregational 35.7%, Assembly of God 13%, Nauru Independent Church 9.5%, Baptist 1.5%, and Seventh Day Adventist 0.7%), Roman Catholic 33%, other 3.7%, none 1.8%, unspecified 1.1% (2011 est.)

Languages: Nauruan 93% (official, a distinct Pacific Island language), English 2% (widely understood, spoken, and used for most government and commercial purposes), other 5% (includes I-Kiribati 2% and Chinese 2%)

Literacy: N/A

Government Type: republic

Capital: no official capital; government offices in Yaren District

Independence: 31 January 1968 (from the Australia-, NZ-, and United Kingdom-administered UN trusteeship)

GDP Per Capita: $14,800

Occupations: employed in mining phosphates, public administration, education, and transportation

Currency: Australian dollar (AUD)

Nepál

Long Name: Federal Democratic Republic of Nepál

Location: Southern Asia, between China and India

Area: 56,827 sq. mi. (147,181 sq. km)

Climate: varies from cool summers and severe winters in north to subtropical summers and mild winters in south

Terrain: Tarai or flat river plain of the Ganges in south, central hill region, rugged Himalayas in north

Population: 31,551,305

Population Growth Rate: 1.79%

Birth Rate (per 1,000): 20.64

Death Rate (per 1,000): 6.56

Life Expectancy: 67.52 years

Ethnic Groups: Chhettri 16.6%, Brahman-Hill 12.2%, Magar 7.1%, Tharu 6.6%, Tamang 5.8%, Newar 5%, Kami 4.8%, Muslim 4.4%, Yadav 4%, Rai 2.3%, Gurung 2%, Damai/Dholii 1.8%, Thakuri 1.6%, Limbu 1.5%, Sarki 1.4%, Teli 1.4%, Chamar/Harijan/Ram 1.3%, Koiri/Kushwaha 1.2%, other 19%

Religion: Hindu 81.3%, Buddhist 9%, Muslim 4.4%, Kirant 3.1%, Christian 1.4%, other 0.5%, unspecifed 0.2% (2011 est.)

Languages: Nepali (official) 44.6%, Maithali 11.7%, Bhojpuri 6%, Tharu 5.8%, Tamang 5.1%, Newar 3.2%, Magar 3%, Bajjika 3%, Urdu 2.6%, Avadhi 1.9%, Limbu 1.3%, Gurung 1.2%, other 10.4%, unspecified 0.2%

Literacy: 63.9%

Government Type: democratic republic

Capital: Kathmandu

Independence: 1768 (unified by Prithvi Narayan Shah)

GDP Per Capita: $2,500

Occupations: agriculture 69%, industry 12%, services 19%

Currency: Nepalese rupee (NPR)

The Netherlands

Long Name: Kingdom of the Netherlands

Location: Western Europe, bordering the North Sea, between Belgium and Germany

Area: 16,040 sq. mi. (41,543 sq. km)

Climate: temperate; marine; cool summers and mild winters

Terrain: mostly coastal lowland and reclaimed land (polders); some hills in southeast

Population: 16,947,904

Population Growth Rate: 0.41%

Birth Rate (per 1,000): 10.83

Death Rate (per 1,000): 8.66

Life Expectancy: 81.23 years

Ethnic Groups: Dutch 78.6%, EU 5.8%, Turkish 2.4%, Indonesian 2.2%, Moroccan 2.2%, Surinamese 2.1%, Bonairian, Saba Islander, Sint Eustatian 0.8%, other 5.9% (2014 est.)

Religion: Roman Catholic 28%, Protestant 19% (includes Dutch Reformed 9%, Protestant Church of The Netherlands, 7%, Calvinist 3%), other 11% (includes about 5% Muslim and lesser numbers of Hindu, Buddhist, Jehovah's Witness, and Orthodox), none 42% (2009 est.)

Languages: Dutch (official)

Literacy: 99%

Government Type: constitutional monarchy

Capital: Amsterdam

Independence: 23 January 1579 (the northern provinces of the Low Countries conclude the Union of Utrecht breaking with Spain; on 26 July 1581 they formally declared their independence with an Act of Abjuration; however, it was not until 30 January 1648 and the Peace of Westphalia that Spain recognized this independence)

GDP Per Capita: $49,300

Occupations: agriculture 1.8%, industry 17%, services 81.2%

Currency: euro (EUR)

New Zealand

Long Name: New Zealand

Location: Oceania, islands in the South Pacific Ocean, southeast of Australia

Area: 103,363 sq. mi. (267,710 sq. km)

Climate: temperate with sharp regional contrasts

Terrain: predominately mountainous with some large coastal plains

Population: 4,438,393

Population Growth Rate: 0.82%

Birth Rate (per 1,000): 13.33

Death Rate (per 1,000): 7.36

Life Expectancy: 81.05 years

Ethnic Groups: European 71.2%, Maori 14.1%, Asian 11.3%, Pacific peoples 7.6%, Middle Eastern, Latin American, African 1.1%, other 1.6%, not stated or unidentified 5.4%

Religion: Christian 44.3% (Catholic 11.6%, Anglican 10.8%, Presbyterian and Congregational 7.8%, Methodist, 2.4%, Pentecostal 1.8%, other 9.9%), Hindu 2.1%, Buddhist 1.4%, Maori Christian 1.3%, Islam 1.1%, other religion 1.4% (includes Judaism, Spiritualism and New Age religions, Baha'i, Asian religions other than Buddhism), no religion 38.5%, not stated or unidentified 8.2%, objected to answering 4.1%

Languages: English (de facto official) 89.8%, Maori (de jure official) 3.5%, Samoan 2%, Hindi 1.6%, French

1.2%, Northern Chinese 1.2%, Yue 1%, Other or not stated 20.5%, New Zealand Sign Language (de jure official)

Literacy: 99%

Government Type: parliamentary democracy

Capital: Wellington

Independence: 26 September 1907 (from United Kingdom)

GDP Per Capita: $36,400

Occupations: agriculture 7%, industry 19%, services 74%

Currency: New Zealand dollar (NZD)

Nicaragua

Long Name: Republic of Nicaragua

Location: Central America, bordering both the Caribbean Sea and the North Pacific Ocean, between Costa Rica and Honduras

Area: 50,336 sq. mi. (130,370 sq. km)

Climate: tropical in lowlands, cooler in highlands

Terrain: extensive Atlantic coastal plains rising to central interior mountains; narrow Pacific coastal plain interrupted by volcanoes

Population: 5,907,881

Population Growth Rate: 1%

Birth Rate (per 1,000): 18.03

Death Rate (per 1,000): 5.08

Life Expectancy: 72.98 years

Ethnic Groups: mestizo (mixed Amerindian and white) 69%, white 17%, black 9%, Amerindian 5%

Religion: Roman Catholic 58.5%, Protestant 23.2% (Evangelical 21.6%, Moravian 1.6%), Jehovah's Witnesses 0.9%, other 1.6%, none 15.7% (2005 est.)

Languages: Spanish (official) 95.3%, Miskito 2.2%, mestizo of the Caribbean coast 2%, other 0.5%

Literacy: 82.8%

Government Type: republic

Capital: Managua

Independence: 15 September 1821 (from Spain)

GDP Per Capita: $5,000

Occupations: agriculture 31%, industry 18%, services 50%

Currency: gold cordoba (NIO)

Niger

Long Name: Republic of Niger

Location: Western Africa, southeast of Algeria

Area: 489,189 sq. mi. (1,267,000 sq. km)

Climate: desert; mostly hot, dry, dusty; tropical in extreme south

Terrain: predominately desert plains and sand dunes; flat to rolling plains in south; hills in north

Population: 18,045,729

Population Growth Rate: 3.25%

Birth Rate (per 1,000): 45.45

Death Rate (per 1,000): 20.42

Life Expectancy: 55.13 years

Ethnic Groups: Hausa 53.1%, Zarma/Songhai 21.2%, Tuareg 11%, Fulani (Peul) 6.5%, Kanuri 5.9%, Gurma 0.8%, Arab 0.4%, Tubu 0.4%, other/unavailable 0.9% (2006 est.)

Religion: Muslim 80%, other (includes indigenous beliefs and Christian) 20%

Languages: French (official), Hausa, Djerma

Literacy: 19.1%

Government Type: republic

Capital: Niamey

Independence: 3 August 1960 (from France)

GDP Per Capita: $1,100

Occupations: agriculture 90%, industry 6%, services 4%

Currency: Communaute Financiere Africaine franc (XOF)

Nigeria

Long Name: Federal Republic of Nigeria

Location: Western Africa, bordering the Gulf of Guinea, between Benin and Cameroon

Area: 356,667 sq. mi. (923,768 sq. km)

Climate: varies; equatorial in south, tropical in center, arid in north

Terrain: southern lowlands merge into central hills and plateaus; mountains in southeast, plains in north

Population: 181,562,056

Population Growth Rate: 2.45%

Birth Rate (per 1,000): 37.64

Death Rate (per 1,000): 12.9

Life Expectancy: 53.02 years

Ethnic Groups: Nigeria, Africa's most populous country, is composed of more than 250 ethnic groups; the most populous and politically influential are: Hausa and the Fulani 29%, Yoruba 21%, Igbo (Ibo) 18%, Ijaw 10%, Kanuri 4%, Ibibio 3.5%, Tiv 2.5%

Religion: Muslim 50%, Christian 40%, indigenous beliefs 10%

Languages: English (official), Hausa, Yoruba, Igbo (Ibo), Fulani, over 500 additional indigenous languages

Literacy: 59.6%

Government Type: federal republic

Capital: Abuja

Independence: 1 October 1960 (from United Kingdom)

GDP Per Capita: $6,400

Occupations: agriculture 70%, industry 10%, services: 20%

Currency: naira (NGN)

North Korea

Long Name: Democratic People's Republic of Korea

Location: Eastern Asia, northern half of the Korean Peninsula bordering the Korea Bay and the Sea of Japan, between China and South Korea

Area: 46,540 sq. mi. (120,538 sq. km)

Climate: temperate with rainfall concentrated in summer

Terrain: mostly hills and mountains separated by deep, narrow valleys; coastal plains wide in west, discontinuous in east

Population: 24,983,205

Population Growth Rate: 0.53%

Birth Rate (per 1,000): 14.52

Death Rate (per 1,000): 9.21

Life Expectancy: 70.11 years

Ethnic Groups: racially homogeneous; there is a small Chinese community and a few ethnic Japanese

Religion: traditionally Buddhist and Confucianist, some Christian and syncretic Chondogyo (Religion of the Heavenly Way)

Languages: Korean

Literacy: 100%

Government Type: Communist state one-man dictatorship

Capital: Pyongyang

Independence: 15 August 1945 (from Japan)

GDP Per Capita: $1,800

Occupations: agriculture 37%, industry and services 63%

Currency: North Korean won (KPW)

Norway

Long Name: Kingdom of Norway

Location: Northern Europe, bordering the North Sea and the North Atlantic Ocean, west of Sweden

Area: 125,020 sq. mi. (323,802 sq. km)

Climate: temperate along coast, modified by North Atlantic Current; colder interior with increased precipitation and colder summers; rainy year-round on west coast

Terrain: glaciated; mostly high plateaus and rugged mountains broken by fertile valleys; small, scattered plains; coastline deeply indented by fjords; arctic tundra in north

Population: 5,207,689

Population Growth Rate: 1.13%

Birth Rate (per 1,000): 12.14

Death Rate (per 1,000): 8.12

Life Expectancy: 81.7 years

Ethnic Groups: Norwegian 94.4% (includes Sami, about 60,000), other European 3.6%, other 2% (2007 est.)

Religion: Church of Norway (Evangelical Lutheran-official) 82.1%, other Christian 3.9%, Muslim 2.3%, Roman Catholic 1.8%, other 2.4%, unspecified 7.5% (2011 est.)

Languages: Bokmal Norwegian (official), Nynorsk Norwegian (official), small Sami- and Finnish-speaking minorities; note-Sami is official in six municipalities

Literacy: 100%

Government Type: constitutional monarchy

Capital: Oslo

Independence: 7 June 1905 (Norway declared the union with Sweden dissolved); 26 October 1905 (Sweden agreed to the repeal of the union)

GDP Per Capita: $68,400

Occupations: agriculture 2.2%, industry 20.2%, services 77.6%

Currency: Norwegian krone (NOK)

Oman

Long Name: Sultanate of Oman

Location: Middle East, bordering the Arabian Sea, Gulf of Oman, and Persian Gulf, between Yemen and UAE

Area: 119,499 sq. mi. (309,500 sq. km)

Climate: dry desert; hot, humid along coast; hot, dry interior; strong southwest summer monsoon (May to September) in far south

Terrain: central desert plain, rugged mountains in north and south

Population: 3,286,936

Population Growth Rate: 2.07%

Birth Rate (per 1,000): 24.44

Death Rate (per 1,000): 3.36

Life Expectancy: 75.21 years

Ethnic Groups: Arab, Baluchi, South Asian (Indian, Pakistani, Sri Lankan, Bangladeshi), African

Religion: Muslim (official; majority are Ibadhi, lesser numbers of Sunni and Shia) 85.9%, Christian 6.5%, Hindu 5.5%, Buddhist 0.8%, Jewish <0.1, other 1%, unaffiliated 0.2%

Languages: Arabic (official), English, Baluchi, Urdu, Indian dialects

Literacy: 91.1%

Government Type: monarchy

Capital: Muscat

Independence: 1650 (expulsion of the Portuguese)

GDP Per Capita: $46,200

Occupations: N/A

Currency: Omani rial (OMR)

Pakistan

Long Name: Islamic Republic of Pakistan

Location: Southern Asia, bordering the Arabian Sea, between India on the east and Iran and Afghanistan on the west and China in the north

Area: 307,374 sq. mi. (796,095 sq. km)

Climate: mostly hot, dry desert; temperate in northwest; arctic in north

Terrain: flat Indus plain in east; mountains in north and northwest; Balochistan plateau in west

Population: 199,085,847

Population Growth Rate: 1.46%

Birth Rate (per 1,000): 22.58

Death Rate (per 1,000): 6.49

Life Expectancy: 67.39 years

Ethnic Groups: Punjabi 44.68%, Pashtun (Pathan) 15.42%, Sindhi 14.1%, Sariaki 8.38%, Muhajirs 7.57%, Balochi 3.57%, other 6.28%

Religion: Muslim (official) 96.4% (Sunni 85-90%, Shia 10-15%), other (includes Christian and Hindu) 3.6% (2010 est.)

Languages: Punjabi 48%, Sindhi 12%, Saraiki (a Punjabi variant) 10%, Pashto (alternate name, Pashtu) 8%, Urdu (official) 8%, Balochi 3%, Hindko 2%, Brahui 1%, English (official; lingua franca of Pakistani elite and most government ministries), Burushaski, and other 8%

Literacy: 57.9%

Government Type: federal republic

Capital: Islamabad

Independence: 14 August 1947 (from British India)

GDP Per Capita: $4,900

Occupations: agriculture 43.7%, industry 22.4%, services 33.9%

Currency: Pakistani rupee (PKR)

Palau

Long Name: Republic of Palau

Location: Oceania, group of islands in the North Pacific Ocean, southeast of the Philippines

Area: 177 sq. mi. (459 sq. km)

Climate: tropical; hot and humid; wet season May to November

Terrain: varying geologically from the high, mountainous main island of Babelthuap to low, coral islands usually fringed by large barrier reefs

Population: 21,265

Population Growth Rate: 0.38%

Birth Rate (per 1,000): 11.05

Death Rate (per 1,000): 7.99

Life Expectancy: 72.87 years

Ethnic Groups: Palauan (Micronesian with Malayan and Melanesian admixtures) 72.5%, Carolinian 1%, other Micronesian 2.4%, Filipino 16.3%, Chinese 1.6%, Vietnamese 1.6%, other Asian 3.4%, white 0.9%, other 0.3% (2005 est.)

Religion: Roman Catholic 49.4%, Protestant 30.9% (includes Protestant (general) 23.1%, Seventh Day Adventist 5.3%, and other Protestant 2.5%), Modekngei 8.7% (indigenous to Palau), Jehovah's Witnesses 1.1%, other 8.8%, none or unspecified 1.1% (2005 est.)

Languages: Palauan (official on most islands) 66.6%, Carolinian 0.7%, other Micronesian 0.7%, English (official) 15.5%, Filipino 10.8%, Chinese 1.8%, other Asian 2.6%, other 1.3%

Literacy: 99.5%

Government Type: constitutional government in free association with the US; the Compact of Free Association entered into force 1 October 1994

Capital: Melekeok

Independence: 1 October 1994 (from the US-administered UN trusteeship)

GDP Per Capita: $14,800

Occupations: agriculture 20%

Currency: U.S. dollar (USD)

Panamá

Long Name: Republic of Panama

Location: Central America, bordering both the Caribbean Sea and the North Pacific Ocean, between Colombia and Costa Rica

Area: 29,120 sq. mi. (75,420 sq. km)

Climate: tropical maritime; hot, humid, cloudy; prolonged rainy season (May to January), short dry season (January to May)

Terrain: interior mostly steep, rugged mountains and dissected, upland plains; coastal areas largely plains and rolling hills

Population: 3,657,024

Population Growth Rate: 1.32%

Birth Rate (per 1,000): 18.32

Death Rate (per 1,000): 4.81

Life Expectancy: 78.47 years

Ethnic Groups: mestizo (mixed Amerindian and white) 65%, Native American 12.3% (Ngabe 7.6%, Kuna 2.4%, Embera 0.9%, Bugle 0.8%, other 0.4%, unspecified 0.2%), black or African descent 9.2%, mulatto 6.8%, white 6.7% (2010 est.)

Religion: Roman Catholic 85%, Protestant 15%

Languages: Spanish (official), indigenous languages (including Ngabere (or Guaymi), Buglere, Kuna, Embera, Wounaan, Naso (or Teribe), and Bri Bri), Panamanian English Creole (similar to Jamaican English Creole; a mixture of English and Spanish with elements of Ngabere; also known as Guari Guari and Colon Creole), English, Chinese (Yue and Hakka), Arabic, French Creole, other (Yiddish, Hebrew, Korean, Japanese)

Literacy: 95%

Government Type: constitutional democracy

Capital: Panama

Independence: 3 November 1903 (from Colombia; became independent from Spain 28 November 1821)

GDP Per Capita: $20,900

Occupations: agriculture 17%, industry 18.6%, services 64.4%

Currency: balboa (PAB); U.S. dollar (USD)

Papua New Guinea

Long Name: Independent State of Papua New Guinea

Location: Oceania, group of islands including the eastern half of the island of New Guinea between the Coral Sea and the South Pacific Ocean, east of Indonesia

Area: 178,703 sq. mi. (462,840 sq. km)

Climate: tropical; northwest monsoon (December to March), southeast monsoon (May to October); slight seasonal temperature variation

Terrain: mostly mountains with coastal lowlands and rolling foothills

Population: 6,672,429

Population Growth Rate: 1.78%

Birth Rate (per 1,000): 24.38

Death Rate (per 1,000): 6.53

Life Expectancy: 67.03 years

Ethnic Groups: Melanesian, Papuan, Negrito, Micronesian, Polynesian

Religion: Roman Catholic 27%, Protestant 69.4% (Evangelical Lutheran 19.5%, United Church 11.5%, Seventh-Day Adventist 10%, Pentecostal 8.6%, Evangelical Alliance 5.2%, Anglican 3.2%, Baptist 2.5%, other Protestant 8.9%), Baha'i 0.3%, indigenous beliefs and other 3.3% (2000 census)

Languages: Tok Pisin (official), English (official), Hiri Motu (official), some 836 indigenous languages spoken (about 12% of the world's total); most languages have fewer than 1,000 speakers

Literacy: 64.2%

Government Type: constitutional parliamentary democracy

Capital: Port Moresby

Independence: 16 September 1975 (from the Australian-administered UN trusteeship)

GDP Per Capita: $2,800

Occupations: agriculture 85%, other 15%

Currency: kina (PGK)

Paraguay

Long Name: Republic of Paraguay

Location: Central South America, northeast of Argentina

Area: 157,048 sq. mi. (406,752 sq. km)

Climate: subtropical to temperate; substantial rainfall in the eastern portions, becoming semiarid in the far west

Terrain: grassy plains and wooded hills east of Rio Paraguay; Gran Chaco region west of Rio Paraguay mostly low, marshy plain near the river, and dry forest and thorny scrub elsewhere

Population: 6,783,272

Population Growth Rate: 1.16%

Birth Rate (per 1,000): 16.37

Death Rate (per 1,000): 4.68

Life Expectancy: 76.99 years

Ethnic Groups: mestizo (mixed Spanish and Amerindian) 95%, other 5%

Religion: Roman Catholic 89.6%, Protestant 6.2%, other Christian 1.1%, other or unspecified 1.9%, none 1.1% (2002 census)

Languages: Spanish (official), Guarani (official)

Literacy: 93.9%

Government Type: constitutional republic

Capital: Asuncion

Independence: 14 May 1811 (from Spain)

GDP Per Capita: $8,800

Occupations: agriculture 26.5%, industry 18.5%, services 55%

Currency: guarani (PYG)

Perú

Long Name: Republic of Perú

Location: Western South America, bordering the South Pacific Ocean, between Chile and Ecuador

Area: 496,225 sq. mi. (1,285,216 sq. km)

Climate: varies from tropical in east to dry desert in west; temperate to frigid in Andes

Terrain: western coastal plain (costa), high and rugged Andes in center (sierra), eastern lowland jungle of Amazon Basin (selva)

Population: 30,444,999

Population Growth Rate: 0.97%

Birth Rate (per 1,000): 18.28

Death Rate (per 1,000): 6.01

Life Expectancy: 73.48 years

Ethnic Groups: Amerindian 45%, mestizo (mixed Amerindian and white) 37%, white 15%, black, Japanese, Chinese, and other 3%

Religion: Roman Catholic 81.3%, Evangelical 12.5%, other 3.3%, none 2.9% (2007 est.)

Languages: Spanish (official) 84.1%, Quechua (official) 13%, Aymara (official) 1.7%, Ashaninka 0.3%, other native languages (includes a large number of minor Amazonian languages) 0.7%, other (includes foreign languages and sign language) 0.2% (2007 est.)

Literacy: 94.5%

Government Type: constitutional republic

Capital: Lima

Independence: 28 July 1821 (from Spain)

GDP Per Capita: $12,300

Occupations: agriculture 25.8%, industry 17.4%, services 56.8%

Currency: nuevo sol (PEN)

The Philippines

Long Name: Republic of the Philippines

Location: Southeastern Asia, archipelago between the Philippine Sea and the South China Sea, east of Vietnam

Area: 115,830 sq. mi. (300,000 sq. km)

Climate: tropical marine; northeast monsoon (November to April); southwest monsoon (May to October)

Terrain: mostly mountains with narrow to extensive coastal lowlands

Population: 100,998,376

Population Growth Rate: 1.61%

Birth Rate (per 1,000): 24.27

Death Rate (per 1,000): 6.11

Life Expectancy: 68.96 years

Ethnic Groups: Tagalog 28.1%, Cebuano 13.1%, Ilocano 9%, Bisaya/Binisaya 7.6%, Hiligaynon Ilonggo 7.5%, Bikol 6%, Waray 3.4%, other 25.3%

Religion: Catholic 82.9% (Roman Catholic 80.9%, Aglipayan 2%), Muslim 5%, Evangelical 2.8%, Iglesia ni Kristo 2.3%, other Christian 4.5%, other 1.8%, unspecified 0.6%, none 0.1% (2000 census)

Languages: Filipino (official; based on Tagalog) and English (official); eight major dialects-Tagalog, Cebuano, Ilocano, Hiligaynon or Ilonggo, Bicol, Waray, Pampango, and Pangasinan

Literacy: 96.3%

Government Type: republic

Capital: Manila

Independence: 12 June 1898 (independence proclaimed from Spain); 4 July 1946 (from the United States)

GDP Per Capita: $7,500

Occupations: agriculture 30%, industry 16%, services 54%

Currency: Philippine peso (PHP)

Poland

Long Name: Republic of Poland

Location: Central Europe, east of Germany

Area: 120,728 sq. mi. (312,685 sq. km)

Climate: temperate with cold, cloudy, moderately severe winters with frequent precipitation; mild summers with frequent showers and thundershowers

Terrain: mostly flat plain; mountains along southern border

Population: 38,562,189

Population Growth Rate: -0.09%

Birth Rate (per 1,000): 9.74

Death Rate (per 1,000): 10.19

Life Expectancy: 77.4 years

Ethnic Groups: Polish 96.9%, Silesian 1.1%, German 0.2%, Ukrainian 0.1%, other and unspecified 1.7%

Religion: Catholic 87.2% (includes Roman Catholic 86.9% and Greek Catholic, Armenian Catholic, and Byzantine-Slavic Catholic 0.3%), Orthodox 1.3% (almost all are Polish Autocephalous Orthodox), Protestant 0.4% (mainly Augsburg Evangelical and Pentacostal), other 0.4% (includes Jehovah's Witness, Buddhist, Hare Krishna, Gaudiya Vaishnavism, Muslim, Jewish, Mormon), unspecified 10.8% (2012 est.)

Languages: Polish (official) 98.2%, Silesian 1.4%, other 1.1%, unspecified 1.3%

Literacy: 99.8%

Government Type: republic

Capital: Warsaw

Independence: 11 November 1918 (republic proclaimed)

GDP Per Capita: $26,400

Occupations: agriculture 12.6%, industry 30.4%, services 57%

Currency: zloty (PLN)

Portugal

Long Name: Portuguese Republic

Location: Southwestern Europe, bordering the North Atlantic Ocean, west of Spain

Area: 35,556 sq. mi. (92,090 sq. km)

Climate: maritime temperate; cool and rainy in north, warmer and drier in south

Terrain: mountainous north of the Tagus River, rolling plains in south

Population: 10,825,309

Population Growth Rate: 0.9%

Birth Rate (per 1,000): 9.27

Death Rate (per 1,000): 11.02

Life Expectancy: 79.16 years

Ethnic Groups: homogeneous Mediterranean stock; citizens of black African descent who immigrated to mainland during decolonization number less than 100,000; since 1990 East Europeans have entered Portugal

Religion: Roman Catholic 81%, other Christian 3.3%, other (includes Jewish, Muslim, other) 0.6%, none 6.8%, unspecified 8.3%

Languages: Portuguese (official), Mirandese (official-but locally used)

Literacy: 95.7%

Government Type: republic; parliamentary democracy

Capital: Lisbon

Independence: 1143 (Kingdom of Portugal recognized); 5 October 1910 (republic proclaimed)

GDP Per Capita: $27,800

Occupations: agriculture 8.6%, industry 23.9%, services 67.5%

Currency: euro (EUR)

Qatar

Long Name: State of Qatar

Location: Middle East, peninsula bordering the Persian Gulf and Saudi Arabia

Area: 4,473 sq. mi. (11,586 sq. km)

Climate: arid; mild, pleasant winters; very hot, humid summers

Terrain: mostly flat and barren desert covered with loose sand and gravel

Population: 2,194,817

Population Growth Rate: 3.07%

Birth Rate (per 1,000): 9.84

Death Rate (per 1,000): 1.53

Life Expectancy: 78.59 years

Ethnic Groups: Arab 40%, Indian 18%, Pakistani 18%, Iranian 10%, other 14%

Religion: Muslim 77.5%, Christian 8.5%, other (includes mainly Hindu and other Indian religions) 14% (2004 est.)

Languages: Arabic (official), English commonly used as a second language

Literacy: 97.3%

Government Type: emirate

Capital: Doha

Independence: 3 September 1971 (from United Kingdom)

GDP Per Capita: $145,000

Occupations: primarily industry

Currency: Qatari rial (QAR)

Republic of the Congo

Long Name: Republic of the Congo

Location: Western Africa, bordering the South Atlantic Ocean, between Angola and Gabon

Area: 132,046 sq. mi. (342,000 sq. km)

Climate: tropical; rainy season (March to June); dry season (June to October); persistent high temperatures and humidity; particularly enervating climate astride the Equator

Terrain: coastal plain, southern basin, central plateau, northern basin

Population: 4,755,097

Population Growth Rate: 2%

Birth Rate (per 1,000): 35.85

Death Rate (per 1,000): 10

Life Expectancy: 58.79 years

Ethnic Groups: Kongo 48%, Sangha 20%, M'Bochi 12%, Teke 17%, Europeans and other 3%

Religion: Roman Catholic 33.1%, Awakening Churches/Christian Revival 22.3%, Protestant 19.9%, Salutiste 2.2%, Muslim 1.6%, Kimbanguiste 1.5%, other 8.1%, none 11.3%

Languages: French (official), Lingala and Monokutuba (lingua franca trade languages), many local languages and dialects (of which Kikongo is the most widespread)

Literacy: 79.3%

Government Type: republic

Capital: Brazzaville

Independence: 15 August 1960 (from France)

GDP Per Capita: $6,800

Occupations: N/A

Currency: Communaute Financiere Africaine franc (XAF)

România

Long Name: România

Location: Southeastern Europe, bordering the Black Sea, between Bulgaria and Ukraine

Area: 92,043 sq. mi. (238,391 sq. km)

Climate: temperate; cold, cloudy winters with frequent snow and fog; sunny summers with frequent showers and thunderstorms

Terrain: central Transylvanian Basin is separated from the Plain of Moldavia on the east by the Carpathian Mountains and separated from the Walachian Plain on the south by the Transylvanian Alps

Population: 21,666,350

Population Growth Rate: -0.3%

Birth Rate (per 1,000): 9.14

Death Rate (per 1,000): 11.9

Life Expectancy: 74.92 years

Ethnic Groups: Romanian 83.4%, Hungarian 6.1%, Roma 3.1%, Ukrainian 0.3%, German 0.2%, other 0.7%, unspecified 6.1% (2011 est.)

Religion: Eastern Orthodox (including all sub-denominations) 81.9%, Protestant (various denominations including Reformed and Pentecostal) 6.4%, Roman Catholic 4.3%, other (includes Muslim) 0.9%, none or atheist 0.2%, unspecified 6.3% (2011 est.)

Languages: Romanian (official) 85.4%, Hungarian 6.3%, Romany (Gypsy) 1.2%, other 1%, unspecified 6.1% (2011 est.)

Literacy: 98.8%

Government Type: republic

Capital: Bucharest

Independence: 9 May 1877 (independence proclaimed from the Ottoman Empire; independence recognized 13 July 1878 by the Treaty of Berlin); 26 March 1881 (kingdom proclaimed); 30 December 1947 (republic proclaimed)

GDP Per Capita: $20,600

Occupations: agriculture 27.9%, industry 28.2%, services 43.9%

Currency: "new" leu (RON) was introduced in 2005; "old" leu (ROL) was phased out in 2006

Russia

Long Name: Russian Federation

Location: Northern Asia (the area west of the Urals is considered part of Europe), bordering the Arctic Ocean, between Europe and the North Pacific Ocean

Area: 6,601,668 sq. mi. (17,098,242 sq. km)

Climate: ranges from steppes in the south through humid continental in much of European Russia; subarctic in Siberia to tundra climate in the polar north; winters vary from cool along Black Sea coast to frigid in Siberia; summers vary from warm in the steppes to cool along Arctic coast

Terrain: broad plain with low hills west of Urals; vast coniferous forest and tundra in Siberia; uplands and mountains along southern border regions

Population: 142,423,773

Population Growth Rate: -0.04%

Birth Rate (per 1,000): 11.6

Death Rate (per 1,000): 13.69

Life Expectancy: 70.47 years

Ethnic Groups: Russian 77.7%, Tatar 3.7%, Ukrainian 1.4%, Bashkir 1.1%, Chuvash 1%, Chechen 1%, other 10.2%, unspecified 3.9%

Religion: Russian Orthodox 15-20%, Muslim 10-15%, other Christian 2% (2006 est.)

Languages: Russian, many minority languages

Literacy: 99.7%

Government Type: federation

Capital: Moscow

Independence: 24 August 1991 (from Soviet Union)

GDP Per Capita: $23,700

Occupations: agriculture 9.4%, industry 27.6%, services 63%

Currency: Russian ruble (RUB)

Rwanda

Long Name: Republic of Rwanda

Location: Central Africa, east of Democratic Republic of the Congo

Area: 10,169 sq. mi. (26,338 sq. km)

Climate: temperate; two rainy seasons (February to April, November to January); mild in mountains with frost and snow possible

Terrain: mostly grassy uplands and hills; relief is mountainous with altitude declining from west to east

Population: 12,661,733

Population Growth Rate: 2.56%

Birth Rate (per 1,000): 33.75

Death Rate (per 1,000): 8.96

Life Expectancy: 59.67 years

Ethnic Groups: Hutu (Bantu) 84%, Tutsi (Hamitic) 15%, Twa (Pygmy) 1%

Religion: Roman Catholic 49.5%, Protestant 39.4% (includes Adventist 12.2% and other Protestant 27.2%), other Christian 4.5%, Muslim 1.8%, animist 0.1%, other 0.6%, none 3.6% (2001), unspecified 0.5% (2002 est.)

Languages: Kinyarwanda only (official, universal Bantu vernacular) 93.2%, Kinyarwanda and other language(s) 6.2%, French (official) and other language(s) 0.1%, English (official) and other language(s) 0.1%, Swahili (or Kiswahili, used in commercial centers) 0.02%, other 0.03%, unspecified 0.3% (2002 est.)

Literacy: 70.5%

Government Type: republic; presidential, multiparty system

Capital: Kigali

Independence: 1 July 1962 (from Belgium-administered UN trusteeship)

GDP Per Capita: $1,800

Occupations: agriculture 90%, industry and services 10%

Currency: Rwandan franc (RWF)

Saint Kitts and Nevis

Long Name: Federation of Saint Kitts and Nevis

Location: Caribbean, islands in the Caribbean Sea, about one-third of the way from Puerto Rico to Trinidad and Tobago

Area: 101 sq. mi. (261 sq. km)

Climate: tropical, tempered by constant sea breezes; little seasonal temperature variation; rainy season (May to November)

Terrain: volcanic with mountainous interiors

Population: 51,936

Population Growth Rate: 0.76%

Birth Rate (per 1,000): 13.5

Death Rate (per 1,000): 7.09

Life Expectancy: 75.52 years

Ethnic Groups: predominantly black; some British, Portuguese, and Lebanese

Religion: Anglican, other Protestant, Roman Catholic

Languages: English

Literacy: 97.8%

Government Type: parliamentary democracy

Capital: Basseterre

Independence: 19 September 1983 (from United Kingdom)

GDP Per Capita: $22,800

Occupations: primarily services and industry

Currency: East Caribbean dollar (XCD)

Saint Lucia

Long Name: Saint Lucia

Location: Caribbean, island between the Caribbean Sea and North Atlantic Ocean, north of Trinidad and Tobago

Area: 238 sq. mi. (616 sq. km)

Climate: tropical, moderated by northeast trade winds; dry season January to April, rainy season May to August

Terrain: volcanic and mountainous with some broad, fertile valleys

Population: 163,922

Population Growth Rate: 0.34%

Birth Rate (per 1,000): 13.7

Death Rate (per 1,000): 7.42

Life Expectancy: 77.6 years

Ethnic Groups: black/African descent 85.3%, mixed 10.9%, East Indian 2.2%, other 1.6%, unspecified 0.1% (2010 est.)

Religion: Roman Catholic 61.5%, Protestant 25.5% (includes Seventh Day Adventist 10.4%, Pentecostal 8.9%, Baptist 2.2%, Anglican 1.6%, Church of God 1.5%, other Protestant 0.9%), other Christian 3.4% (includes Evangelical 2.3% and Jehovah's Witness 1.1%), Rastafarian 1.9%, other 0.4%, none 5.9%, unspecified 1.4% (2010 est.)

Languages: English (official), French patois

Literacy: 90.1%

Government Type: parliamentary democracy

Capital: Castries

Independence: 22 February 1979 (from United Kingdom)

GDP Per Capita: $12,000

Occupations: agriculture 21.7%, industry 24.7%, services 53.6%

Currency: East Caribbean dollar (XCD)

Saint Vincent and the Grenadines

Long Name: Saint Vincent and the Grenadines

Location: Caribbean, islands between the Caribbean Sea and North Atlantic Ocean, north of Trinidad and Tobago

Area: 150 sq. mi. (389 sq. km)

Climate: tropical; little seasonal temperature variation; rainy season (May to November)

Terrain: volcanic, mountainous

Population: 102,627

Population Growth Rate: –0.28%

Birth Rate (per 1,000): 13.57

Death Rate (per 1,000): 7.18

Life Expectancy: 75.09 years

Ethnic Groups: black 66%, mixed 19%, East Indian 6%, European 4%, Carib Amerindian 2%, other 3%

Religion: Protestant 75% (Anglican 47%, Methodist 28%), Roman Catholic 13%, other (includes Hindu, Seventh-Day Adventist, other Protestant) 12%

Languages: English, French patois

Literacy: 96%

Government Type: parliamentary democracy

Capital: Kingstown

Independence: 27 October 1979 (from United Kingdom)

GDP Per Capita: $11,000

Occupations: agriculture 26%, industry 17%, services 57%

Currency: East Caribbean dollar (XCD)

Samoa

Long Name: Independent State of Samoa

Location: Oceania, group of islands in the South Pacific Ocean, about half way between Hawaii and New Zealand

Area: 1,093 sq. mi. (2,831 sq. km)

Climate: tropical; rainy season (November to April), dry season (May to October)

Terrain: two main islands (Savaii, Upolu) and several smaller islands and uninhabited islets; narrow coastal plain with volcanic, rocky, rugged mountains in interior

Population: 197,773

Population Growth Rate: 0.58%

Birth Rate (per 1,000): 20.87

Death Rate (per 1,000): 5.32

Life Expectancy: 73.46 years

Ethnic Groups: Samoan 92.6%, Euronesians (persons of European and Polynesian blood) 7%, Europeans 0.4%

Religion: Protestant 57.4% (Congregationalist 31.8%, Methodist 13.7%, Assembly of God 8%, Seventh-Day Adventist 3.9%), Roman Catholic 19.4%, Mormon 15.2%, Worship Centre 1.7%, other Christian 5.5%, other 0.7%, none 0.1%, unspecified 0.1% (2011 est.)

Languages: Samoan (Polynesian), English

Literacy: 99%

Government Type: parliamentary democracy

Capital: Apia

Independence: 1 January 1962 (from New Zealand-administered UN trusteeship)

GDP Per Capita: $5,400

Occupations: agriculture 65%, industry N/A, services N/A

Currency: tala (SAT)

San Marino

Long Name: Most Serene Republic of San Marino

Location: Southern Europe, an enclave in central Italy

Area: 24 sq. mi. (61 sq. km)

Climate: Mediterranean; mild to cool winters; warm, sunny summers

Terrain: rugged mountains

Population: 33,020

Population Growth Rate: 0.82%

Birth Rate (per 1,000): 8.63

Death Rate (per 1,000): 8.45

Life Expectancy: 83.24 years

Ethnic Groups: Sammarinese, Italian

Religion: Roman Catholic

Languages: Italian

Literacy: 96%

Government Type: republic

Capital: San Marino

Independence: 3 September AD 301

GDP Per Capita: $62,100

Occupations: agriculture 0.2%, industry 33.5%, services 66.3%

Currency: euro (EUR)

São Tomé and Príncipe

Long Name: Democratic Republic of São Tomé and Príncipe

Location: Western Africa, islands in the Gulf of Guinea, straddling the Equator, west of Gabon

Area: 372 sq. mi. (964 sq. km)

Climate: tropical; hot, humid; one rainy season (October to May)

Terrain: volcanic, mountainous

Population: 194,006

Population Growth Rate: 1.84%

Birth Rate (per 1,000): 34.23

Death Rate (per 1,000): 7.24

Life Expectancy: 64.58 years

Ethnic Groups: mestico, angolares (descendants of Angolan slaves), forros (descendants of freed slaves), servicais (contract laborers from Angola, Mozambique, and Cabo Verde), tongas (children of servicais born on the islands), Europeans (primarily Portuguese), Asians (mostly Chinese)

Religion: Catholic 55.7%, Adventist 4.1%, Assembly of God 3.4%, New Apostolic 2.9%, Mana 2.3%, Universal Kingdom of God 2%, Jehovah's Witness 1.2%, other 6.2%, none 21.2%, unspecified 1% (2012 est.)

Languages: Portuguese 98.4% (official), Forro 36.2%, Cabo Verdian 8.5%, French 6.8%, Angolar 6.6%, English 4.9%, Lunguie 1%, other (including sign language) 2.4%

Literacy: 74.9%

Government Type: republic

Capital: São Tomé

Independence: 12 July 1975 (from Portugal)

GDP Per Capita: $3,400

Occupations: population mainly engaged in subsistence agriculture and fishing; shortages of skilled workers

Currency: dobra (STD)

Saudi Arabia

Long Name: Kingdom of Saudi Arabia

Location: Middle East, bordering the Persian Gulf and the Red Sea, north of Yemen

Area: 829,996 sq. mi. (2,149,690 sq. km)

Climate: harsh, dry desert with great temperature extremes

Terrain: mostly uninhabited, sandy desert

Population: 27,752,316

Population Growth Rate: 1.6%

Birth Rate (per 1,000): 18.51

Death Rate (per 1,000): 3.33

Life Expectancy: 75.05 years

Ethnic Groups: Arab 90%, Afro-Asian 10%

Religion: Muslim (official; citizens are 85-90% Sunni and 10-15% Shia), other (includes Eastern Orthodox, Protestant, Roman Catholic, Jewish, Hindu, Buddhist, and Sikh) (2012 est.)

Languages: Arabic

Literacy: 94.7%

Government Type: monarchy

Capital: Riyadh

Independence: 23 September 1932 (unification of the kingdom)

GDP Per Capita: $54,600

Occupations: agriculture 6.7%, industry 21.4%, services 71.9%

Currency: Saudi riyal (SAR)

Sénégal

Long Name: Republic of Sénégal

Location: Western Africa, bordering the North Atlantic Ocean, between Guinea-Bissau and Mauritania

Area: 75,955 sq. mi. (196,722 sq. km)

Climate: tropical; hot, humid; rainy season (May to November) has strong southeast winds; dry season (December to April) dominated by hot, dry, harmattan wind

Terrain: generally low, rolling, plains rising to foothills in southeast

Population: 13,975,834

Population Growth Rate: 2.45%

Birth Rate (per 1,000): 34.52

Death Rate (per 1,000): 8.46

Life Expectancy: 61.32 years

Ethnic Groups: Wolof 38.7%, Pular 26.5%, Serer 15%, Mandinka 4.2%, Jola 4%, Soninke 2.3%, other 9.3% (includes Europeans and persons of Lebanese descent) (2010-11 est.)

Religion: Muslim 95.4% (most adhere to one of the four main Sufi brotherhoods), Christian 4.2% (mostly Roman Catholic), animist 0.4% (2010-11 est.)

Languages: French (official), Wolof, Pulaar, Jola, Mandinka

Literacy: 57.7%

Government Type: republic

Capital: Dakar

Independence: 4 April 1960 (from France); complete independence achieved upon dissolution of federation with Mali on 20 August 1960

GDP Per Capita: $2,500

Occupations: agriculture 77.5%, industry and services 22.5%

Currency: Communaute Financiere Africaine franc (XOF)

Serbia

Long Name: Republic of Serbia

Location: Southeastern Europe, between Macedonia and Hungary

Area: 29,913 sq. mi. (77,474 sq. km)

Climate: in the north, continental climate (cold winters and hot, humid summers with well distributed rainfall); in other parts, continental and Mediterranean climate (relatively cold winters with heavy snowfall and hot, dry summers and autumns)

Terrain: extremely varied; to the north, rich fertile plains; to the east, limestone ranges and basins; to the southeast, ancient mountains and hills

Population: 7,176,794

Population Growth Rate: –0.46

Birth Rate (per 1,000): 9.08

Death Rate (per 1,000): 13.66

Life Expectancy: 75.26 years

Ethnic Groups: Serb 83.3%, Hungarian 3.5%, Romany 2.1%, Bosniak 2%, other 5.7%, undeclared or unknown 3.4% (2011 est.)

Languages: Serbian (official) 88.1%, Hungarian 3.4%, Bosnian 1.9%, Romany 1.4%, other 3.4%, undeclared or unknown 1.8%

Literacy: 98.1%

Government Type: republic

Capital: Belgrade

Independence: 5 June 2006 (from Serbia and Montenegro)

GDP Per Capita: $13,600

Occupations: agriculture 21.9%, industry 15.6%, services 62.5%

Currency: Serbian dinar (RSD)

Seychelles

Long Name: Republic of Seychelles

Location: archipelago in the Indian Ocean, northeast of Madagascar

Area: 176 sq. mi. (455 sq. km)

Climate: tropical marine; humid; cooler season during southeast monsoon (late May to September); warmer season during northwest monsoon (March to May)

Terrain: Mahe Group is granitic, narrow coastal strip, rocky, hilly; others are coral, flat, elevated reefs

Population: 92,430

Population Growth Rate: 0.83%

Birth Rate (per 1,000): 14.19

Death Rate (per 1,000): 6.89

Life Expectancy: 74.49 years

Ethnic Groups: mixed French, African, Indian, Chinese, and Arab

Religion: Roman Catholic 76.2%, Protestant 10.6% (Anglican 6.1%, Pentecoastal Assembly 1.5%, Seventh-Day Adventist 1.2%, other Protestant 1.6), other Christian 2.4%, Hindu 2.4%, Muslim 1.6%, other non-Christian 1.1%, unspecified 4.8%, none 0.9% (2010 est.)

Languages: Seychellois Creole (official) 89.1%, English (official) 5.1%, French (official) 0.7%, other 3.8%, unspecified 1.4% (2010 est.)

Literacy: 91.8%

Government Type: republic

Capital: Victoria

Independence: 29 June 1976 (from United Kingdom)

GDP Per Capita: $27,000

Occupations: agriculture 3%, industry 23%, services 74%

Currency: Seychelles rupee (SCR)

Sierra Leone

Long Name: Republic of Sierra Leone

Location: Western Africa, bordering the North Atlantic Ocean, between Guinea and Liberia

Area: 27,699 sq. mi. (71,740 sq. km)

Climate: tropical; hot, humid; summer rainy season (May to December); winter dry season (December to April)

Terrain: coastal belt of mangrove swamps, wooded hill country, upland plateau, mountains in east

Population: 5,879,098

Population Growth Rate: 2.35%

Birth Rate (per 1,000): 37.03

Death Rate (per 1,000): 10.81

Life Expectancy: 57.79 years

Ethnic Groups: Temne 35%, Mende 31%, Limba 8%, Kono 5%, Kriole 2% (descendants of freed Jamaican slaves who were settled in the Freetown area in the late-18th century; also known as Krio), Mandingo 2%, Loko 2%, other 15% (includes refugees from Liberia's recent civil war, and small numbers of Europeans, Lebanese, Pakistanis, and Indians) (2008 census)

Religion: Muslim 60%, Christian 10%, indigenous beliefs 30%

Languages: English (official, regular use limited to literate minority), Mende (principal vernacular in the south), Temne (principal vernacular in the north), Krio (English-based Creole, spoken by the descendants of freed Jamaican slaves who were settled in the Freetown area, a lingua franca and a first language for 10% of the population but understood by 95%)

Literacy: 48.1%

Government Type: constitutional democracy

Capital: Freetown

Independence: 27 April 1961 (from United Kingdom)

GDP Per Capita: $1,600

Occupations: N/A

Currency: leone (SLL)

Singapore

Long Name: Republic of Singapore

Location: Southeastern Asia, islands between Malaysia and Indonesia

Area: 269 sq. mi. (697 sq. km)

Climate: tropical; hot, humid, rainy; two distinct monsoon seasons-Northeastern monsoon (December to March) and Southwestern monsoon (June to September); inter-monsoon-frequent afternoon and early evening thunderstorms

Terrain: lowland; gently undulating central plateau contains water catchment area and nature preserve

Population: 5,674,472

Population Growth Rate: 1.89%

Birth Rate (per 1,000): 8.27

Death Rate (per 1,000): 3.43

Life Expectancy: 84.68 years

Ethnic Groups: Chinese 74.2%, Malay 13.3%, Indian 9.2%, other 3.3% (2013 est.)

Religion: Buddhist 33.9%, Muslim 14.3%, Taoist 11.3%, Catholic 7.1%, Hindu 5.2%, other Christian 11%, other 0.7%, none 16.4% (2010 est.)

Languages: Mandarin (official) 36.3%, English (official) 29.8%, Malay (official) 11.9%, Hokkien 8.1%, Tamil (official) 4.4%, Cantonese 4.1%, Teochew 3.2%, other Indian languages 1.2%, other Chinese dialects 1.1%, other 1.1% (2010 est.)

Literacy: 96.8%

Government Type: parliamentary republic

Capital: Singapore

Independence: 9 August 1965 (from Malaysian Federation)

GDP Per Capita: $85,700

Occupations: agriculture 1.3%, industry 14.8%, services 83.9%

Currency: Singapore dollar (SGD)

Slovakia

Long Name: Slovak Republic

Location: Central Europe, south of Poland

Area: 18,933 sq. mi. (49,035 sq. km)

Climate: temperate; cool summers; cold, cloudy, humid winters

Terrain: rugged mountains in the central and northern part and lowlands in the south

Population: 5,445,027

Population Growth Rate: 0.02%

Birth Rate (per 1,000): 9.91

Death Rate (per 1,000): 9.74

Life Expectancy: 76.88 years

Ethnic Groups: Slovak 80.7%, Hungarian 8.5%, Roma 2%, other and unspecified 8.8% (2011 est.)

Religion: Roman Catholic 62%, Protestant 8.2%, Greek Catholic 3.8%, other or unspecified 12.5%, none 13.4% (2011 est.)

Languages: Slovak (official) 78.6%, Hungarian 9.4%, Roma 2.3%, Ruthenian 1%, other or unspecified 8.8% (2011 est.)

Literacy: 99.6%

Government Type: parliamentary democracy

Capital: Bratislava

Independence: 1 January 1993 (Czechoslovakia split into the Czech Republic and Slovakia)

GDP Per Capita: $29,500

Occupations: agriculture 3.5%, industry 25.9%, services 70.6%

Currency: Slovak koruna (SKK)

Slovenia

Long Name: Republic of Slovenia

Location: Central Europe, eastern Alps bordering the Adriatic Sea, between Austria and Croatia

Area: 7,827 sq. mi. (20,273 sq. km)

Climate: Mediterranean climate on the coast, continental climate with mild to hot summers and cold winters in the plateaus and valleys to the east

Terrain: a short coastal strip on the Adriatic, an alpine mountain region adjacent to Italy and Austria, mixed mountains and valleys with numerous rivers to the east

Population: 1,983,412

Population Growth Rate: -0.26%

Birth Rate (per 1,000): 8.42

Death Rate (per 1,000): 11.37

Life Expectancy: 78.01 years

Ethnic Groups: Slovene 83.1%, Serb 2%, Croat 1.8%, Bosniak 1.1%, other or unspecified 12% (2002 census)

Religion: Catholic 57.8%, Muslim 2.4%, Orthodox 2.3%, other Christian 0.9%, unaffiliated 3.5%, other or unspecified 23%, none 10.1% (2002 census)

Languages: Slovenian (official) 91.1%, Serbo-Croatian 4.5%, other or unspecified 4.4%, Italian (official, only in municipalities where Italian national communities reside), Hungarian (official, only in municipalities where Hungarian national communities reside) (2002 census)

Literacy: 99.7%

Government Type: parliamentary republic

Capital: Ljubljana

Independence: 25 June 1991 (from Yugoslavia)

GDP Per Capita: $30,900

Occupations: agriculture 8.3%, industry 30.8%, services 60.9%

Currency: euro (EUR)

Solomon Islands

Long Name: Solomon Islands

Location: Oceania, group of islands in the South Pacific Ocean, east of Papua New Guinea

Area: 11,157 sq. mi. (28,896 sq. km)

Climate: tropical monsoon; few extremes of temperature and weather

Terrain: mostly rugged mountains with some low coral atolls

Population: 622,469

Population Growth Rate: 2.02%

Birth Rate (per 1,000): 25.77

Death Rate (per 1,000): 3.85

Life Expectancy: 75.12 years

Ethnic Groups: Melanesian 95.3%, Polynesian 3.1%, Micronesian 1.2%, other 0.3% (2009 est.)

Religion: Protestant 73.4% (Church of Melanesia 31.9%, South Sea Evangelical 17.1%, Seventh Day Adventist 11.7%, United Church 10.1%, Christian Fellowship Church 2.5%), Roman Catholic 19.6%, other Christian 2.9%, other 4%, none 0.03%, unspecified 0.1% (2009 est.)

Languages: Melanesian pidgin (in much of the country is lingua franca), English (official but spoken by only 1%-2% of the population), 120 indigenous languages

Literacy: 84.1%

Government Type: parliamentary democracy

Capital: Honiara

Independence: 7 July 1978 (from United Kingdom)

GDP Per Capita: $2,000

Occupations: agriculture 75%, industry 5%, services 20%

Currency: Solomon Islands dollar (SBD)

Somalia

Long Name: Somalia

Location: Eastern Africa, bordering the Gulf of Aden and the Indian Ocean, east of Ethiopia

Area: 246,200 sq. mi. (637,657 sq. km)

Climate: principally desert; northeast monsoon (December to February), moderate temperatures in north and hot in south; southwest monsoon (May to October), torrid in the north and hot in the south, irregular rainfall, hot and humid periods (tangambili) between monsoons

Terrain: mostly flat to undulating plateau rising to hills in north

Population: 10,616,380

Population Growth Rate: 1.83%

Birth Rate (per 1,000): 40.45

Death Rate (per 1,000): 13.62

Life Expectancy: 51.96 years

Ethnic Groups: Somali 85%, Bantu and other non-Somali 15% (including 30,000 Arabs)

Religion: Sunni Muslim

Languages: Somali (official), Arabic, Italian, English

Literacy: N/A

Government Type: no permanent national government; transitional, parliamentary federal government

Capital: Mogadishu

Independence: 1 July 1960 (from a merger of British Somaliland, which became independent from the United Kingdom on 26 June 1960, and Italian Somaliland, which became independent from the Italian-administered UN trusteeship on 1 July 1960, to form the Somali Republic)

GDP Per Capita: $400

Occupations: agriculture 71%, industry and services 29%

Currency: Somali shilling (SOS)

South Africa

Long Name: Republic of South Africa

Location: Southern Africa, at the southern tip of the continent of Africa

Area: 470,693 sq. mi. (1,219,090 sq. km)

Climate: mostly semiarid; subtropical along east coast; sunny days, cool nights

Terrain: vast interior plateau rimmed by rugged hills and narrow coastal plain

Population: 53,675,563

Population Growth Rate: 1.33%

Birth Rate (per 1,000): 20.75

Death Rate (per 1,000): 9.91

Life Expectancy: 62.34 years

Ethnic Groups: black African 80.2%, white 8.4%, colored 8.8%, Indian/Asian 2.5%

Religion: Protestant 36.6% (Zionist Christian 11.1%, Pentecostal/Charismatic 8.2%, Methodist 6.8%, Dutch Reformed 6.7%, Anglican 3.8%), Catholic 7.1%, Muslim 1.5%, other Christian 36%, other 2.3%, unspecified 1.4%, none 15.1% (2001 census)

Languages: IsiZulu (official) 22.7%, IsiXhosa (official) 16%, Afrikaans (official) 13.5%, English (official) 9.6%, Sepedi (official) 9.1%, Setswana (official) 8%, Sesotho (official) 7.6%, Xitsonga (official) 4.5%, siSwati (official) 2.5%, Tshivenda (official) 2.4%, isiNdebele (official) 2.1%, sign language 0.5%, other 1.6% (2011 est.)

Literacy: 94.3%

Government Type: republic

Capital: Pretoria

Independence: 31 May 1910 (Union of South Africa formed from four British colonies: Cape Colony, Natal, Transvaal, and Orange Free State); 31 May 1961 (republic declared) 27 April 1994 (majority rule)

GDP Per Capita: $13,400

Occupations: agriculture 4%, industry 18%, services 66%

Currency: rand (ZAR)

South Korea

Long Name: Republic of Korea

Location: Eastern Asia, southern half of the Korean Peninsula bordering the Sea of Japan and the Yellow Sea

Area: 38,502 sq. mi. (99,720 sq. km)

Climate: temperate, with rainfall heavier in summer than winter

Terrain: mostly hills and mountains; wide coastal plains in west and south

Population: 49,115,196

Population Growth Rate: 0.14%

Birth Rate (per 1,000): 8.19

Death Rate (per 1,000): 6.75

Life Expectancy: 80.04 years

Ethnic Groups: homogeneous (except for about 20,000 Chinese)

Religion: Christian 31.6% (Protestant 24%, Roman Catholic 7.6%), Buddhist 24.2%, other or unknown 0.9%, none 43.3% (2010 survey)

Languages: Korean, English widely taught in junior high and high school

Literacy: N/A

Government Type: republic

Capital: Seoul

Independence: 15 August 1945 (from Japan)

GDP Per Capita: $36,700

Occupations: agriculture 5.7%, industry 24.2%, services 70.2%

Currency: South Korean won (KRW)

Spain

Long Name: Kingdom of Spain

Location: Southwestern Europe, bordering the Bay of Biscay, Mediterranean Sea, North Atlantic Ocean, and Pyrenees Mountains, southwest of France

Area: 195,124 sq. mi. (505,370 sq. km)

Climate: temperate; clear, hot summers in interior, more moderate and cloudy along coast; cloudy, cold winters in interior, partly cloudy and cool along coast

Terrain: large, flat to dissected plateau surrounded by rugged hills; Pyrenees in north

Population: 48,146,134

Population Growth Rate: 0.89%

Birth Rate (per 1,000): 9.64

Death Rate (per 1,000): 9.04

Life Expectancy: 81.57 years

Ethnic Groups: composite of Mediterranean and Nordic types

Religion: Roman Catholic 94%, other 6%

Languages: Castilian Spanish (official nationwide) 74%, Catalan (official in Catalonia, the Balearic Islands, and the Valencian Community [where it is known as Valencian]) 17%, Galician (official in Galicia) 7%, Basque (official in the Basque Country and in the Basque-speaking area of Navarre) 2%, Aranese (official in the northwest corner of Catalonia (Vall d'Aran) along with Catalan; <5,000 speakers)

Literacy: 98.1%

Government Type: parliamentary monarchy

Capital: Madrid

Independence: the Iberian peninsula was characterized by a variety of independent kingdoms prior to the Muslim occupation that began in the early eighth century C.E. and lasted nearly seven centuries; the small Christian redoubts of the north began the reconquest almost immediately, culminating in the seizure of Granada in 1492; this event completed the unification of several kingdoms and is traditionally considered the forging of present-day Spain

GDP Per Capita: $35,200

Occupations: agriculture 2.9%, industry 15%, services 58.4%

Currency: euro (EUR)

Sri Lanka

Long Name: Democratic Socialist Republic of Sri Lanka

Location: Southern Asia, island in the Indian Ocean, south of India

Area: 25,332 sq. mi. (65,610 sq. km)

Climate: tropical monsoon; northeast monsoon (December to March); southwest monsoon (June to October)

Terrain: mostly low, flat to rolling plain; mountains in south-central interior

Population: 22,053,488

Population Growth Rate: 0.84%

Birth Rate (per 1,000): 15.85

Death Rate (per 1,000): 6.11

Life Expectancy: 76.56 years

Ethnic Groups: Sinhalese 74.9%, Sri Lankan Tamil 11.2%, Sri Lankan Moors 9.2%, Indian Tamil 4.2%, other 0.5% (2012 est.)

Religion: Buddhist (official) 70.2%, Hindu 12.6%, Muslim 9.7%, Roman Catholic 6.1%, other Christian 1.3%, other 0.05% (2012 est.)

Languages: Sinhala (official and national language) 74%, Tamil (national language) 18%, other 8%, English is commonly used in government and is spoken competently by about 10% of the population

Literacy: 92.6%

Government Type: republic

Capital: Colombo

Independence: 4 February 1948 (from United Kingdom)

GDP Per Capita: $11,200

Occupations: agriculture 28.4%, industry 25.7%, services 45.9%

Currency: Sri Lankan rupee (LKR)

The Súdán

Long Name: Republic of the Súdán

Location: Northern Africa, bordering the Red Sea, between Egypt and Eritrea

Area: 718,723 sq. mi. (1,861,484 sq. km)

Climate: tropical in south; arid desert in north; rainy season varies by region (April to November)

Terrain: generally flat, featureless plain; mountains in far south, northeast and west; desert dominates the north

Population: 36,108,853

Population Growth Rate: 1.72%

Birth Rate (per 1,000): 29.19

Death Rate (per 1,000): 7.66

Life Expectancy: 63.68 years

Ethnic Groups: Sudanese Arab (approximately 70%), Fur, Beja, Nuba, Fallata

Religion: Sunni Muslim, small Christian minority

Languages: Arabic (official), English (official), Nubian, Ta Bedawie, Fur

Literacy: 75.9%

Government Type: Government of National Unity (GNU)-the National Congress Party (NCP) and Sudan People's Liberation Movement (SPLM) formed a power-sharing government under the 2005 Comprehensive Peace Agreement (CPA); the NCP, which came to power by military coup in 1989, is the majority partner; the agreement stipulates national elections in 2009

Capital: Khartoum

Independence: 1 January 1956 (from Egypt and United Kingdom)

GDP Per Capita: $4,500

Occupations: agriculture 80%, industry 7%, services 13%

Currency: Sudanese pounds (SDG)

Suriname

Long Name: Republic of Suriname

Location: Northern South America, bordering the North Atlantic Ocean, between French Guiana and Guyana

Area: 63,251 sq. mi. (163,820 sq. km)

Climate: tropical; moderated by trade winds

Terrain: mostly rolling hills; narrow coastal plain with swamps

Population: 579,633

Population Growth Rate: 1.08%

Birth Rate (per 1,000): 16.34

Death Rate (per 1,000): 6.13

Life Expectancy: 71.97 years

Ethnic Groups: Hindustani (also known locally as "East Indians"; their ancestors emigrated from northern India in the latter part of the 19th century) 37%, Creole (mixed white and black) 31%, Javanese 15%, "Maroons" (their African ancestors were brought to the country in the 17th and 18th centuries as slaves and escaped to the interior) 10%, Amerindian 2%, Chinese 2%, white 1%, other 2%

Religion: Hindu 27.4%, Protestant 25.2% (predominantly Moravian), Roman Catholic 22.8%, Muslim 19.6%, indigenous beliefs 5%

Literacy: 95.6%

Government Type: constitutional democracy

Capital: Paramaribo

Independence: 25 November 1975 (from the Netherlands)

GDP Per Capita: $16,700

Occupations: agriculture 11.2%, industry 19.5%, services 69.3%

Currency: Surinam dollar (SRD)

Swaziland

Long Name: Kingdom of Swaziland

Location: Southern Africa, between Mozambique and South Africa

Area: 6,704 sq. mi. (17,364 sq. km)

Climate: varies from tropical to near temperate

Terrain: mostly mountains and hills; some moderately sloping plains

Population: 1,435,613

Population Growth Rate: 1.11%

Birth Rate (per 1,000): 24.67

Death Rate (per 1,000): 13.56

Life Expectancy: 51.05 years

Ethnic Groups: African 97%, European 3%

Religion: Zionist 40% (a blend of Christianity and indigenous ancestral worship), Roman Catholic 20%, Muslim 10%, other (includes Anglican, Bahai, Methodist, Mormon, Jewish) 30%

Languages: English (official, government business conducted in English), siSwati (official)

Literacy: 87.5%

Government Type: monarchy

Capital: Mbabane

Independence: 6 September 1968 (from United Kingdom)

GDP Per Capita: $9,800

Occupations: agriculture 70%, industry N/A, services N/A

Currency: lilangeni (SZL)

Sweden

Long Name: Kingdom of Sweden

Location: Northern Europe, bordering the Baltic Sea, Gulf of Bothnia, Kattegat, and Skagerrak, between Finland and Norway

Area: 173,860 sq. mi. (450,295 sq. km)

Climate: temperate in south with cold, cloudy winters and cool, partly cloudy summers; subarctic in north

Terrain: mostly flat or gently rolling lowlands; mountains in west

Population: 9,801,616

Population Growth Rate: 0.8%

Birth Rate (per 1,000): 11.99

Death Rate (per 1,000): 9.4

Life Expectancy: 81.98 years

Ethnic Groups: indigenous population: Swedes with Finnish and Sami minorities; foreign-born or first-generation immigrants: Finns, Yugoslavs, Danes, Norwegians, Greeks, Turks

Religion: Lutheran 87%, other (includes Roman Catholic, Orthodox, Baptist, Muslim, Jewish, and Buddhist) 13%

Languages: Swedish, small Sami- and Finnish-speaking minorities

Literacy: N/A

Government Type: constitutional monarchy

Capital: Stockholm

Independence: 6 June 1523 (Gustav Vasa elected king)

GDP Per Capita: $48,000

Occupations: agriculture 2%, industry 12%, services 86%

Currency: Swedish krona (SEK)

Switzerland

Long Name: Swiss Confederation

Location: Central Europe, east of France, north of Italy

Area: 15,937 sq. mi. (41,277 sq. km)

Climate: temperate, but varies with altitude; cold, cloudy, rainy/snowy winters; cool to warm, cloudy, humid summers with occasional showers

Terrain: mostly mountains (Alps in south, Jura in northwest) with a central plateau of rolling hills, plains, and large lakes

Population: 8,121,830

Population Growth Rate: 0.71%

Birth Rate (per 1,000): 10.5

Death Rate (per 1,000): 8.13

Life Expectancy: 82.5 years

Ethnic Groups: German 65%, French 18%, Italian 10%, Romansch 1%, other 6%

Religion: Roman Catholic 38.2%, Protestant 26.9%, other Christian 5.6%, Muslim 5%, other 1.6%, none 21.4%, unspecified 1.3% (2013 est.)

Languages: German (official) 63.5%, French (official) 22.5%, Italian (official) 8.1%, English 4.4%, Portuguese 3.4%, Albanian 3.1%, Serbo-Croatian 2.5%, Spanish 2.2%, Romansch (official) 0.5%, other 6.6%

Literacy: N/A

Government Type: formally a confederation but similar in structure to a federal republic

Capital: Bern

Independence: 1 August 1291 (founding of the Swiss Confederation)

GDP Per Capita: $59,300

Occupations: agriculture 3.4%, industry 23.4%, services 73.2%

Currency: Swiss franc (CHF)

Syria

Long Name: Syrian Arab Republic

Location: Middle East, bordering the Mediterranean Sea, between Lebanon and Turkey

Area: 71,498 sq. mi. (185,180 sq. km)

Climate: mostly desert; hot, dry, sunny summers (June to August) and mild, rainy winters (December to February) along coast; cold weather with snow or sleet periodically in Damascus

Terrain: primarily semiarid and desert plateau; narrow coastal plain; mountains in west

Population: 17,064,854

Population Growth Rate: –0.16%

Birth Rate (per 1,000): 22.17

Death Rate (per 1,000): 4

Life Expectancy: 74.69 years

Ethnic Groups: Arab 90.3%, Kurds, Armenians, and other 9.7%

Religion: Muslim 87% (official; includes Sunni 74% and Alawi, Ismaili, and Shia 13%), Christian 10% (includes Orthodox, Uniate, and Nestorian), Druze 3%, Jewish (few remaining in Damascus and Aleppo)

Languages: Arabic (official); Kurdish, Armenian, Aramaic, Circassian widely understood; French, English somewhat understood

Literacy: 86.4%

Government Type: republic under an authoritarian military-dominated regime

Capital: Damascus

Independence: 17 April 1946 (from League of Nations mandate under French administration)

GDP Per Capita: $5,100

Occupations: agriculture 17%, industry 16%, services 67%

Currency: Syrian pound (SYP)

Taiwan

Long Name: Taiwan

Location: Eastern Asia, islands bordering the East China Sea, Philippine Sea, South China Sea, and Taiwan Strait, north of the Philippines, off the southeastern coast of China

Area: 13,892 sq. mi. (35,980 sq. km)

Climate: tropical; marine; rainy season during southwest monsoon (June to August); cloudiness is persistent and extensive all year

Terrain: eastern two-thirds mostly rugged mountains; flat to gently rolling plains in west

Population: 23,415,126

Population Growth Rate: 0.23%

Birth Rate (per 1,000): 8.47

Death Rate (per 1,000): 7.11

Life Expectancy: 79.98 years

Ethnic Groups: Taiwanese (including Hakka) 84%, mainland Chinese 14%, indigenous 2%

Religion: mixture of Buddhist and Taoist 93%, Christian 4.5%, other 2.5%

Languages: Mandarin Chinese (official), Taiwanese (Min), Hakka dialects

Literacy: 98.5%

Government Type: multiparty democracy

Capital: Taipei

Independence: Following the Communist victory on the mainland in 1949, two million Nationalists fled to Taiwan and established a government using the 1946 constitution drawn up for all of China. Over the next five decades, the ruling authorities gradually democratized and incorporated the local population within the governing structure. Taiwan is functionally independent of China and has diplomatic relations with many countries, but China still claims it as officially part of the People's Republic of China.

GDP Per Capita: $47,500

Occupations: agriculture 5%, industry 36.1%, services 58.9%

Currency: New Taiwan dollar (TWD)

Tajikistan

Long Name: Republic of Tajikistan

Location: Central Asia, west of China

Area: 55,637 sq. mi. (144,100 sq. km)

Climate: mid-latitude continental, hot summers, mild winters; semiarid to polar in Pamir Mountains

Terrain: Pamir and Alay Mountains dominate landscape; western Fergana Valley in north, Kofarnihon and Vakhsh Valleys in southwest

Population: 8,191,958

Population Growth Rate: 1.71%

Birth Rate (per 1,000): 24.38

Death Rate (per 1,000): 6.18

Life Expectancy: 67.39 years

Ethnic Groups: Tajik 84.3%, Uzbek 13.8% (includes Lakai, Kongrat, Katagan, Barlos, Yuz), other 2% (includes Kyrgyz, Russian, Turkmen, Tatar, Arab) (2010 est.)

Religion: Sunni Muslim 85%, Shia Muslim 5%, other 10% (2003 est.)

Languages: Tajik (official), Russian widely used in government and business

Literacy: 99.8%

Government Type: republic

Capital: Dushanbe

Independence: 9 September 1991 (from Soviet Union)

GDP Per Capita: $2,800

Occupations: agriculture 46.5%, industry 10.8%, services 42.8%

Currency: somoni (TJS)

Tanzania

Long Name: United Republic of Tanzania

Location: Eastern Africa, bordering the Indian Ocean, between Kenya and Mozambique

Area: 365,755 sq. mi. (947,300 sq. km)

Climate: varies from tropical along coast to temperate in highlands

Terrain: plains along coast; central plateau; highlands in north, south

Population: 51,045,882

Population Growth Rate: 2.79%

Birth Rate (per 1,000): 36.39

Death Rate (per 1,000): 8

Life Expectancy: 61.71 years

Ethnic Groups: mainland-African 99% (of which 95% are Bantu consisting of more than 130 tribes), other 1% (consisting of Asian, European, and Arab); Zanzibar-Arab, African, mixed Arab and African

Religion: Christian 61.4%, Muslim 35.2%, folk religion 1.8%, other 0.2%, unaffiliated 1.4%

Languages: Kiswahili or Swahili (official), Kiunguja (name for Swahili in Zanzibar), English (official, primary language of commerce, administration, and higher education), Arabic (widely spoken in Zanzibar), many local languages

Literacy: 70.68%

Government Type: republic

Capital: Dar es Salaam

Independence: 26 April 1964; Tanganyika became independent 9 December 1961 (from United Kingdom-administered UN trusteeship); Zanzibar became independent 19 December 1963 (from United Kingdom); Tanganyika united with Zanzibar 26 April 1964 to form the United Republic of Tanganyika and Zanzibar; renamed United Republic of Tanzania 29 October 1964

GDP Per Capita: $3,000

Occupations: agriculture 80%, industry and services 20%

Currency: Tanzanian shilling (TZS)

Thailand

Long Name: Kingdom of Thailand

Location: Southeastern Asia, bordering the Andaman Sea and the Gulf of Thailand, southeast of Burma

Area: 198,117 sq. mi. (513,120 sq. km)

Climate: tropical; rainy, warm, cloudy southwest monsoon (mid-May to September); dry, cool northeast monsoon (November to mid-March); southern isthmus always hot and humid

Terrain: central plain; Khorat Plateau in the east; mountains elsewhere

Population: 67,976,405

Population Growth Rate: 0.34%

Birth Rate (per 1,000): 11.19

Death Rate (per 1,000): 7.8

Life Expectancy: 74.43 years

Ethnic Groups: Thai 95.9%, Burmese 2%, other 1.3%, unspecified 0.9% (2010 est.)

Religion: Buddhist (official) 93.6%, Muslim 4.9%, Christian 1.2%, other 0.2%, none 0.1% (2010 est.)

Languages: Thai (official) 90.7%, Burmese 1.3%, other 8%

Literacy: 92.6%

Government Type: constitutional monarchy

Capital: Bangkok

Independence: 1238 (traditional founding date; never colonized)

GDP Per Capita: $16,100

Occupations: agriculture 32.2%, industry 16.7%, services 51.1%

Currency: baht (THB)

Timór-Leste

Long Name: Democratic Republic of Timór-Leste

Location: Southeastern Asia, northwest of Australia in the Lesser Sunda Islands at the eastern end of the Indonesian archipelago; note-Timor-Leste includes the eastern half of the island of Timor, the Oecussi (Ambeno) region on the northwest portion of the island of Timor, and the islands of Pulau Atauro and Pulau Jaco

Area: 5,743 sq. mi. (14,874 sq. km)

Climate: tropical; hot, humid; distinct rainy and dry seasons

Terrain: mountainous

Population: 1,231,116

Population Growth Rate: 2.42%

Birth Rate (per 1,000): 34.16

Death Rate (per 1,000): 6.1

Life Expectancy: 67.72 years

Ethnic Groups: Austronesian (Malayo-Polynesian), Papuan, small Chinese minority

Religion: Roman Catholic 96.9%, Protestant / Evangelical 2.2%, Muslim 0.3%, other 0.6% (2005)

Languages: Tetum (official), Portuguese (official), Indonesian, English, there are about 16 indigenous languages (Tetum, Galole, Mambae, and Kemak are spoken by significant numbers of people)

Literacy: 67.5%

Government Type: republic

Capital: Dili

Independence: 28 November 1975 (independence proclaimed from Portugal); 20 May 2002 is the official date of international recognition of Timor-Leste's independence from Indonesia

GDP Per Capita: $5,800

Occupations: agriculture 64%, industry 10%, services 26%

Currency: U.S. dollar (USD)

Togo

Long Name: Togolese Republic

Location: Western Africa, bordering the Bight of Benin, between Benin and Ghana

Area: 21,925 sq. mi. (56,785 sq. km)

Climate: tropical; hot, humid in south; semiarid in north

Terrain: gently rolling savanna in north; central hills; southern plateau; low coastal plain with extensive lagoons and marshes

Population: 7,552,318

Population Growth Rate: 2.69%

Birth Rate (per 1,000): 34.13

Death Rate (per 1,000): 7.26

Life Expectancy: 64.51 years

Ethnic Groups: African (37 tribes; largest and most important are Ewe, Mina, and Kabre) 99%, European and Syrian-Lebanese less than 1%

Religion: Christian 29%, Muslim 20%, indigenous beliefs 51%

Languages: French (official and the language of commerce), Ewe and Mina (the two major African languages in the south), Kabye (sometimes spelled Kabiye) and Dagomba (the two major African languages in the north)

Literacy: 66.5%

Government Type: republic under transition to multiparty democratic rule

Capital: Lome

Independence: 27 April 1960 (from French-administered UN trusteeship)

GDP Per Capita: $1,500

Occupations: agriculture 65%, industry 5%, services 30%

Currency: Communaute Financiere Africaine franc (XOF)

Tonga

Long Name: Kingdom of Tonga

Location: Oceania, archipelago in the South Pacific Ocean, about two-thirds of the way from Hawaii to New Zealand

Area: 288 sq. mi. (747 sq. km)

Climate: tropical; modified by trade winds; warm season (December to May), cool season (May to December)

Terrain: most islands have limestone base formed from uplifted coral formation; others have limestone overlying volcanic base

Population: 106,501

Population Growth Rate: 0.03%

Birth Rate (per 1,000): 23

Death Rate (per 1,000): 4.85

Life Expectancy: 76.04 years

Ethnic Groups: Tongan 96.6%, part-Tongan 1.7%, other 1.7%, unspecified 0.03% (2006 est.)

Religion: Protestant 64.9% (includes Free Wesleyan Church 37.3%, Free Church of Tonga 11.4%, Church of Tonga 7.2%, Tokaikolo Christian Church 2.6%, Assembly of God 2.3% Seventh Day Adventist 2.2%, Constitutional Church of Tonga 0.9%, Anglican 0.8% and Full Gospel Church 0.2%), Mormon 16.8%, Roman Catholic 15.6%, other 1.1%, none 0.03%, unspecified 1.7% (2006 est.)

Languages: English and Tongan 87%, Tongan (official) 10.7%, English (official) 1.2%, other 1.1%, unspecified 0.03% (2006 est.)

Literacy: 99.4%

Government Type: constitutional monarchy

Capital: Nukmu'alofa

Independence: 4 June 1970 (from United Kingdom protectorate)

GDP Per Capita: $5,100

Occupations: agriculture 27.5%, industry 27.5%, services 20.06%

Currency: pa'anga (TOP)

Trinidad and Tobago

Long Name: Republic of Trinidad and Tobago

Location: Caribbean, islands between the Caribbean Sea and the North Atlantic Ocean, northeast of Venezuela

Area: 1,980 sq. mi. (5,128 sq. km)

Climate: tropical; rainy season (June to December)

Terrain: mostly plains with some hills and low mountains

Population: 1,222,363

Population Growth Rate: -0.13%

Birth Rate (per 1,000): 13.46

Death Rate (per 1,000): 8.56

Life Expectancy: 72.59 years

Ethnic Groups: East Indian 35.4%, African 34.2%, mixed-other 15.3%, mixed African/East Indian 7.7%, other 1.3%, unspecified 6.2% (2011 est.)

Religion: Protestant 32.1% (Pentecostal/Evangelical/Full Gospel 12%, Baptist 6.9%, Anglican 5.7%, Seventh-Day Adventist 4.1%, Presbyterian/Congretational 2.5, other Protestant 0.9), Roman Catholic 21.6%, Hindu 18.2%, Muslim 5%, Jehovah's Witness 1.5%, other 8.4%, none 2.2%, unspecified 11.1% (2011 est.)

Languages: English (official), Caribbean Hindustani (a dialect of Hindi), French, Spanish, Chinese

Literacy: 99%

Government Type: parliamentary democracy

Capital: Port-of-Spain

Independence: 31 August 1962 (from United Kingdom)

GDP Per Capita: $32,800

Occupations: agriculture 3.8%, manufacturing, mining, and quarrying 12.8%, construction and utilities 20.4%, services 62.9%

Currency: Trinidad and Tobago dollar (TTD)

Túnisia

Long Name: Túnisian Republic

Location: Northern Africa, bordering the Mediterranean Sea, between Algeria and Libya

Area: 63,170 sq. mi. (163,610 sq. km)

Climate: temperate in north with mild, rainy winters and hot, dry summers; desert in south

Terrain: mountains in north; hot, dry central plain; semiarid south merges into the Sahara

Population: 11,037,225

Population Growth Rate: 0.89%

Birth Rate (per 1,000): 16.64

Death Rate (per 1,000): 5.98

Life Expectancy: 75.89 years

Ethnic Groups: Arab 98%, European 1%, Jewish and other 1%

Religion: Muslim (official; Sunni) 99.1%, other (includes Christian, Jewish, Shia Muslim, and Baha'i) 1%

Languages: Arabic (official, one of the languages of commerce), French (commerce), Berber (Tamazight)

Literacy: 81.8%

Government Type: republic

Capital: Tunis

Independence: 20 March 1956 (from France)

GDP Per Capita: $11,600

Occupations: agriculture 14.8%, industry 33.2%, services 51.7%

Currency: Tunisian dinar (TND)

Turkey

Long Name: Republic of Turkey

Location: Southeastern Europe and Southwestern Asia (that portion of Turkey west of the Bosporus is geographically part of Europe), bordering the Black Sea, between Bulgaria and

Georgia, and bordering the Aegean Sea and the Mediterranean Sea, between Greece and Syria

Area: 302,535 sq. mi. (783,562 sq. km)

Climate: temperate; hot, dry summers with mild, wet winters; harsher in interior

Terrain: high central plateau (Anatolia); narrow coastal plain; several mountain ranges

Population: 79,414,269

Population Growth Rate: 1.26%

Birth Rate (per 1,000): 16.33

Death Rate (per 1,000): 5.88

Life Expectancy: 74.57 years

Ethnic Groups: Turkish 70-75%, Kurdish 18%, other minorities 7-12% (2008 est.)

Religion: Muslim 99.8% (mostly Sunni), other 0.2% (mostly Christians and Jews)

Languages: Turkish (official), Kurdish, other minority languages

Literacy: 95%

Government Type: republican parliamentary democracy

Capital: Ankara

Independence: 29 October 1923

GDP Per Capita: $20,500

Occupations: agriculture 25.5%, industry 26.2%, services 48.4%

Currency: Turkish lira (TRY)

Türkmenistan

Long Name: Türkmenistan

Location: Central Asia, bordering the Caspian Sea, between Iran and Kazakhstan

Area: 188,456 sq. mi. (488,100 sq. km)

Climate: subtropical desert

Terrain: flat-to-rolling sandy desert with dunes rising to mountains in the south; low mountains along border with Iran; borders Caspian Sea in west

Population: 5,231,422

Population Growth Rate: 1.14%

Birth Rate (per 1,000): 19.4

Death Rate (per 1,000): 6.13

Life Expectancy: 69.78 years

Ethnic Groups: Turkmen 85%, Uzbek 5%, Russian 4%, other 6% (2003)

Religion: Muslim 89%, Eastern Orthodox 9%, unknown 2%

Languages: Turkmen 72%, Russian 12%, Uzbek 9%, other 7%

Literacy: 99.7%

Government Type: republic; authoritarian presidential rule, with little power outside the executive branch

Capital: Ashgabat

Independence: 27 October 1991 (from Soviet Union)

GDP Per Capita: $15,600

Occupations: agriculture 48.2%, industry 14%, services 37.8%

Currency: Turkmen manat (TMM)

Tuvalu

Long Name: Tuvalu

Location: Oceania, island group consisting of nine coral atolls in the South Pacific Ocean, about one-half of the way from Hawaii to Australia

Area: 10 sq. mi. (26 sq. km)

Climate: tropical; moderated by easterly trade winds (March to November); westerly gales and heavy rain (November to March)

Terrain: very low-lying and narrow coral atolls

Population: 10,869

Population Growth Rate: 0.82%

Birth Rate (per 1,000): 23.74

Death Rate (per 1,000): 8.74

Life Expectancy: 66.16 years

Ethnic Groups: Polynesian 96%, Micronesian 4%

Religion: Protestant 98.4% (Church of Tuvalu (Congregationalist) 97%, Seventh-Day Adventist 1.4%), Baha'i 1%, other 0.6%

Languages: Tuvaluan, English, Samoan, Kiribati (on the island of Nui)

Literacy: N/A

Government Type: constitutional monarchy with a parliamentary democracy

Capital: Funafuti

Independence: 1 October 1978 (from United Kingdom)

GDP Per Capita: $3,400

Occupations: people make a living mainly through exploitation of the sea, reefs, and atolls and from wages sent home by those abroad (mostly workers in the phosphate industry and sailors)

Currency: Australian dollar (AUD)

Uganda

Long Name: Republic of Uganda

Location: Eastern Africa, west of Kenya

Area: 93,065 sq. mi. (241,038 sq. km)

Climate: tropical; generally rainy with two dry seasons (December to February, June to August); semiarid in northeast

Terrain: mostly plateau with rim of mountains

Population: 37,101,745

Population Growth Rate: 3.24%

Birth Rate (per 1,000): 43.79

Death Rate (per 1,000): 10.69

Life Expectancy: 54.93 years

Ethnic Groups: Baganda 16.9%, Banyankole 9.5%, Basoga 8.4%, Bakiga 6.9%, Iteso 6.4%, Langi 6.1%, Acholi 4.7%, Bagisu 4.6%, Lugbara 4.2%, Bunyoro 2.7%, other 29.6% (2002 census)

Religion: Roman Catholic 41.9%, Protestant 42% (Anglican 35.9%, Pentecostal 4.6%, Seventh Day Adventist 1.5%), Muslim 12.1%, other 3.1%, none 0.9%

Languages: English (official national language, taught in grade schools, used in courts of law and by most newspapers and some radio broadcasts), Ganda or Luganda (most widely used of the Niger-Congo languages, preferred for native language publications in the capital and may be taught in school), other Niger-Congo languages, Nilo-Saharan languages, Swahili, Arabic

Literacy: 78.4%

Government Type: republic

Capital: Kampala

Independence: 9 October 1962 (from United Kingdom)

GDP Per Capita: $2,100

Occupations: agriculture 82%, industry 5%, services 13%

Currency: Ugandan shilling (UGX)

Ukraine

Long Name: Ukraine

Location: Eastern Europe, bordering the Black Sea, between Poland, Romania, and Moldova in the west and Russia in the east

Area: 233,032 sq. mi. (603,550 sq. km)

Climate: temperate continental; Mediterranean only on the southern Crimean coast; precipitation disproportionately distributed, highest in west and north, lesser in east and southeast; winters vary from cool along the Black Sea to cold farther inland; summers are warm across the greater part of the country, hot in the south

Terrain: most of Ukraine consists of fertile plains (steppes) and plateaus, mountains being found only in the west (the Carpathians), and in the Crimean Peninsula in the extreme south

Population: 44,429,471

Population Growth Rate: -0.6%

Birth Rate (per 1,000): 10.72

Death Rate (per 1,000): 14.46

Life Expectancy: 71.57 years

Ethnic Groups: Ukrainian 77.8%, Russian 17.3%, Belarusian 0.6%, Moldovan 0.5%, Crimean Tatar 0.5%, Bulgarian 0.4%, Hungarian 0.3%, Romanian 0.3%, Polish 0.3%, Jewish 0.2%, other 1.8% (2001 est.)

Religion: Orthodox (includes Ukrainian Autocephalous Orthodox (UAOC), Ukrainian Orthodox-Kyiv Patriarchate (UOC-KP), Ukrainian Orthodox-Moscow Patriarchate (UOC-MP), Ukrainian Greek Catholic, Roman Catholic, Protestant, Muslim, Jewish

Languages: Ukrainian (official) 67.5%, Russian (regional language) 29.6%, other (includes small Crimean Tatar-, Moldavian-, and Hungarian-speaking minorities) 2.9% (2001 est.)

Literacy: 99.8%

Government Type: republic

Capital: Kyiv

Independence: 24 August 1991 (from Soviet Union)

GDP Per Capita: $8,000

Occupations: agriculture 5.6%, industry 26%, services 68.4%

Currency: hryvnia (UAH)

United Arab Emirates

Long Name: United Arab Emirates

Location: Middle East, bordering the Gulf of Oman and the Persian Gulf, between Oman and Saudi Arabia

Area: 32,278 sq. mi. (83,600 sq. km)

Climate: desert; cooler in eastern mountains

Terrain: flat, barren coastal plain merging into rolling sand dunes of vast desert wasteland; mountains in east

Population: 5,779,760

Population Growth Rate: 2.58%

Birth Rate (per 1,000): 15.43

Death Rate (per 1,000): 1.97

Life Expectancy: 77.29 years

Ethnic Groups: Emirati 19%, other Arab and Iranian 23%, South Asian 50%, other expatriates (includes Westerners and East Asians) 8%

Religion: Muslim (Islam; official) 76%, Christian 9%, other (primarily Hindu and Buddhist, less than 5% of the population consists of Parsi, Baha'i, Druze, Sikh, Ahmadi, Ismaili, Dawoodi Bohra Muslim, and Jewish) 15%

Languages: Arabic (official), Persian, English, Hindi, Urdu

Literacy: 93.8%

Government Type: federation with specified powers delegated to the UAE federal government and other powers reserved to member emirates

Capital: Abu Dhabi

Independence: 2 December 1971 (from United Kingdom)

GDP Per Capita: $67,000

Occupations: agriculture 7%, industry 15%, services 78%

Currency: Emirati dirham (AED)

United Kingdom

Long Name: United Kingdom of Great Britain and Northern Ireland

Location: Western Europe, islands including the northern one-sixth of the island of Ireland between the North Atlantic Ocean and the North Sea, northwest of France

Area: 94,058 sq. mi. (243,610 sq. km)

Climate: temperate; moderated by prevailing southwest winds over the North Atlantic Current; more than one-half of the days are overcast

Terrain: mostly rugged hills and low mountains; level to rolling plains in east and southeast

Population: 64,088,222

Population Growth Rate: 0.54%

Birth Rate (per 1,000): 12.17

Death Rate (per 1,000): 9.35

Life Expectancy: 80.54 years

Ethnic Groups: white 87.2%, black/African/Caribbean/black British 3%, Asian/Asian British: Indian 2.3%, Asian/Asian British: Pakistani 1.9%, mixed 2%, other 3.7% (2011 est.)

Religion: Christian (includes Anglican, Roman Catholic, Presbyterian, Methodist) 59.5%, Muslim 4.4%, Hindu 1.3%, other 2%, unspecified 7.2%, none 25.7% (2011 est.)

Languages: English

Literacy: N/A

Government Type: constitutional monarchy

Capital: London

Independence: England has existed as a unified entity since the 10th century; the union between England and Wales, begun in 1284 with the Statute of Rhuddlan, was not formalized until 1536 with an Act of Union; in another Act of Union in 1707, England and Scotland agreed to permanently join as Great Britain; the legislative union of Great Britain and Ireland was implemented in 1801, with the adoption of the name the United Kingdom of Great Britain and Ireland; the Anglo-Irish treaty of 1921 formalized a partition of Ireland; six northern Irish counties remained part of the United Kingdom as Northern Ireland and the current name of the country, the United Kingdom of Great Britain and Northern Ireland, was adopted in 1927

GDP Per Capita: $41,200

Occupations: agriculture 1.3%, industry 15.2%, services 83.5%

Currency: British pound (GBP)

United States

Long Name: United States of America

Location: North America, bordering both the North Atlantic Ocean and the North Pacific Ocean, between Canada and Mexico

Area: 3,796,742 sq. mi. (9,833,517 sq. km)

Climate: mostly temperate, but tropical in Hawaii and Florida, arctic in Alaska, semiarid in the great plains west of the Mississippi River, and arid in the Great Basin of the southwest; low winter temperatures in the northwest are ameliorated occasionally in January and February by warm chinook winds from the eastern slopes of the Rocky Mountains

Terrain: vast central plain, mountains in west, hills and low mountains in east; rugged mountains and broad river valleys in Alaska; rugged, volcanic topography in Hawaii

Population: 321,368,864

Population Growth Rate: 0.78%

Birth Rate (per 1,000): 12.49

Death Rate (per 1,000): 8.15

Life Expectancy: 79.68 years

Ethnic Groups: white 79.96%, black 12.85%, Asian 4.43%, Amerindian and Alaska native 0.97%, native Hawaiian and other Pacific islander 0.18%, two or more races 1.61% (July 2007 estimate)

Religion: Protestant 51.3%, Roman Catholic 23.9%, Mormon 1.7%, other Christian 1.6%, Jewish 1.7%, Buddhist 0.7%, Muslim 0.6%, other or unspecified 2.5%, unaffiliated 12.1%, none 4% (2007 est.)

Languages: English 79.2%, Spanish 12.9%, other Indo-European 3.8%, Asian and Pacific island 3.3%, other 0.9% (2011 est.)

Literacy: N/A

Government Type: Constitution-based federal republic; strong democratic tradition

Capital: Washington, DC

Independence: 4 July 1776 (from Great Britain)

GDP Per Capita: $56,300

Occupations: farming, forestry, and fishing 0.7%, manufacturing, extraction, transportation, and crafts 20.3%, managerial, professional, and technical 37.3%, sales and office 24.2%, other services 17.6%

Currency: U.S. dollar (USD)

Uruguay

Long Name: Oriental Republic of Uruguay

Location: Southern South America, bordering the South Atlantic Ocean, between Argentina and Brazil

Area: 68,037 sq. mi. (176,215 sq. km)

Climate: warm temperate; freezing temperatures almost unknown

Terrain: mostly rolling plains and low hills; fertile coastal lowland

Population: 3,341,893

Population Growth Rate: 0.27%

Birth Rate (per 1,000): 13.07

Death Rate (per 1,000): 9.45

Life Expectancy: 77 years

Ethnic Groups: white 88%, mestizo 8%, black 4%, Amerindian (practically nonexistent)

Religion: Roman Catholic 47.1%, non-Catholic Christians 11.1%, nondenominational 23.2%, Jewish 0.3%, atheist or agnostic 17.2%, other 1.1%

Languages: Spanish, Portunol, or Brazilero (Portuguese-Spanish mix on the Brazilian frontier)

Literacy: 98.5%

Government Type: constitutional republic

Capital: Montevideo

Independence: 25 August 1825

GDP Per Capita: $21,800

Occupations: agriculture 13%, industry 14%, services 73%

Currency: Uruguayan peso (UYU)

Uzbekistan

Long Name: Republic of Uzbekistan

Location: Central Asia, north of Afghanistan

Area: 172,741 sq. mi. (447,400 sq. km)

Climate: mostly mid-latitude desert, long, hot summers, mild winters; semiarid grassland in east

Terrain: mostly flat-to-rolling sandy desert with dunes; broad, flat intensely irrigated river valleys along course of Amu Darya, Syr Darya (Sirdaryo), and Zarafshon; Fergana Valley in east surrounded by mountainous Tajikistan and Kyrgyzstan; shrinking Aral Sea in west

Population: 29,199,942

Population Growth Rate: 0.93%

Birth Rate (per 1,000): 17

Death Rate (per 1,000): 5.3

Life Expectancy: 73.55 years

Ethnic Groups: Uzbek 80%, Russian 5.5%, Tajik 5%, Kazakh 3%, Karakalpak 2.5%, Tatar 1.5%, other 2.5%

Religion: Muslim 88% (mostly Sunnis), Eastern Orthodox 9%, other 3%

Languages: Uzbek 74.3%, Russian 14.2%, Tajik 4.4%, other 7.1%

Literacy: 99.6%

Government Type: republic; authoritarian presidential rule, with little power outside the executive branch

Capital: Tashkent

Independence: 1 September 1991 (from Soviet Union)

GDP Per Capita: $6,100

Occupations: agriculture 25.9%, industry 13.2%, services 60.9%

Currency: soum (UZS)

Vanuatu

Long Name: Republic of Vanuatu

Location: Oceania, group of islands in the South Pacific Ocean, about three-quarters of the way from Hawaii to Australia

Area: 4,706 sq. mi. (12,189 sq. km)

Climate: tropical; moderated by southeast trade winds from May to October; moderate rainfall from November to April; may be affected by cyclones from December to April

Terrain: mostly mountainous islands of volcanic origin; narrow coastal plains

Population: 272,264

Population Growth Rate: 1.95%

Birth Rate (per 1,000): 25.04

Death Rate (per 1,000): 4.09

Life Expectancy: 73.06 years

Ethnic Groups: Ni-Vanuatu 97.6%, part Ni-Vanuatu 1.1%, other 1.3% (2009 est.)

Religion: Protestant 70% (includes Presbyterian 27.9%, Anglican 15.1%, Seventh Day Adventist 12.5%, Assemblies of God 4.7%, Church of Christ 4.5%, Neil Thomas Ministry 3.1%, and Apostolic 2.2%), Roman Catholic 12.4%, customary beliefs 3.7% (including Jon Frum cargo cult), other 12.6%, none 1.1%, unspecified 0.2% (2009 est.)

Languages: local languages (more than 100) 63.2%, Bislama (official; creole) 33.7%, English (official) 2%, French (official) 0.6%, other 0.5% (2009 est.)

Literacy: 85.2%

Government Type: parliamentary republic

Capital: Port-Vila

Independence: 30 July 1980 (from France and United Kingdom)

GDP Per Capita: $2,600

Occupations: agriculture 65%, industry 5%, services 30%

Currency: vatu (VUV)

Vatican City

Long Name: State of the Vatican City, The Holy See

Location: Southern Europe, an enclave of Rome (Italy)

Area: 0.17 sq. mi. (0.44 sq. km)

Climate: temperate; mild, rainy winters (September to May) with hot, dry summers (May to September)

Terrain: urban; low hill

Population: 824

Population Growth Rate: 0.003%

Birth Rate (per 1,000): N/A

Death Rate (per 1,000): N/A

Life Expectancy: N/A

Ethnic Groups: Italians, Swiss, other

Religion: Roman Catholic

Languages: Italian, Latin, French, various other languages

Literacy: 100%

Government Type: ecclesiastical

Capital: Vatican City

Independence: 11 February 1929 (from Italy)

GDP Per Capita: N/A

Occupations: essentially services with a small amount of industry; nearly all dignitaries, priests, nuns, guards, and the approximately 3,000 lay workers live outside the Vatican

Currency: euro (EUR)

Venezuela

Long Name: Bolivarian Republic of Venezuela

Location: Northern South America, bordering the Caribbean Sea and the North Atlantic Ocean, between Colombia and Guyana

Area: 352,143 sq. mi. (912,050 sq. km)

Climate: tropical; hot, humid; more moderate in highlands

Terrain: Andes Mountains and Maracaibo Lowlands in northwest; central plains (llanos); Guiana Highlands in southeast

Population: 29,275,460

Population Growth Rate: 1.39%

Birth Rate (per 1,000): 19.16

Death Rate (per 1,000): 5.31

Life Expectancy: 74.54 years

Ethnic Groups: Spanish, Italian, Portuguese, Arab, German, African, indigenous people

Religion: nominally Roman Catholic 96%, Protestant 2%, other 2%

Languages: Spanish (official), numerous indigenous dialects

Literacy: 93%

Government Type: federal republic

Capital: Caracas

Independence: 5 July 1811 (from Spain)

GDP Per Capita: $16,100

Occupations: agriculture 7.3%, industry 21.8%, services 70.9%

Currency: bolivar (VEB)

Vietnam

Long Name: Socialist Republic of Vietnam

Location: Southeastern Asia, bordering the Gulf of Thailand, Gulf of Tonkin, and South China Sea, alongside China, Laos, and Cambodia

Area: 127,881 sq. mi. (331,210 sq. km)

Climate: tropical in south; monsoonal in north with hot, rainy season (May to September) and warm, dry season (October to March)

Terrain: low, flat delta in south and north; central highlands; hilly, mountainous in far north and northwest

Population: 94,348,835

Population Growth Rate: 0.97%

Birth Rate (per 1,000): 15.96

Death Rate (per 1,000): 5.93

Life Expectancy: 73.16 years

Ethnic Groups: Kinh (Viet) 85.7%, Tay 1.9%, Thai 1.8%, Muong 1.5%, Khmer 1.5%, Mong 1.2%, Nung 1.1%, others 5.3% (1999 census)

Religion: Buddhist 9.3%, Catholic 6.7%, Hoa Hao 1.5%, Cao Dai 1.1%, Protestant 0.5%, Muslim 0.1%, none 80.8% (1999 census)

Languages: Vietnamese (official), English (increasingly favored as a second language), some French, Chinese, and Khmer, mountain area languages (Mon-Khmer and Malayo-Polynesian)

Literacy: 94.5%

Government Type: Communist state

Capital: Hanoi

Independence: 2 September 1945 (from France)

GDP Per Capita: $6,100

Occupations: agriculture 48%, industry 21%, services: 31%

Currency: dong (VND)

Yemen

Long Name: Republic of Yemen

Location: Middle East, bordering the Arabian Sea, Gulf of Aden, and Red Sea, between Oman and Saudi Arabia

Area: 203,850 sq. mi. (527,968 sq. km)

Climate: mostly desert; hot and humid along west coast; temperate in western mountains affected by seasonal monsoon; extraordinarily hot, dry, harsh desert in east

Terrain: narrow coastal plain backed by flat-topped hills and rugged mountains; dissected upland desert plains in center slope into the desert interior of the Arabian Peninsula

Population: 26,737,317

Population Growth Rate: 2.47%

Birth Rate (per 1,000): 29.98

Death Rate (per 1,000): 6.28

Life Expectancy: 65.18 years

Ethnic Groups: predominantly Arab; but also Afro-Arab, South Asians, Europeans

Religion: Muslim 99.1% (official; virtually all are citizens, an estimated 65% are Sunni and 35% are Shia), other 0.9% (includes Jewish, Baha'i, Hindu, and Christian; many are refugees or temporary foreign residents) (2010 est.)

Languages: Arabic

Literacy: 70.1%

Government Type: republic

Capital: Sanaa

Independence: 22 May 1990 (Republic of Yemen was established with the merger of the Yemen Arab Republic [Yemen (Sanaa) or North Yemen] and the Marxist-dominated People's Democratic Republic of Yemen [Yemen (Aden) or South Yemen])

GDP Per Capita: $2,800

Occupations: most people are employed in agriculture and herding; services, construction, industry, and commerce account for less than one-fourth of the labor force

Currency: Yemeni rial (YER)

Zambia

Long Name: Republic of Zambia

Location: Southern Africa, east of Angola

Area: 290,587 sq. mi. (752,618 sq. km)

Climate: tropical; modified by altitude; rainy season (October to April)

Terrain: mostly high plateau with some hills and mountains

Population: 15,066,266

Population Growth Rate: 2.88%

Birth Rate (per 1,000): 42.13

Death Rate (per 1,000): 12.67

Life Expectancy: 52.15 years

Ethnic Groups: Bemba 21%, Tonga 13.6%, Chewa 7.4%, Lozi 5.7%, Nsenga 5.3%, Tumbuka 4.4%, Ngoni 4%, Lala 3.1%, Kaonde 2.9%, Namwanga 2.8%, Lunda (north Western) 2.6%, Mambwe 2.5%, Luvale 2.2%, Lamba 2.1%, Ushi 1.9%, Lenje 1.6%, Bisa 1.6%, Mbunda 1.2%, other 13.8%, unspecified 0.4% (2010 est.)

Religion: Protestant 75.3%, Roman Catholic 20.2%, other 2.7% (includes Muslim Buddhist, Hindu, and Baha'i), none 1.8% (2010 est.)

Languages: Bembe 33.4%, Nyanja 14.7%, Tonga 11.4%, Lozi 5.5%, Chewa 4.5%, Nsenga 2.9%, Tumbuka 2.5%, Lunda (North Western) 1.9%, Kaonde 1.8%, Lala 1.8%, Lamba 1.8%, English (official) 1.7%, Luvale 1.5%, Mambwe 1.3%, Namwanga 1.2%, Lenje 1.1%, Bisa 1%, other 9.2%, unspecified 0.4%

Literacy: 63.4%

Government Type: republic

Capital: Lusaka

Independence: 24 October 1964 (from United Kingdom)

GDP Per Capita: $4,300

Occupations: agriculture 85%, industry 6%, services 9%

Currency: Zambian kwacha (ZMK)

Zimbabwe

Long Name: Republic of Zimbabwe

Location: Southern Africa, between South Africa and Zambia

Area: 150,872 sq. mi. (390,757 sq. km)

Climate: tropical; moderated by altitude; rainy season (November to March)

Terrain: mostly high plateau with higher central plateau (high veld); mountains in east

Population: 14,229,541

Population Growth Rate: 2.21%

Birth Rate (per 1,000): 32.26

Death Rate (per 1,000): 10.13

Life Expectancy: 57.05 years

Ethnic Groups: African 99.4% (predominantly Shona; Ndebele is the second largest ethnic group), other 0.4%, unspecified 0.2% (2012 est.)

Religion: Protestant 75.9% (includes Apostolic 38%, Pentecostal 21.1%, other 16.8%), Roman Catholic 8.4%, other Christian 8.4%, other 1.2% (includes traditional, Muslim), none 6.1% (2011 est.)

Languages: Shona (official; most widely spoken), Ndebele (official, second most widely spoken), English (official; traditionally used for official business), 13 minority languages (official; includes

Chewa, Chibarwe, Kalanga, Koisan, Nambya, Ndau, Shangani, sign language, Sotho, Tonga, Tswana, Venda, and Xhosa)

Literacy: 86.5%

Government Type: parliamentary democracy

Capital: Harare

Independence: 18 April 1980 (from United Kingdom)

GDP Per Capita: $2,100

Occupations: agriculture 66%, industry 10%, services 24%

Currency: Zimbabwean dollar (ZWD)

Websites for Further Reading

About.com, History of Africa: http://africanhistory.about.com
Accuweather: http://www.accuweather.com
Aeroflot Russian Airlines: http://www.aeroflot.com
African Union: http://www.au.int
Al Bawaba Middle East News and Arab World Headlines: http://www.albawaba.com
Alaska Volcano Observatory: https://www.avo.alaska.edu
Algemeiner Newspaper: http://www.algemeiner.com
AllTrips West Yellowstone, Montana: http://www.westyellowstonenet.com
American Foundation for AIDS Research: http://www.amfar.org
American Museum of Natural History: http://www.amnh.org
American Planning Association: https://www.planning.org
American Society of Civil Engineers: http://www.asce.org
Any Latitude Blog: http://www.anylatitude.com
Arctic Centre University of Lapland: http://www.arcticcentre.org
Arirang TV/Radio: http://www.arirang.co.kr
Asia and the Pacific Policy Society: http://www.policyforum.net
The Atlantic: http://www.theatlantic.com
Basilique Notre-Dame de la Paix de Yamoussoukro: http://www.ndpbasilique.org
Biofuels Digest: http://www.biofuelsdigest.com
Bloomberg: http://www.bloomberg.com
Borderplex Alliance: http://www.borderplexalliance.org
Breaking Energy: http://breakingenergy.com
Brenthurst Foundation: http://www.thebrenthurstfoundation.org
The Brisbane Times: http://www.brisbanetimes.com.au
British Broadcasting Corporation: http://bbc.com
Brookings Institution: http://www.brookings.edu
Business Insider: http://www.businessinsider.com.au

BuzzFeed: http://www.buzzfeed.com
CBS News: http://www.cbsnews.com
Center for Strategic and International Studies: http://csis.org
Centers for Disease Control and Prevention: http://www.cdc.gov
Centre for Aviation: http://centreforaviation.com
Charities Aid Foundation: https://www.cafonline.org
City Mayors Foundation: http://www.citymayors.com
City of Boston, Massachusetts: http://www.cityofboston.gov
City of New York, New York: http://www.nyc.gov
City of Truth or Consequences, New Mexico: http://www.torcnm.org
CNBC: http://www.cnbc.com
CNN: http://edition.cnn.com
CNN-Money: http://money.cnn.com
Conservation Institute: http://www.conservationinstitute.org
Convention on Biological Diversity: https://www.cbd.int
Convert Units: http://www.convertunits.com
Cooperative Institute for Research In Environmental Sciences: http://cires.colorado.edu
Council on Foreign Relations: http://www.cfr.org
County of Inyo, California: http://www.inyocounty.us
Current: http://www.currentresults.com
The Daily Mail: http://www.dailymail.co.uk
Data Center: http://www.datacenterresearch.org
Detroit Historical Society: http://detroithistorical.org
Dictionnaire Biographique Du Canada: http://www.biographi.ca
Discovering Lewis & Clark: http://www.lewis-clark.org
Discovery Channel: http://news.discovery.com
Economic Times: http://articles.economictimes.indiatimes.com
Economist: http://www.economist.com
Embassy of the United States Dili, Timor-Leste: http://timor-leste.gov.tl
Emporia State University: http://academic.emporia.edu
Enchanted Learning: http://www.enchantedlearning.com
Encyclopaedia Britannica: http://www.britannica.com
Encyclopedia of Chicago: http://www.encyclopedia.chicagohistory.org
Encyclopedia of the Nations: http://www.nationsencyclopedia.com
Ethanol Producer Magazine: http://ethanolproducer.com
Euronews: http://www.euronews.com
European Commission: http://ec.europa.eu

European Space Agency: http://www.esa.int
Evergreen State College: http://academic.evergreen.edu
Examiner.com: http://www.examiner.com
Export.gov: http://www.export.gov
Express: http://www.express.co.uk
Eyewitness to History.com: http://www.eyewitnesstohistory.com
Federal Communications Commission: http://fcc.gov
FEMA Flood Map Service Center: https://msc.fema.gov
Finfacts Ireland: http://www.finfacts.ie
First Flight Centennial: http://www.firstflightcentennial.org
Florida Center for Instructional Technology: http://fcit.usf.edu
Focus Migration: http://focus-migration.hwwi.de
Forbes Magazine: http://www.forbes.com
Foreign Policy: https://foreignpolicy.com
Fox News: http://www.foxnews.com
Free Dictionary: http://www.thefreedictionary.com
Fukuoka Properties: http://www.fukuokaproperties.com
Gallup.com: http://www.gallup.com
Gallup-Healthways Well-Being Index: http://www.well-beingindex.com
Garmin, Inc.: http://www8.garmin.com
Genome Sciences Centre: http://mkweb.bcgsc.ca
GeoCarta: http://geocarta.blogspot.com
Geohive-Population Statistics: http://www.geohive.com
Geology.com: http://geology.com
Gilbert's Potoroo Action Group: http://www.potoroo.org
Global Water Intel: http://www.globalwaterintel.com
Government of Japan, Ministry of Land, Infrastructure, Transport and Tourism: http://www.mlit.go.jp
Greenpeace USA: http://www.greenpeace.org
Groupe Eurotunnel: http://www.eurotunnelgroup.com
The Guardian: http://www.theguardian.com
Guide to Iceland: https://guidetoiceland.is
Harvard Business Review: https://hbr.org
Harvard University: http://www.harvard.edu
Harvard University Summer School: http://www.summer.harvard.edu
Hawaiian Islands: http://www.gohawaii.com
Highest Lake: http://www.highestlake.com

The Hindu: http://www.thehindu.com
Historic UK: http://www.historic-uk.com
History.com: http://www.history.com
Huffington Post: http://www.huffingtonpost.com
Humanity in Action: http://www.humanityinaction.org
IFLSCIENCE: http://www.iflscience.com
India Today: http://indiatoday.intoday.in
Industry Tap: http://www.industrytap.com
Inside Gov: http://divorce-laws.insidegov.com
Institution of Mechanical Engineers: http://www.imeche.org
International Business Times: http://www.ibtimes.com
International Desalination Association: http://idadesal.org
International Permafrost Association: http://ipa.arcticportal.org
International Rice Research Institute: http://irri.org
Internet Encyclopedia of Philosophy: http://www.iep.utm.edu
Internet Live Stats: http://www.internetlivestats.com
Jagran Josh Exam Prep: http://www.jagranjosh.com
Japan Meteorological Agency: http://www.jma.go.jp
Jewish Virtual Library: http://www.jewishvirtuallibrary.org
JR Central Japan Rail Company: http://english.jr-central.co.jp
Kauffman Foundation: http://www.kauffman.org
Korea.net: http://www.korea.net
Kuensel Online: http://www.kuenselonline.com
Lake Baikal: http://lakebaikal.org
Lee Scharich blog: https://sperglord.wordpress.com
Legatum Institute: http://www.li.com
Legatum Prosperity Index: http://www.prosperity.com
Lewis and Clark Journals/University of Nebraska: http://lewisandclarkjournals.unl.edu
Liberian Registry: http://www.liscr.com
Live Science: http://www.livescience.com
The Local: http://www.thelocal.fr
The Los Angeles Times: http://www.latimes.com
Mall of the Emirates: http://www.malloftheemirates.com
Map Happy: http://maphappy.org
Maps of the World: http://www.mapsofworld.com
MarketWatch: http://www.marketwatch.com
 McDonald's Corporation: http://www.aboutmcdonalds.com

Medium: https://medium.com

The Mercury News: http://www.mercurynews.com

Metric Conversions: http://www.metric-conversions.org

Metropolitan Museum of Art: http://www.metmuseum.org

Migration Policy Institute: http://www.migrationpolicy.org

MIT Technology Review: http://www.technologyreview.com

Montanakids.com: http://montanakids.com

Moody's Investors Service: https://www.moodys.com

Mother Nature Network: http://www.mnn.com

Museum of Unnatural Mystery: http://www.unmuseum.org

The National: http://www.thenational.ae

National Aeronautics and Space Administration Global Climate Change: http://climate.nasa.gov

National Aeronautics and Space Administration Science Mission Directorate: http://science.nasa.gov

National Aeronautics and Space Administration Near Earth Object Program: http://neo.jpl.nasa.gov

National Association of Colleges and Employers: https://www.naceweb.org

National Center for Education Statistics: https://nces.ed.gov

National Council for Geographic Education: http://www.ncge.org

National Geographic Magazine: http://education.nationalgeographic.com

National Geographic Society: http://www.nationalgeographic.org

National Oceanic and Atmospheric Administration: http://www.noaa.gov

National Oceanic and Atmospheric Administration Arctic Theme Page: http://www.arctic.noaa.gov

National Oceanic and Atmospheric Administration Centers for Environmental Info: http://www.ncdc.noaa.gov

National Oceanic and Atmospheric Administration National Data Buoy Center: http://www.ndbc.noaa.gov

National Oceanic and Atmospheric Administration National Ocean Service: http://oceanservice.noaa.gov

National Oceanic and Atmospheric Administration National Weather Service: http://www.crh.noaa.gov

National Oceanic and Atmospheric Administration National Weather Service Southern: http://www.srh.noaa.gov

National Oceanic and Atmospheric Administration Pacific Tsunami Warning Center: http://ptwc.weather.gov

National Oceanic and Atmospheric Administration Satellites and Information: http://ols.nndc.noaa.gov

National Oceanic and Atmospheric Administration Storm Prediction Center: http://www.spc.noaa.gov

National Park Service: http://www.nps.gov

National Parks Conservation Association: http://www.npca.org

National Philanthropic Trust: http://www.nptrust.org

National Public Radio: http://www.npr.org

National Snow and Ice Data Center: https://nsidc.org

National Weather Service Lightning Safety: http://www.lightningsafety.noaa.gov

Nationmaster: http://www.nationmaster.com

Natural History Museum: http://www.nhm.ac.uk

Nelson Mandela Foundation: https://www.nelsonmandela.org

New Scientist: http://www.newscientist.com

New York Public Library: http://exhibitions.nypl.org

New York Times: http://www.nytimes.com

New Yorker: http://www.newyorker.com

News 12 Connecticut: http://connecticuthistory.org

Newsweek: http://www.newsweek.com

National Marine Manufacturers Association: http://www.nmma.org

NK News: http://www.nknews.org

Nobelprize.org: http://www.nobelprize.org

Number Sleuth: http://www.numbersleuth.org

Observatory of Economic Complexity: https://atlas.media.mit.edu

Office of the United States Trade Representative: https://ustr.gov

OilPrice.com: http://oilprice.com

Olympic.org: http://registration.olympic.org

Open Culture: http://www.openculture.com

Organization of American States Foreign Trade Information System: http://www.sice.oas.org

Organization of the Petroleum Exporting Countries: http://www.opec.org

Pennsylvania State University: http://sites.psu.edu

Peter G. Peterson Foundation: http://pgpf.org

Peterson Institute for International Economics: http://blogs.piie.com

Pew Research Center: http://www.pewforum.org

Photius.com: http://www.photius.com

Phys.org: http://phys.org

Physics.org: http://www.physics.org

Population Institute: http://www.populationinstitute.org

PR Newswire: http://www.prnewswire.com

Princeton University Library: http://libweb5.princeton.edu
Project Amazonas: http://www.projectamazonas.org
Prospects: http://www.prospects.ac.uk
Public Broadcasting Service: http://www.pbs.org
Public Radio International: http://www.pri.org
Quartz: http://qz.com
Quatr.us: http://quatr.us
Reuters: http://www.reuters.com
Rick Steves Europe: https://www.ricksteves.com
Rio Times: http://riotimesonline.com
Rolling Stone Magazine: http://www.rollingstone.com
Romantic Asheville.com: http://www.romanticasheville.com
Royal Flying Doctor Service: https://www.flyingdoctor.org.au
S7 Airlines: http://www.s7.ru
Sahara Question: http://sahara-question.com
Sarah Woodbury blog: http://www.sarahwoodbury.com
Science Blogs: http://scienceblogs.com
Seawater Desalination Huntington Beach Facility: http://hbfreshwater.com
Sidem-Veolia Company: http://www.sidem-desalination.com
Slate Magazine: http://www.slate.com
Smashing Lists: http://www.smashinglists.com
Smith College: http://www.smith.edu
Smithsonian.com: http://www.smithsonianmag.com
Softschools.com: http://www.softschools.com
South African Government: http://www.gov.za
South African History Online: http://www.sahistory.org.za
South Tyrol Museum of Archaeology: http://www.iceman.it
Space.com: http://www.space.com
State of Alaska: http://www.alaska.org
Stars and Stripes: http://www.stripes.com
State of Arkansas Tourism: http://www.arkansas.com
State of Michigan: http://www.michigan.gov
State of Nevada: http://nv.gov
State of Oregon: http://www.oregon.gov
Statistics New Zealand: http://www.stats.govt.nz
The Sydney Morning Herald: http://www.smh.com.au
System Dynamics Society: http://www.systemdynamics.org

Tech Crunch: http://techcrunch.com

Tech Insider: http://www.techinsider.io

Tech Republic: http://www.techrepublic.com

The Telegraph: http://www.telegraph.co.uk

Theodore's Royalty & Monarchy Site: http://www.royaltymonarchy.com

Time and Date.com: http://www.timeanddate.com

Top 10 List Land: http://www.top10listland.com

Top Five of Anything: http://top5ofanything.com

Tour de France: http://www.letour.com

Tourneau, LLC: https://www.tourneau.com

Trans Siberian Express: http://www.transsiberianexpress.net

Treasury of the Government of New Zealand: http://www.treasury.govt.nz

UAEinteract: http://www.uaeinteract.com

U.N. Educational, Scientific and Cultural Organization Institute for Statistics: http://www.uis.unesco.org

U.N. Educational, Scientific and Cultural Organization World Heritage Centre: http://whc.unesco.org

United Nations: http://www.un.org

United Nations Children's Emergency Fund: http://www.unicef.org

United Nations Department of Economic and Social Affairs: http://esa.un.org

United Nations Food and Agriculture Organization: http://www.fao.org

United Nations Office of the High Commissioner for Refugees: http://www.unhcr.org

United Nations Office on Drugs and Crime: http://www.unodc.org

United Nations World Health Organization: http://www.who.int

United Nations World Tourism Organization: http://www.unwto.org

United Press International: http://www.upi.com

Universe Today: http://www.universetoday.com

University of California Santa Barbara Center for Spatial Studies: http://spatial.ucsb.edu

University of Illinois Urbana-Champaign, Department of English: http://www.english.illinois.edu

University of Ouagadougou, Burkina Faso: http://www.univ-ouaga.bf

University of Wisconsin Green Bay: https://www.uwgb.edu

U.S. Antarctic Program: http://www.usap.gov

U.S. Bureau of Reclamation: http://www.usbr.gov

U.S. Census Bureau: http://censtats.census.gov

U.S. Census Bureau Quick Facts: http://quickfacts.census.gov

U.S. Central Intelligence Agency: https://www.cia.gov

U.S. Climate Data: http://www.usclimatedata.com

U.S. Department of Agriculture Economic Research Service: http://www.ers.usda.gov

U.S. Department of Agriculture Foreign Agriculture Service: http://gain.fas.usda.gov

U.S. Department of Agriculture Natural Resources Conservation Service: http://www.nrcs.usda.gov

U.S. Department of Commerce International Trade Administration: http://www.trade.gov

U.S. Department of Education: http://www2.ed.gov

U.S. Department of Homeland Security Federal Emergency Management Administration: https://www.fema.gov

U.S. Department of Labor Bureau of Labor Statistics: http://www.bls.gov

U.S. Department of the Interior Bureau of Indian Affairs: http://www.bia.gov

U.S. Department of the Navy: http://www.navy.mil

U.S. Department of Transportation Assistant Secretary for Research and Technology: http://www.rita.dot.gov

U.S. Energy Information Administration: http://www.eia.gov

U.S. Environmental Protection Agency: http://www.epa.gov

U.S. Geological Survey: http://www.usgs.gov

U.S. Geological Survey Earthquake Hazards Program: http://earthquake.usgs.gov

U.S. Geological Survey Publications Warehouse: http://pubs.usgs.gov

U.S. Geological Survey Water Resources: https://water.usgs.gov

U.S. Immigration and Customs Enforcement: http://www.ice.gov

U.S. News and World Report: http://www.usnews.com

USA for UNHCR: http://www.unrefugees.org

U.S.A.gov: http://search.usa.gov

USA Today: http://www.usatoday.com

USA Today-Travel Tips: http://traveltips.usatoday.com

Vanguard News Nigeria: http://www.vanguardngr.com

Visual Arts Encyclopedia: http://www.visual-arts-cork.com

Volcanoes: http://www.volcanoes.org.uk

Wall Street Journal: http://www.wsj.com

Washington Post: http://www.washingtonpost.com

Washington Times: http://www.washingtontimes.com

Water Education Foundation: http://www.watereducation.org

Water-technology.net: http://www.water-technology.net

Weather Channel: http://www.weather.com

Weather Underground: http://www.wunderground.com

WFMZ-TV: http://www.wfmz.com

Wiley Online Library: http://onlinelibrary.wiley.com
Wired: http://www.wired.com
Wonderopolis: http://wonderopolis.org
World Atlas: http://www.worldatlas.com
World Bank: http://www.worldbank.org
World Cocoa Foundation: http://worldcocoafoundation.org
World Data Center for Geomagnetism, Kyoto: http://wdc.kugi.kyoto-u.ac.jp
World Meteorological Organization Arizona State University: http://wmo.asu.edu
World Meteorological Society: http://globalcryospherewatch.org
World Nuclear Association: http://www.world-nuclear.org
World Policy Institute: http://www.worldpolicy.org
World Time Server: http://www.worldtimeserver.com
World Time Zone: http://www.worldtimezone.com
XE: http://www.xe.com
Yale University Malaysian and Singaporean Association: http://masa.sites.yale.edu

Glossary

absolute location—describes the location of a place by using grid coordinates, most commonly latitude and longitude.

acid rain—precipitation of sulfuric and nitric acids caused by pollutants of motor vehicles and industrial activity released into the atmosphere.

air pollution—particles in the air such as dust, smoke, volcanic ash, and pollens, as well as chemicals and particulates from combustion and industrial activity.

a.m.—an abbreviation for the Latin words "ante meridiem," meaning "before midday."

Andes—a mountain chain that runs along the entire west coast of South America, from Panama (at the southern tip of Central America) to the Strait of Magellan (at the southern tip of South America).

Antarctic Circle—a line of latitude at 66.5 degrees south of the equator.

anthropogenic—man-made.

Apennines—a mountain range extending from northern to southern Italy for approximately 750 miles (1,200 km).

aqueducts—channels that transport water hundreds of miles.

aquifer—an underground collection of water that is surrounded by rock.

archipelago—a chain (or group) of islands that are close to one another.

Arctic Circle—a line of latitude at 66.5 degrees north of the equator.

Asia Minor—the term used for the larger part of Turkey that lies in Asia, east of the Strait of Bosporus.

Atacama Desert—the driest non-polar desert on Earth, located along the Pacific coast of northern Chile.

atoll—tiny island formed when a volcano, around which coral often grows, erodes away, leaving a circular wall of coral with a lagoon at the center.

atomic clock—uses measurements of energy released from atoms to precisely measure time.

axis—the imaginary line that passes through the north and south poles about which the earth revolves.

ayatollah—a high-ranking Muslim Shi'i religious scholar, religious leader or cleric.

azimuth—another method for stating compass direction.

balkanization—the fragmentation of a country into ethnic, language, or cultural divisions by territory.

bar scale—a scale that uses a graphic to show the relationship between distance on the map to distance in the area represented.

basins and ranges—are sets of valleys and mountains that are spaced close together.

bay—smaller bodies of water partially surrounded by land.

B.C.E.—abbreviation for "before the common/current/Christian era."

Bedouins—a semi-nomadic group that have inhabited areas of the Middle East.

boiling point—the temperature when water boils. At sea level, it is 212° Fahrenheit (100° Celsius).

brain drain—when highly educated or highly skilled individuals leave their home countries to go to countries where opportunities are better.

canal—connects two bodies of water that lie at different elevations.

capital—a city in which is the seat of government.

capitol—a building housing the seat of government.

Caprivi Strip—the narrow strip of land that protrudes from the northeast corner of Namibia, surrounded on three sides by the countries of Angola, Zambia, and Botswana.

cartel—an organization made up of businesses that band together to eliminate competition, collude to maintain high prices, and control supply and production of a product or service.

cartographer—someone who contributes to the scientific, technological, and artistic components in the making of maps.

C.E.—abbreviation for "common/current/Christian era."

census—an enumeration, or counting, of a population.

Central America—an area that includes the countries that connect North and South America and are located between Mexico and Colombia.

central business district—an area of a city, often located downtown, that is the primary concentration of commercial buildings.

CFCs—chlorofluorocarbons.

choke point—a narrow waterway between two larger bodies of water that can be easily closed or blocked to control water transportation routes.

city—a legal entity with a delegated power by a state and county to govern and provide services to its citizens.

climate—the long-term (usually 30-year) average weather for a particular place.

Continental Divide—the line that divides the flow of water in North America, either toward the Pacific Ocean or Atlantic Ocean.

continental drift—when tectonic plates slowly move, crashing into each other, or sliding side by side, or separating apart, all of which activity can lead to the creation of mountain ranges, volcanoes, earthquakes, or new seas and oceans.

continentality—areas of a continent that are distant from an ocean that experience greater extremes in temperature than do places that are closer to an ocean.

continents—the six or seven large land masses on the planet. If you count seven continents, these include Europe, Asia, Africa, Australia, Antarctica, North America, and South America.

coral reefs—formed by the accumulation of calcium carbonate that comes from the external skeletons of tiny animals called coral polyps.

coriolis effect—due to the rotation of the earth, any object on or near the earth's surface will veer to the right in the Northern Hemisphere and to the left in the Southern Hemisphere

country—the equivalent of a state, a political entity.

Crimea—a diamond-shaped peninsula attached by the Isthmus of Perekop to southern Ukraine, and protruding into the Black Sea.

cryosphere—the area of the earth where water has solidified, and includes ice, floating ice, glaciers, permafrost, and snow.

cultivation—the deliberate attempt to sow and manage wild plants and seed.

dam—structures that block the flow of a river, allowing a reservoir of water to build up.

DART—Deep-Ocean Assessment and Reporting of Tsunami.

delta—a low-lying area where a river meets the sea.

desalination—the process of removing some salt and minerals from ocean water in order to produce consumable freshwater for a population.

desert—an area of light rainfall that usually has little plant or animal life due to the dry conditions.

desertification—the process that turns viable agricultural land into desert because of overgrazing, inefficient irrigation systems, and deforestation.

dike—magma that has risen up through a crack between layers of rock.

domestication—when people experiment with, and consciously select, the right seeds to grow for various conditions; when people tame formerly wild animals for use in agriculture or as pets.

drainage basin—the area that includes all of the tributaries for an individual stream or river.

dust devil—columns of brown, dust-filled air, which can rise dozens of feet, caused by warm air rising on dry, clear days.

El Niño—a large body of warm water that moves between the eastern and western Pacific Ocean near the equator.

epicenter—the point on the earth's surface that is directly above the hypocenter, where earthquakes actually occur.

equator—a line that divides the earth into northern and southern hemispheres.

evapotranspiration—the process of water moving into the atmosphere.

fault—a fracture or a collection of fractures in the earth's surface where movement has occurred.

FEMA—Federal Emergency Management Agency.

fjords—canyons with high cliffs hanging over a thin bay of water.

floodplain—the area surrounding a river that, when unmodified by human structures, would normally be flooded during a river flood.

fossil—the outline of the remains of a plant or animal embedded in rock.

fossil fuels—underground fuels such as natural gas, oil, and coal that are encased in rocks, just like fossils.

freeways—multi-lane highways that use on- and off-ramps, rather than intersections,

Fujita scale—measures the strength of a tornado based on observed damage and effects.

Gaza Strip—The area of land along the Mediterranean Sea at the border of Israel and Egypt.

gazetteer—an index that lists the latitude and longitude of places within a specific region or across the entire world.

GDP—gross domestic product, the value of all goods and services produced in a country in a year.

geographic literacy—when someone understands the interactions and interconnections between people and our physical world, and the implications that these interconnections may cause.

geography—writing about the earth.

geologic time—a time scale that divides the history of the planet Earth into eras, periods, and epochs from the birth of the planet to the present.

glacier—a mass of ice that stays frozen throughout the year and flows downhill.

global warming—the gradual increase of the earth's average temperature.

GNP—gross national product, the total value of GDP plus all income from investments around the world.

GPS—global positioning system is a network of satellites in orbit around the earth that may give us a precise location of either ourselves or a point of interest.

green revolution—began in the 1960s as an effort by international organizations (especially the United Nations) to help increase the agricultural production of less developed nations.

greenhouse effect—a natural process of the atmosphere that traps some of the sun's heat near the earth.

greenhouse gases—chemical compounds that absorb and emit thermal radiation, are present in our atmosphere and allow direct sunlight to reach the earth's surface.

gulf—large bodies of water partially surrounded by land.

gypsies—nomadic tribes that travel throughout Europe.

hail—water droplets (raindrops) that have turned to ice inside of clouds.

hemisphere—half of the earth.

highway—a paved road connecting distant towns.

Horn of Africa—the eastern peninsula of continental Africa that includes the countries of Djibouti, Eritrea, Ethiopia, and Somalia.

horse latitudes—high-pressure regions, more formally known as sub-tropic highs, which are warm and don't have much wind.

hot springs—created by underground water that is heated and percolates to the earth's surface.

hydrologic cycle—the movement of water from the atmosphere to the land, rivers, oceans, and plants and then back into the atmosphere.

Ice Age—period of geologic time when the climate of the planet has cooled, and sheets of ice covered large portions of land.

ice core sample—a thick column of ice, sometimes hundreds of feet long, that is produced by drilling a circular pipe-like device into thick ice and then pulling out the cylindrical piece.

igneous rocks—formed when liquid magma under the surface of the earth, or lava on the surface of the earth, cools and hardens.

incidence map—a map that plots where and how people have been infected or exposed to potentially harmful viruses and infectious diseases.

Indochina—the name of the peninsula in Southeast Asia composed of Myanmar (Burma), Thailand, Cambodia, Laos, Vietnam, and the mainland portion of Malaysia.

Industrial Revolution—began in the eighteenth century in England with the transformation from an agriculture-based economy to an industry-based economy.

irredentism—a term used to describe a situation in which a minority group in one country shares the culture and heritage of people within another country.

irrigation—the process of artificially watering crops.

isotherms—lines of equal temperature.

jet stream—a band of swiftly moving air located high in the atmosphere.

jungle—a forest that is composed of very dense vegetation.

Katabatic wind—high density air that moves from a higher elevation down a slope because of the force of gravity.

knot—equivalent to one nautical mile per hour.

Kurds—an indigenous ethnic group who live in southeastern Turkey, northwestern Iran, northeastern Iraq, Syria, and Armenia.

Latin America—an area that encompasses Central America as well as Mexico and all of the countries of South America.

lava—when magma erupts or flows from a volcano onto the earth's surface.

legend—usually found in a box on the map, contains information that explains the symbols used on a map.

lines of latitude and longitude—make up a grid system that was developed to help determine the location of points on the earth.

lingua franca—a language used between people who do not share a common language.

locks—used to gradually move the ships from one elevation to another within a canal.

long lots—long and narrow pieces of property.

Maghreb—a region of Northern Africa composed of the Atlas Mountains and coasts of Morocco, Algeria, Tunisia, and Libya farther east.

magma—hot, liquefied rock that lies underneath the surface of the earth.

Magnetic North—the direction that corresponds to the magnetic field lines to which compasses point.

Manifest Destiny—the phrase used to describe the assumption that American expansion to the Pacific Ocean was inevitable and ordained by God.

maquiladoras—Mexican factories owned by foreign (usually U.S.) corporations.

meander—streams and rivers that have carved a flat floodplain and flow in "s"-shaped curves.

Mediterranean climate—a climate similar to the one found along the Mediterranean Sea: warm, hot, and dry in the summer and mild, cool, and wet in the winter.

megacity—cities with populations greater than ten million people.

megalopolis—a huge metropolitan area connecting large urban areas.

Melanesia—a small region that is located northeast of Australia, south of the equator, and west of 180 degrees longitude, but excludes New Zealand. It includes the countries of Vanuatu, Fiji, Papua New Guinea, and the Solomon Islands, as well as the French overseas territory of New Caledonia, but excludes New Zealand.

Mercalli scale—measures the power of an earthquake as felt by humans and structures.

MERCOSUR—Southern Cone Common Market, a regional trade group that includes Brazil, Argentina, Paraguay, and Uruguay.

Mesopototamia—an ancient region situated between the Tigris and Euphrates Rivers, from contemporary southern Turkey to the Persian Gulf, and includes modern-day Turkey, Syria, and Kuwait.

metamorphic rocks—rocks that were once sedimentary, igneous, or even another metamorphic rock, but have been altered by heat, pressure, or other outside force without going through a liquid phase.

Micronesia—an area of the world that consists of islands east of the Philippines, west of the International Date Line, north of the equator, and south of the Tropic of Cancer.

monsoon—winds that flow from the ocean to the continent during the summer and from the continent to the ocean in the winter.

NAFTA—North American Free Trade Agreement.

nation—a group of people who share a common heritage and culture.

nautical mile—equivalent to approximately 6,076 feet (1,852 meters) or 1.15 miles (1.85 km).

near earth object—any object in space that is relatively close to Earth and is of any size.

nomads—tribes that move from place to place in a seasonal circuit over a large region.

North Magnetic Pole—where compass needles around the world point.

nuclear winter—what would follow a large-scale nuclear war when radioactive particles, dust, and smoke are released into the atmosphere, creating a large cloud over the planet and blocking out sunlight, reducing temperatures worldwide.

ocean—a large body of salt water unobstructed by continents or land.

ocean current—water that is constantly moving in giant circles.

Oceania—a region in the central and southern Pacific that includes Australia, New Zealand, Papua New Guinea, and the islands that compose Polynesia, Melanesia, and Micronesia.

Ogallala Aquifer—a huge aquifer that spans an area from western Texas to South Dakota, including parts of Colorado, Kansas, Nebraska, Oklahoma, and New Mexico.

OPEC—Organization of Petroleum Exporting Countries.

orogeny—what happens when two tectonic plates collide forming a mountain, related to continental drift.

oxbow lake—a crescent-shaped lake that is formed when the meander, or curve, of a river is cut off from the rest of the river during a flood, or when the curve of the meander becomes so large that the river begins flowing along a new path.

ozone layer—part of the stratosphere, a layer of the earth's atmosphere that lies about 10 to 30 miles (16 to 48 km) above the surface of the earth.

Pangea—the continent that existed when all of the land on Earth was lumped together into one large continent 250 million years ago.

permafrost—ground (including soil, rock, organic material, and ice) that is permanently at or below 32° Fahrenheit (0° Celsius) for a minimum of two consecutive years.

perpetual resources—natural resources, such as solar energy, wind, and tidal energy, that have no chance of being used in excess of their availability.

PHA—Potentially Hazardous Asteroid, an object that is 3.28 feet (1 meter) in length or bigger, which is large enough to be detected by astronomers on Earth, and has an orbital path that could bring it in close proximity to Earth.

physical map—shows natural features of the land such as mountains, rivers, lakes, streams, and deserts.

pidgin—a language that has a small vocabulary and is a combination and distortion of two or more languages.

Pinyin—a system for transliterating Chinese characters into the Roman alphabet.

plaza—an open public square at the center of the downtown of many Latin American cities.

p.m.—an abbreviation for the Latin words "post meridiem," meaning "after midday."

political map—shows human-made features and boundaries such as cities, highways, and countries.

Polynesia—an area of the world that consists of more than 1,000 islands in a great, triangular region bounded by Hawaii to the north, New Zealand to the southwest, and Easter Island to the southeast.

Prime Meridian—the longitudinal line that passes through the Royal Observatory in Greenwich, England.

primogeniture—the system of inheritance in which all inheritable land and property is passed on to the first-born son.

rain forest—any densely vegetated area that receives over 40 inches (100 centimeters) of rain a year.

refugee—person living somewhere outside of his/her home country, usually because of deplorable conditions at home, warfare, or government and societal discrimination and economic oppression.

relative location—a description of location using the relation of one place to another.

relief map—portrays various elevations in different colors.

renewable resource—one that can be replenished within a generation.

Richter scale—the scale used to measure the energy released by an earthquake.

ridge—a crack between tectonic plates, where new ocean floor is being created as magma flows up from under the earth.

Ring of Fire—dense accumulation of earthquakes and volcanoes in and around the Pacific Ocean.

Saffir-Simpson Hurricane Wind Scale—a measurement used to rank the intensity and destructive capacity of hurricanes.

satellite images—accurate photographs of the earth's surface, allowing cartographers to precisely determine the location of roads, cities, rivers, and other features on the earth.

scale—indicates the level of detail and defines the distances between objects on a map.

sea—any body of salt water, partially or completely enclosed by land.

sedimentary rocks—rocks formed by the accumulation and squeezing together of layers of sediment (particles of rock or remains of plant and animal life) at the bottom of rivers, lakes, and oceans or even on land.

seismic sea wave—another name for a tsunami.

sextant—an instrument used to measure the angle between the horizon and a celestial body.

Sherpa—an indigenous ethnic group in Tibet and Nepal.

snow—water vapor that freezes in clouds before falling to the earth.

sounding—a method once used for determining the depth of the ocean

State—a country.

state—spelled with a lower-case "s," it is a division within a country.

strait—a narrow body of water between islands or continents that connects two larger bodies of water.

subcontinent—a landmass that has its own continental shelf and its own continental plate.

subduction—when crust from one plate slides under the crust of another.

sundial—an instrument that uses the sun to measure time.

surface—upper layer of the earth composed of three layers: crust, mantle (upper and lower), and core (outer and inner).

theocracy—a country where the civil leader is believed to receive direct guidance from God.

third world—a term that once referred to those countries that did not align themselves with the United States (first world) or the Soviet Union (second world) during the Cold War, or non-aligned nations.

thunderstorms—localized, atmospheric phenomena that produce heavy rain, thunder and lightning, and sometimes hail, which are formed in cumulonimbus clouds (clouds that are big and bulbous) that rise many miles into the sky.

topographic map—shows human and physical features of the earth and can be distinguished from other maps by its great detail and by its contour lines indicating elevation.

tornadoes—very powerful, yet tiny storms that have destructive winds capable of leveling buildings and other structures.

tree line—the point of elevation above which trees can no longer grow because of low temperatures and frozen ground (permafrost).

Trenches—deep, "v"-shaped depressions on the earth's surface, lying in the deepest parts of the earth's oceans, mostly found in the Pacific Ocean.

tributary—any stream that flows into another stream.

tropical glaciers—masses of ice found high in the mountains of tropical regions in the world.

tropical rain forest—a rain forest that lies between the Tropic of Cancer in the north (northernmost latitude around the earth, where the sun appears directly overhead during the Northern Solstice) and the Tropic of Capricorn in the south (the southernmost latitude around the earth, where the sun appears directly overhead during the Southern Solstice) or within the "tropics."

tropics—the lines of latitude where the sun is directly overhead on the summer solstices.

True north—the direction that one can map along the surface of the earth to the geographic North Pole.

tsunami—forceful movement of water caused by an earthquake that occurs under the ocean or near the coast of a land mass.

tundra—dry, barren plains that have significant areas of frozen soil or permafrost.

turnpike—toll road.

UNESCO—United Nations Educational, Scientific and Cultural Organization.

urban area—an area that consists of a central city and its surrounding suburbs.

volcano—the result of hot, liquid magma rising or being pushed to the surface of the earth.

wadi—the Arabic word for a gully or other stream bed that is dry for most of the year, but turns into a stream during a short rainy season.

warning—means that a hazard is already occurring or is imminent.

watch—means that such a hazardous event is likely to occur or is predicted to occur, such as a tornado watch or a flood watch.

water clock—a clock that is operated by dripping water from containers at measured intervals and then measuring it.

watershed—the boundary between drainage basins.

weather—the current condition of the atmosphere.

westerlies—winds that flow at mid-latitudes (30 to 60 degrees north and south of the equator) from west to east around the earth.

Willy-willy—indigenous Australian word for dust devil.

Index

Note: (ill.) indicates photos and illustrations.

NUMBERS

A

B

C

D

E

F

G

H

I

J

K

L

M

N

O

P

Q

R

S

T

U

V

W

X, Y

Z